D1237436

MOLECULAR NEUROPHARMACOLOGY

MOLECULAR
NEUROPHARMACOLOGY

Strategies and Methods

Edited by

ARNE SCHOUSBOE, PhD, DSc

*Department of Pharmacology,
The Danish University of Pharmaceutical Sciences,
Copenhagen, Denmark*

and

HANS BRÄUNER-OSBORNE, PhD, DSc

*Department of Medical Chemistry,
The Danish University of Pharmaceutical Sciences,
Copenhagen, Denmark*

Humana Press ✳ Totowa, New Jersey

© 2004 Humana Press Inc.
999 Riverview Drive, Suite 208
Totowa, New Jersey 07512

humanapress.com

Production Editor: Tracy Catanese

Cover Illustration: Figures 2 and 10 from Chapter 1, "Computational Studies of Ligand–Receptor Interactions in Ionotropic Glutamate Receptors" by Jeremy R. Greenwood and Tommy Liljefors.

Cover design by Patricia F. Cleary.

For additional copies, pricing for bulk purchases, and/or information about other Humana titles, contact Humana at the above address or at any of the following numbers: Tel.: 973-256-1699; Fax: 973-256-8341; E-mail: humana@humanapr.com or visit our website at www.humanapress.com

The opinions expressed herein are the views of the authors and may not necessarily reflect the official policy of the National Institute on Drug Abuse or any other parts of the US Department of Health and Human Services. The US Government does not endorse or favor any specific commercial product or company. Trade, proprietary, or company names appearing in this publication are used only because they are considered essential in the context of the studies reported herein.

This publication is printed on acid-free paper. ∞
ANSI Z39.48-1984 (American National Standards Institute) Permanence of Paper for Printed Library Materials.

Printed in the United States of America. 10 9 8 7 6 5 4 3 2 1

1-59259-672-X (e-book)

Library of Congress Cataloging-in-Publication Data

Molecular neuropharmacology : strategies and methods / edited by Arne Schousboe and Hans Bräuner-Osborne.
    p. ; cm.
  Includes bibliographical references and index.
  ISBN 1-58829-143-X (alk. paper)
    1. Neurotransmitter receptors. 2. Neuropharmacology. 3. Molecular pharmacology. I. Schousboe, Arne. II. Bräuner-Osborne, Hans.
    [DNLM: 1. Neuopharmacology--methods. 2. Receptors, Neurotransmitter. 3. Neurotransmitters. 4. Receptors, GABA. 5. Receptors, Glutamate. WL 102.8 M7179 2004]
  QP364.7.M64 2004
  615'.78--cd22

2003056600

PREFACE

The advent of the cloning of a large number of receptors and transporters for neurotransmitters and the simultaneous increase in the sophistication of tools available to produce specific mutations and chimeras of these proteins have provided scientists with the tools to understand the pharmacological and functional properties of such receptors and transporters at an hitherto unattained level. When this knowledge is combined with expertise in medicinal and computational chemistry, the basis for understanding the interactions between ligands and their corresponding macromolecules is greatly facilitated.

Molecular Neuropharmacology: Strategies and Methods is intended to bridge the gap between molecular biology and advanced chemistry. In addition, it attempts to include information about x-ray crystallographic analyses whenever available. This book discusses interdisciplinary interactions for monoamine transporters, amino acid transporters, ionotropic receptors, metabotropic glutamate receptors, GABA receptors, and other G protein-coupled receptors.

Arne Schousboe, PhD, DSc
Hans Bräuner-Osborne, PhD, DSc

CONTENTS

PART I. MOLECULAR PHARMACOLOGY OF GLUTAMATE RECEPTORS

PART II. MOLECULAR PHARMACOLOGY OF GABA RECEPTORS

PART III. MOLECULAR PHARMACOLOGY
OF TRANSMITTER TRANSPORTERS

CONTRIBUTORS

BERNHARD BETTLER • *Department of Clinical-Biological Sciences, Pharmacenter, University of Basel, Basel, Switzerland*

THIJS BEUMING • *Department of Physiology and Biophysics, Mount Sinai School of Medicine, New York, NY*

LARS BORRE • *Department of Biochemistry, Hadassah Medical School, The Hebrew University, Jerusalem, Israel*

HANS BRÄUNER-OSBORNE • *Department of Medicinal Chemistry, The Danish University of Pharmaceutical Sciences, Copenhagen, Denmark*

FABIEN CAMPAGNE • *Department of Physiology and Biophysics and Institute for Computational Biomedicine, Mount Sinai School of Medicine, New York, NY*

VERNON R. J. CLARKE • *The MRC Centre for Synaptic Plasticity, Department of Anatomy, Medical School, University of Bristol, Bristol, UK*

GRAHAM L. COLLINGRIDGE • *The MRC Centre for Synaptic Plasticity, Department of Anatomy, Medical School, University of Bristol, Bristol, UK*

BJARKE EBERT • *Department of Molecular Pharmacology, H. Lundbeck A/S, Valby, Denmark*

MARTA FILIZOLA • *Department of Physiology and Biophysics, Mount Sinai School of Medicine, New York, NY*

JEAN-MARC FRITSCHY • *Institute of Pharmacology and Toxicology, University of Zurich, Zurich, Switzerland*

BENTE FRØLUND • *Department of Medicinal Chemistry, The Danish University of Pharmaceutical Sciences, Copenhagen, Denmark*

ULRIK GETHER • *Molecular Neuropharmacology Group, Department of Pharmacology, The Panum Institute, University of Copenhagen, Copenhagen, Denmark*

NAOMI R. GOLDBERG • *Center for Molecular Recognition, Columbia University, New York, NY*

JEREMY R. GREENWOOD • *Department of Medicinal Chemistry, The Danish University of Pharmaceutical Sciences, Copenhagen, Denmark*

JONATHAN A. JAVITCH • *Center for Molecular Recognition, Columbia University, New York, NY*

ANDERS A. JENSEN • *Department of Medicinal Chemistry, The Danish University of Pharmaceutical Sciences, Copenhagen, Denmark*

ANNE T. JØRGENSEN • *Department of Medicinal Chemistry, The Danish University of Pharmaceutical Sciences, Copenhagen, Denmark*

BARUCH I. KANNER • *Department of Biochemistry, Hadassah Medical School, The Hebrew University, Jerusalem, Israel*

MICHAEL P. KAVANAUGH • *Vollum Institute, Oregon Health Sciences University, Portland, OR*

POVL KROGSGAARD-LARSEN • *Department of Medicinal Chemistry, The Danish University of Pharmaceutical Sciences, Copenhagen, Denmark*

ix

ORLA MILLER LARSSON • *Department of Pharmacology, The Danish University of Pharmaceutical Sciences, Copenhagen, Denmark*

SARI E. LAURI • *Neuroscience Center and Department of Biosciences, University of Helsinki, Helsinki, Finland*

TOMMY LILJEFORS • *Department of Medicinal Chemistry, The Danish University of Pharmaceutical Sciences, Copenhagen, Denmark*

CLAUS JUUL LOLAND • *Molecular Neuropharmacology Group, Department of Pharmacology, The Panum Institute, University of Copenhagen, Copenhagen, Denmark*

MARTIN MORTENSEN • *Department of Pharmacology, The Danish University of Pharmaceutical Sciences, Copenhagen, Denmark*

RENAE M. RYAN • *Department of Pharmacology, University of Sydney, Sydney, Australia*

ALAN SARUP • *Department of Pharmacology, The Danish University of Pharmaceutical Sciences, Copenhagen, Denmark*

ARNE SCHOUSBOE • *Department of Pharmacology, The Danish University of Pharmaceutical Sciences, Copenhagen, Denmark*

LUCY SKRABANEK • *Department of Physiology and Biophysics and Institute for Computational Biomedicine, Mount Sinai School of Medicine, New York, NY*

SIGNE Í STÓRUSTOVU • *Department of Molecular Pharmacology, H. Lundbeck A/S, Valby, Denmark*

SALLY ANNE THOMPSON • *Department of Pharmacology, MSD Neuroscience Research Centre, Harlow, UK*

ROBERT J. VANDENBERG • *Department of Pharmacology, University of Sydney, Sydney, Australia*

IRACHE VISIERS • *Department of Physiology and Biophysics, Mount Sinai School of Medicine, New York, NY, and Millenium Pharmaceuticals, Cambridge, MA*

KEITH A. WAFFORD • *Department of Pharmacology, MSD Neuroscience Research Centre, Harlow, UK*

HAREL WEINSTEIN • *Department of Physiology and Biophysics, and Institute for Computational Biomedicine, Weil Cornell College of Medicine, New York, NY*

COLOR PLATES

Color Plates 1–11 appear in an insert following p. 84.

I

MOLECULAR PHARMACOLOGY OF GLUTAMATE RECEPTORS

1

Computational Studies of Ligand–Receptor Interactions in Ionotropic Glutamate Receptors

Jeremy R. Greenwood and Tommy Liljefors

1. INTRODUCTION

The determination by Gouaux and co-workers of the three-dimensional (3D) structure of a construct corresponding to the ligand-binding domain (S1S2) of the iGluR2 ionotropic glutamate receptor subunit constitutes a breakthrough in glutamate receptor research. Since the first publication of an iGluR2 construct (S1S2I) in complex with kainate *(1)*, a number of X-ray structures based on a slightly modified construct (S1S2J) have been reported, including the ligand-free *apo* form of the protein *(2)* and ligand–iGlur2 complexes involving antagonists as well as agonists. The ligands in these complexes include the agonists *(S)*-glutamate (Glu) *(2)*, *(S)*-2-amino-3-hydroxy-5-methyl-4-isoxazolyl propionic acid (AMPA) *(2)*, kainate (2), *(S)*-2-amino-3-(3-carboxy-5-methyl-4-isoxazolyl)-propionic acid (ACPA), *(3)*, *(S)*-2-amino-3-[3-hydroxy-5-(2-methyl-2*H*-tetrazol-5-yl)isoxazol-4-yl]propionic acid (2-methyltetrazolyl AMPA) *(3)*, *(S)*-2-amino-3-(4-bromo-3-hydroxy-5-isoxazolyl) propionic acid (Br-HIBO) *(3);* and the competitive antagonists 6,7-dinitro-2,3-quinoxalinedione (DNQX) *(2)* and *(S)*-2-amino-3-[5-*tert*-butyl-3-(phosphonomethoxy)-4-isoxazolyl]-propionic acid (ATPO) *(4)*. The structures of these ligands are shown in Fig. 1.

In order to study the structural features responsible for the observed selectivity of Br-HIBO for the iGluR1 subunit over the iGluR3 subunit *(5,6)*, ACPA and Br-HIBO have also been co-crystallized with the mutant (Y702F)iGluR2-S1S2J and the X-ray structures solved *(3)*.

Recently, our understanding of the mechanism of glutamate receptor desensitization has greatly expanded following an investigation of a series of co-crystallized iGluR2-S1S2J mutants: two complexes of the nondesensitizing (L483Y)iGluR2 binding core, one with DNQX and one with AMPA; the rapidly desensitizing (N754D)iGluR2 mutant in complex with kainate; and a ternary complex of both Glu and the allosteric desensitization blocker cyclothiazide (CTZ) with (N754S)iGluR2, a mutation conferring enhanced sensitivity to CTZ *(7)*.

These X-ray structures provide a wealth of information about the interactions of agonists, antagonists, and allosteric modulators with the glutamate receptor subunit

From: *Molecular Neuropharmacology: Strategies and Methods*
Edited by: A. Schousboe and H. Bräuner-Osborne © Humana Press Inc., Totowa, NJ

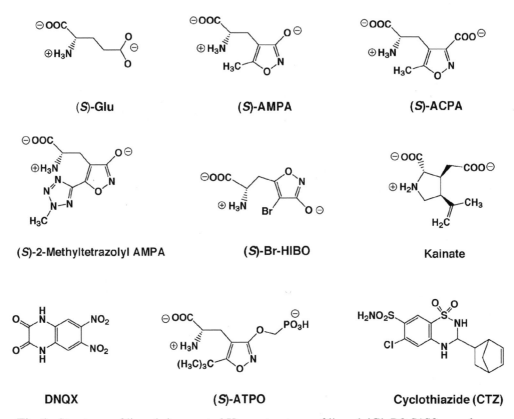

Fig. 1. Structures of ligands in reported X-ray structures of ligand–iGluR2-S1S2 complexes.

iGluR2. Considering the high degree of amino acid sequence identity between ligand binding domains of iGluR subunits in general and between the AMPA-selective subunits iGluR1-4 in particular, much of this information may also be used for analyzing ligand–receptor interactions in subunits other than iGluR2.

In order to fully exploit the information contained in the X-ray structures, interdisciplinary research is required, involving such disciplines as computational chemistry, medicinal chemistry, and molecular pharmacology in partnership with X-ray crystallography. In this context, the X-ray structures provide an excellent springboard for studying structure-activity/selectivity relationships, as well as the structure-based design of new ligands.

This chapter provides examples of the contributions of computational chemistry to such an interdisciplinary research approach. These include:

- Studies of the structural dynamics of ligand-free protein or ligand-protein complexes by molecular dynamics
- The use of homology modeling to develop 3D-structures of subunits for which an experimental structure is not available
- Calculations of optimal hydrogen bonding and hydrogen bond networks
- Identification of energetically favorable sites for water molecules and molecular fragments within the binding cavity, facilitating the design of novel ligands
- Automatic flexible docking of molecules into the binding pocket, e.g., in order to study structure-activity/selectivity relationships
- Computer-aided design of novel ligands

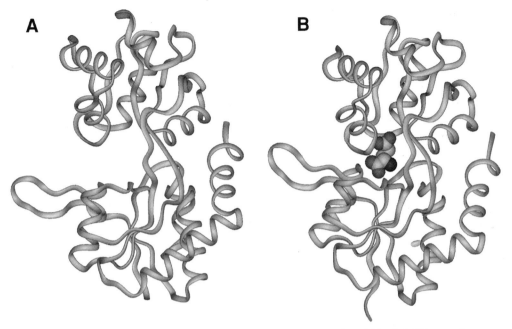

Fig. 2. (**A**) The *apo* structure of iGluR1-S1S2 and (**B**) the Glu–iGlur2-S1S2 complex.

The focus of the present chapter is on the AMPA receptor subunits iGluR1-4, but examples taken from studies of the low-affinity kainate receptor subunits iGluR5-7 are also included.

2. MOLECULAR DYNAMICS SIMULATIONS

A particularly important feature revealed by the ligand-protein complexes of the iGluR2-S1S2 construct is the response of the receptor to the ligand. The domains of the ligand-binding core are in an open state in the *apo* structure, whereas different ligands induce different degrees of domain closure (Fig. 2). The degree of domain closure has been correlated successfully with ligand efficacy, as measured in electrophysiology experiments using homomers of the nondesensitizing (L483Y)iGluR2i(Q) subunit *(3)*. This provides strong evidence that receptor activation occurs as a result of domain closure. Studies on the structural dynamics of the protein and of ligand–protein complexes are thus of great importance for gaining a detailed understanding of receptor activation.

An X-ray structure provides a static picture of a dynamic process. However, the dynamic behavior of the protein may be studied by computational molecular dynamics (MD) simulations. The structural dynamics of the iGlur2-S1S2J construct have recently been studied by Arinaminpathy et al. *(8)* using MD simulations of 2–5 ns duration, including a water box of 11000 explicit water molecules surrounding the protein. The aim of the study was to explore how the motion of the domains is influenced by the ligand, as compared with the motion of the ligand-free *apo* state. The study included the Glu–iGluR2 and kainate–iGluR2 ligand–protein complexes. In addition, they simulated Glu–iGluR2 with the ligand replaced by five water molecules in order to study the feasibility of reproducing the conformational change from an agonist-induced domain-closed state to the open *apo* state.

The simulations (2 ns timescale) show that the motion of domain 2 (surrounding the γ-carboxylate group of Glu) is much larger than that of domain 1 (surrounding the α-amino acid moiety of bound Glu) in agreement with an analysis of the nuclear magnetic resonance (NMR) relaxation dynamics of Glu–iGluR2 *(9)*. Furthermore, the simulations show that the motion of domain 2 relative to domain 1 is smallest for the complex including the full agonist Glu and significantly larger for that including the partial agonist kainate. The *apo* state produced the largest domain motion. These results indicate that by comparison with partial agonists (such as kainate), full agonists (such as Glu) not only induce a larger degree of domain closure but also lower the degree of domain mobility in the closed state.

MD simulations offer the particularly interesting possibility of simulating the process of domain closure (or domain opening). However, in a 5 ns MD simulation of the domain-closed protein structure of the glutamate complex with the ligand replaced by five water molecules, the protein remained in the closed state throughout the simulation. This indicates that simulating the change in conformation from a closed to an open (or open to closed) state requires much longer simulation times than 5 ns.

3. HOMOLOGY MODELING OF iGluR SUBTYPES

Given the ongoing interest in developing subtype-specific ligands for iGluRs, both as pharmacological tools and in the case of antagonists as potential therapeutic agents (some of which have reached clinical trials), it is highly desirable to develop good homology models of the binding sites of all subtypes, with a view to docking studies and structure-based design.

Producing useful homology models of the binding domains of iGluR1,3-7 from the crystal structures of iGluR2 is both easy and difficult. It is easy because the overall homology is high; particularly within the binding sites and particularly between the iGluR1-4 subunits to which AMPA binds selectively (Figs. 3 and 4). It is difficult because the binding site is highly polar and contains ionized residues, which means that water molecules play a role in determining structure, as well as being involved in ligand binding, and such water molecules have been shown to shift substantially depending on the subtype and the ligand. Difficult also because the size and shape of the receptor binding cavity varies dramatically depending on the ligand. The solution here is to have a wide range of structures available displaying different degrees of domain closure, from which to build homology models. Recently, such a range of complexes has become available for iGluR2-S1S2J: from the open *apo* state *(2),* slight closure with the antagonists ATPO *(4)* and DNQX *(2)* (0–4%), substantial closure with the nondesensitising partial agonist kainate *(1,2)* (12%), further closure with partially desensitizing full agonists such as thio-ATPA (18%) *(10),* up to full closure with rapidly desensitizing full agonists such as Glu *(2)* and 2-methyltetrazolyl AMPA *(3)* (20–21%). Aligning iGluR1-7 using the alignment algorithm BLAST *(11)* is straightforward (*see* Fig. 4), and constructing a binding site model of a given subtype is little more than a threading exercise, and can be readily accomplished using such tools as SWISSPROT *(12)*. However, if one wishes to model a particular ligand bound to a particular subtype, a choice of domain closure must first be made. For example, kainate is a partial agonist at iGluR2 but a full agonist at iGluR5-7. Thus, homology models built from kainate–iGluR2 are certainly unsuitable for modeling antagonist binding, and less than ideal for modeling the binding of full ago-

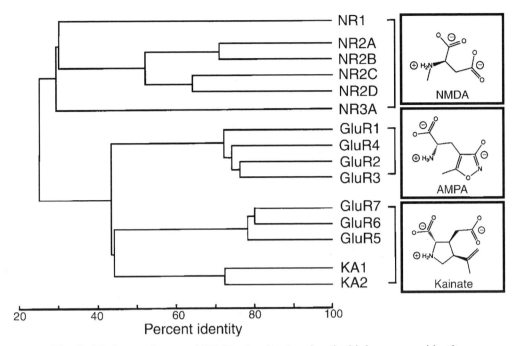

Fig. 3. Phylogenetic tree of iGluR subunits showing the high sequence identity.

nists; a model of kainate-iGluR5 built from kainate-iGluR2 can be expected to underestimate the degree of domain closure induced by kainate at iGluR5 (13).

Having chosen an initial structure and built a homology model, the next challenge is to determine whether the side-chain arrangement in the binding site is appropriate. Typically, side-chain conformations are chosen from libraries (14), with minimal conformational searching and energy minimization. If a flexible side-chain occludes the binding site, preventing the docking of known tight-binding ligands, it has obviously been modeled in the incorrect conformation. Various strategies can be used to improve the side-chain conformations: restrained minimization with or without ligand/waters, restrained Monte Carlo searching, or MD. Though the latter approach has become popular for refining homology models, it is also computationally expensive and perhaps unnecessary for the iGluR binding site where the homology is so high. The first approach was used to model the iGluR3/4-preferring ligand CPW399 bound to iGluR3 (15) and the second was used to improve a homology model of iGluR5, Fig. 5 (13).

The final challenge for producing a good homology model of the binding site is determining the hydrogen-bond network, including water molecules. The binding site can be analyzed using the program GRID (16) to determine the likely location of tightly bound water molecules. Water molecules can also be copied across from the X-ray structure and energy-minimized. A hydrogen-bonding network can then be built manually, or with the help of tools such as WHATIF (17), but ultimately this process is one of educated guesswork. Using a Monte Carlo search to explore the orientation of all rotatable hydroxyls and waters in the vicinity of the binding site is more time-consuming, but gives a clearer picture of the possible hydrogen bond networks and their relative energies. Again, MD simulations with explicit water might be employed to determine average water occupancies and hydrogen bond directions, but given the high

	402	450	478	485	650	654	686	702	705	723
GluR1	L E D	K Y G	A P L T I	V R E	T L E A	G S T K	T T E	A Y L	L E S T M N	G Y G
GluR2	L E S	K Y G	A P L T I	V R E	T L D S	G S T K	T T A	A Y L	L E S T M N	G Y G
GluR3	L E S	K Y G	A P L T I	V R E	T L D S	G S T K	T T A	A F L	L E S T M N	G Y G
GluR4	M E S	K Y G	A P L T I	V R E	T L D S	G S T K	T T A	A F L	L E S T M N	G Y G
GluR5	L E E	K Y G	A P L T I	V R E	A V R D	G G S T M	N S D	A L L	M E S T S I	G Y G
GluR6	L E E	K Y G	A P L T I	V R E	A V E D	G A T M	S N E	A F L	M E S T T I	G Y G
GluR7	L E E	K Y G	A P L T I	V R E	A V K D	G A T M	N N E	A L L	M E S T T I	G Y G
Function	"lock"	Pkt	$-NH_3^+$	$-COO^-$	Pkt	distal anion	"lock"	Wat	$-NH_3^+$	Pkt

Fig. 4. Alignment of amino acid residues surrounding the binding sites of iGluR1-7. Function: "lock," interdomain hydrogen bond; Pkt, amphiphilic or hydrophobic residues bordering the pocket; $-NH_3^+$, COO^-, and distal anion bind the respective charged moieties of glutamate; Wat, binds to a binding-site water molecule in iGluR2.

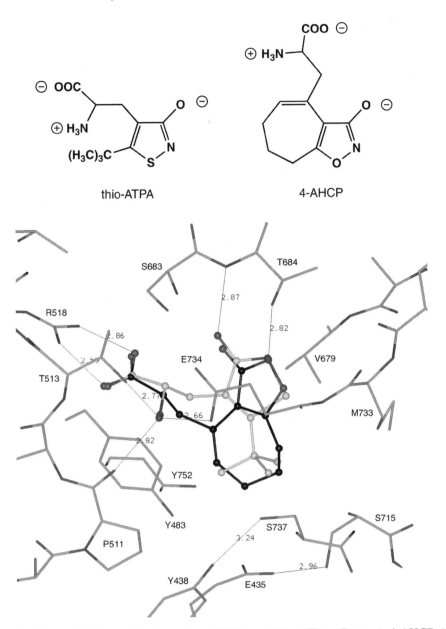

Fig. 5. (Top) iGluR5-specific ligands 4-AHCP and thio-ATPA. (Bottom) 4-AHCP (dark) and thio-ATPA (cyan) docked to a homology model of iGluR5 by Glide *(13)*.

homology of the binding site and the amount of information that can be gained using simpler techniques, as yet no authors have resorted to this technique for improving iGluR homology models.

A number of groups have published homology models of various iGluR binding domains. Kizelsztein et al. *(18)* modeled iGluR1 from iGluR2, finding negligible difference between the two binding sites. Lampinen et al. *(19)* modeled iGluR4 in conjunction with a mutation study, noting the importance of F724 (F702Y in iGluR2). Banke et al. and Campiani et al. modeled and compared iGluR1 with iGluR3 *(6,15)*,

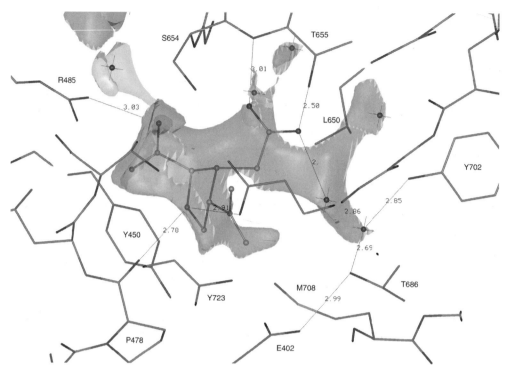

Fig. 6. Kainate–iGluR2-S1S2 complex, including isoenergy contours according to GRID *(16)* – methyl probe (beige, –3.0 kcal/mol), water probe (cyan, –10.0 kcal/mol), and anionic oxygen probe (magenta, –11.0 kcal/mol).

while Brehm et al. compared iGluR2 with iGluR5 *(13)*. Figure 4 shows the alignment of binding site residues and high sequence identity within the binding site as has been generally noted for iGluR1-7.

4. CHARACTERIZATION OF THE BINDING SITE

4.1. The Binding Site of Agonized iGluR2

The GRID map shown in Fig. 6 was calculated when the first structure of the extra-cellular binding domain of iGluR2 (kainate–iGluR-S1S2I) became available *(1)*. The calculation was performed on the empty cavity without ligand or water molecules. There are several features to note: first, the binding site is quite narrow, and kainate fills out most of the available space. In fact, the cavity becomes even more narrow when full agonists such as AMPA bind. Second, the water and anion probes clearly highlight the region occupied by the α-amino acid group, leaving no doubt that this highly conserved receptor motif is optimized for amino acid recognition. Third, any space in the cavity not occupied by kainate is occupied by water molecules, and for a larger ligand to bind as an agonist, it must displace one or more waters with moieties that can out-compete water. In this context, the grid map not only correctly identifies favorable sites for water molecules, it also highlights which ones are more tightly or loosely bound, as indicated here by the size of the blue region surrounding a given water position. For example, a single loosely bound water molecule lies within the region described by the

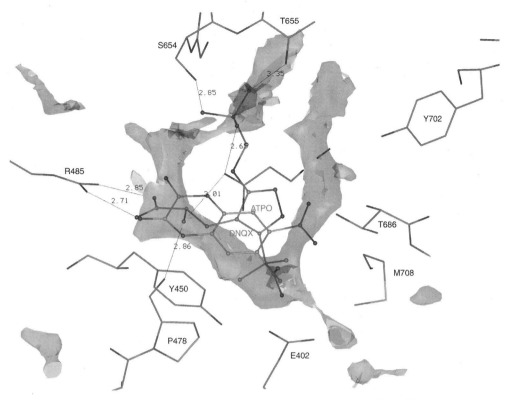

Fig. 7. Overlay of the DNQX and ATPO–iGluR2-S1S2 complexes including isoenergy contours according to GRID-methyl probe (beige, –2.9 kcal/mol), water probe (cyan, –8.5 kcal/mol), and hydrogen phosphate anion probe (magenta, –14.4 kcal/mol).

methyl probe near the distal carboxylate; this is now known to be displaced by the isoxazole heteroatoms of AMPA and the 3-oxyanion of Br-HIBO.

4.2. The Binding Site of Antagonized iGluR2

By contrast with the very tight binding site induced by agonists and partial agonists, with sharply defined pharmacophore requirements and comparatively few opportunities for extension, the *apo* and antagonized states of the receptor allow large ligands from diverse structural classes. To date, crystal structures of only two different antagonists bound to iGluR2 have appeared, in addition to the *apo* structure: the phosphonate AMPA derivative ATPO *(4),* and the quinoxalinedione DNQX. The latter has been co-crystallized with both the native iGluR2 receptor construct *(2)* and a nondesensitizing L483Y mutant *(7);* there are no significant binding site differences between these two, although the ligand and some side chains appear better refined in the latter structure. By comparing the interactions made by these two antagonists, and by conducting GRID analysis of the binding site, a detailed picture of the receptor topography has been established *(4).*

It can be seen at once from the GRID map of the binding site that neither molecule is optimized in terms of occupying the available space in the receptor, nor in terms of making all possible polar contacts (Fig. 7). Neither antagonist approaches nonconserved residues such as Y702 or L650 (below T655 – not shown). By contrast with the role of the 3-isoxazolol of AMPA, which makes strong polar contacts with domain 2 in the AMPA–iGluR2 complex, the isoxazole of ATPO lies in a "dead" zone right in the middle of the receptor. Hence its only contribution is to act as a scaffold for the pharmacophore elements and to provide a slight entropic contribution by excluding disordered water. This strongly suggests that alternative scaffolds could be employed in place of the isoxazole. A synthetic program is in progress to test this hypothesis, with promising preliminary results *(20)*.

5. DOCKING LIGANDS INTO THE iGluR2 BINDING SITE

Flexible automated docking is a computational method for predicting the binding mode of a ligand to a given protein. The protein structure may be experimentally derived (X-ray crystallography, NMR, etc.) or obtained by homology modeling (*see* Subheading 2). By comparison with manual docking, e.g., based on superimposing the ligand on a known template ligand, flexible automated docking is unbiased, in the sense that no assumptions are made about the bioactive conformation of the ligand or its binding orientation. Flexible docking is particularly useful for structure-activity studies, including ligands for which experimental ligand–protein complexes are not available (*see* Subheading 6), for *in silico* screening of potential new ligands, and for studying the effects of mutations on ligand binding, facilitating the interpretation of the results of mutation experiments.

Several computer programs for this purpose are available, including Glide *(21)*, FlexX *(22)*, GOLD *(23)*, and AUTODOCK *(24)*. Docking algorithms identify possible binding poses (a "pose" is a complete specification of the conformation of the ligand, and its the position and orientation relative to the receptor) and then rank (score) the poses using various scoring functions. The aim of a scoring function is to identify the correct binding mode of the ligand by estimating which pose has the most favorable binding energy. For a recent review of ligand docking, *see* ref. *(25)*. It should be noted that a limitation of the majority of the present generation of docking programs is that the geometry of the protein is kept fixed during the docking process.

The developers of docking software generally recommend docking the ligand without including water molecules, even if water molecules are observed in the binding cavity of the experimental structure. The various X-ray structures of ligand–iGluR2 complexes display several water molecules in contact with the ligand. Thus, in the context of docking to iGluR subunits, it is of particular interest to investigate the effect of observed water molecules on the docking results. In the discussion below, only results obtained using the program Glide are discussed.

5.1. Docking DNQX into iGluR2 Excluding Water Molecules

The binding pose of DNQX predicted by Glide to be the best (highest scoring) is extremely close to the experimental one *(26)*, as displayed in Fig. 8. The root mean square deviation (rmsd) between the experimentally observed and docked structures is calculated to be 0.29 Å. The ring structures of the two structures superimpose particularly well, whereas small differences in the conformations about the phenyl-NO_2 bonds are seen. Interestingly, quite some variation is seen in the experimental DNQX conformation among the four crystallographically distinct complexes available, indicating that these bonds in fact show some binding-site flexibility. Because inclusion of water

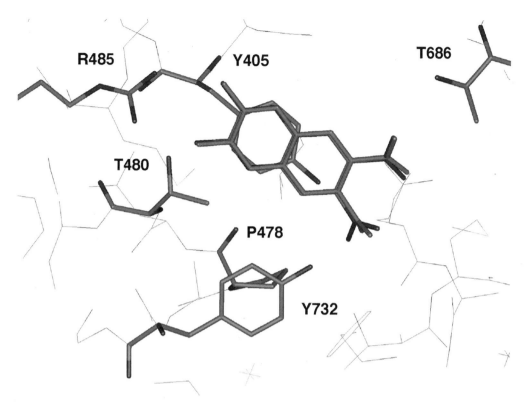

Fig. 8. DNQX docked into the binding pocket of iGluR2 using Glide *(21).* The experimentally observed structure is shown in magenta and the atoms of the docked structure are type-coded.

molecules in the case of DNQX is not required to reproduce the experimental data, the results indicate that the water molecules in the binding cavity of iGluR2 are not major determinants of the binding mode of this antagonist.

5.2. Docking AMPA into iGluR2 With and Without Inclusion of Water Molecules

AMPA provides an interesting test case because it displays a binding mode that differs from those of, e.g., glutamate, kainate, and 2-methyltetrazolyl AMPA at iGluR2.

Figure 9 displays glutamate and AMPA superimposed in their respective binding modes within the iGluR2-S1S2J binding pocket as determined by X-ray crystallography *(2).* Contrary to all expectation, the 3-oxyanion and ring nitrogen atom of AMPA do not superimpose with the carboxylate oxygens of glutamate. A water molecule in the AMPA–iGluR2 complex occupies the position corresponding to the upper carboxylate oxygen of glutamate.

Interestingly, with no water molecules included in the calculations Glide predicts an incorrect binding mode for AMPA that strongly resembles the binding modes displayed by glutamate, kainate, and 2-methyltetrazolyl AMPA, as indicated in Fig. 10A *(26).* The 3-oxyanion and ring nitrogen atom of AMPA are predicted to occupy positions similar to those of the carboxylate oxygens in Glu–iGluR2. The docking algo-

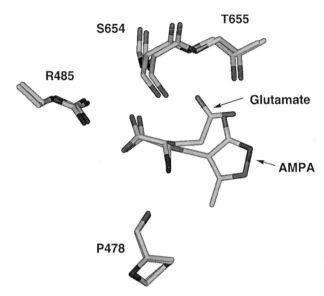

Fig. 9. Superposition of glutamate and AMPA as bound to iGluR2 according to X-ray structures.

rithm identifies the experimentally observed binding mode of AMPA, but only ranks it as number 3.

When all experimentally observed waters in the binding pocket are included in the calculations, the highest-ranked pose superimposes nicely with the experimentally observed binding mode (rmsd = 0.21 Å). This shows that one or more water molecules within the binding pocket are crucial for determining the binding mode of AMPA. When including only the water molecule located between the 3-oxyanion of AMPA and the S654 NH, the pose ranked highest by Glide corresponds closely to the experimentally observed binding mode (rmsd = 0.20 Å) as displayed in Fig. 10B.

Thus, in contrast to the DNQX case previously discussed, it is necessary to take binding-site water molecules into account when docking AMPA into the iGluR2 binding pocket. This implies that for reliable docking of ligands into an iGluR protein, the binding pocket should be carefully analyzed using, e.g., the GRID methodology discussed in Subheading 4, to investigate whether strongly bound water molecules that can a stabilize a particular binding mode may be present.

6. STRUCTURE-ACTIVITY/SELECTIVITY STUDIES

6.1. Structure-Activity Studies

The iGluR2 crystal structure is increasingly being used to dock ligands (Subheading 5) and thus guide synthesis and rationalize observed activities. Figure 11 displays the structures of the ligands discussed in Subheadings 6.1. and 6.2. One of the first examples was the comparison of a series of hydroxyimidazole, 1-hydroxypyrazole, and 1-hydroxytriazole derivatives of AMPA, which are, respectively, inactive, weakly agonistic, and strongly agonistic at AMPA receptors. Simple manual overlay of these near analogs of AMPA with the crystal structure of AMPA bound to GluR2-S1S2 is sufficient to illustrate the source of the structure-activity relationships (SAR) *(27)*.

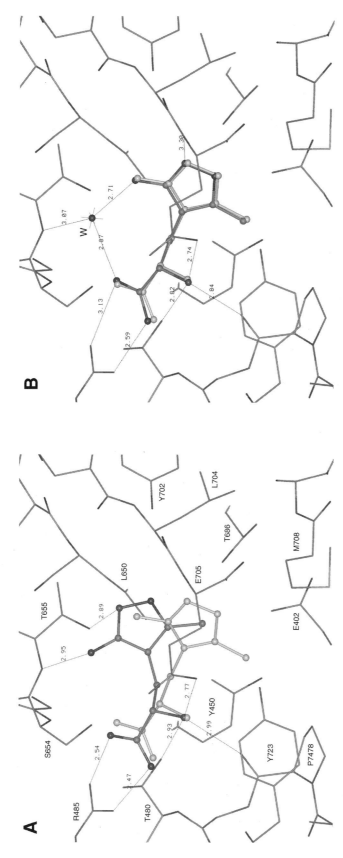

Fig. 10. AMPA docked to iGluR2 using Glide (21). **(A)** Excluding water molecules from the calculations; **(B)** Including one water molecule, marked W. The experimentally observed AMPA structure is shown in magenta and the atoms of the docked structure are type-coded.

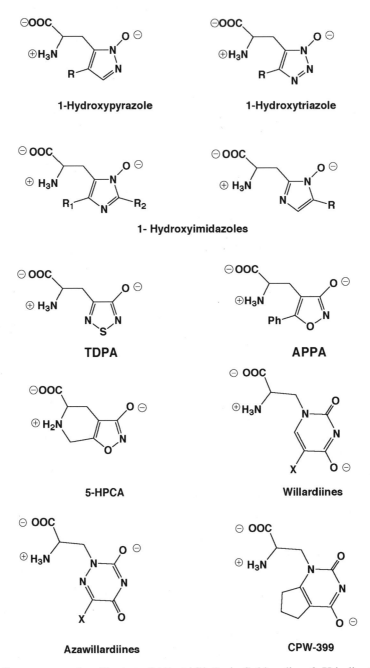

Fig. 11. Structures used to illustrate SAR at iGluRs in Subheading 6. X indicates H, Me, or halogen.

More recently, some AMPA agonists showing unusual stereostructure-activity have been compared using the commercial docking package Glide *(21)*. These include the enantiomers of 2-amino-3-(3-hydroxy-1,2,5-thiadiazol-4-yl)propionic acid (TDPA) and 2-amino-3-hydroxy-5-phenyl-4-isoxazolyl propionic acid (APPA) *(28)*, as well as 3-hydroxy-4,5,6,7-tetrahydroisoxazolo[5,4-*c*]pyridine-5-carboxylic acid (5-HPCA).

TDPA is the thiadiazole analog of AMPA, which is distinguished by the fact that both the *(R)*- and *(S)*-isomers are approximately equipotent *(29)*, whereas only the *(S)*-forms of AMPA and glutamate show significant binding. Conformational analysis reveals that low-energy conformers of both isomers are able to achieve a similar spatial orientation of the charged pharmacophore elements, by dint of the unoccupied 5-position ring nitrogen. Docking illustrates both similarities and differences in the binding modes (Fig. 12).

While *(S)*-TDPA is predicted to occupy the receptor in much the same way as AMPA, including the water molecule between the two anionic moieties and T655, this water is displaced by the β-methylene of *(R)*-TDPA. At the same time, the thiadiazole of the *(R)*-isomer rotates, such that the 2-position ring nitrogen comes nearer to the water interposed between Y702 and the ligand. Approaching Y702 has consequences for AMPA receptor subunit selectivity, discussed in detail in Subheading 6.2., prompting the testing of the enantiomers at cloned homomeric AMPA receptors. Preliminary data indicates that *(R)*-TDPA indeed shows some preference for iGluR1,2, whereas the profile of *(S)*-TDPA is very similar to that of *(S)*-AMPA *(30)*.

The enantiomers of 5-phenyl analog of AMPA, *(S)*-, and *(R)*-APPA have also received attention as displaying interesting stereostructure-activity; the *(S)*-enantiomer is a modestly potent AMPA agonist, whereas the *(R)*-enantiomer displays very weak but measurable antagonism, the originally measured effect of the racemate being one of partial agonism *(31)*. By docking *(S)*- and *(R)*-APPA to the agonized and antagonized forms of iGluR2, respectively *(28)*, the agonism and antagonism of *(S)*- and *(R)*-APPA could be rationalized.

The conformationally constrained bicyclic AMPA analog 5-HPCA presents another interesting case of unusual stereostructure activity. Here, enantioseparation revealed that the *(S)*-isomer was entirely inactive, whereas the *(R)*-form was a weak agonist. This observation was originally met with some scepticism, and a series of crystallographic and spectroscopic experiments were carried out to double check the absolute configuration of the racemates. However the stereoselectivity can be easily validated by docking to iGluR2. Attempting to dock *(S)*-5-HPCA using Glide gives a poor scoring result and an inverted binding mode as the highest-ranked pose, whereas the docked binding mode of *(R)*-5-HPCA is not dissimilar to kainate, despite the opposite configuration of the α-amino acid (Fig. 13). Some parallels can be drawn between the folded binding conformation of *(R)*-TDPA and the conformationally restricted *(R)*-5-HPCA.

6.2. Selectivity Studies

On the basis of homology and pharmacology, we can divide ionotropic glutamate receptors that bind AMPA derivatives into three subclasses: iGluR1/2, iGluR3/4, and iGluR5 (although AMPA itself binds only weakly to iGluR5 and this subunit is normally grouped with iGluR6,7 as a low-affinity kainate receptor). To date, agonists have been discovered with a strong preference for iGluR1/2 (Br-HIBO) *(5)*, for iGluR5 (thio-ATPA) *(32)*, and in a few cases a weak preference for iGluR3/4 (2-methyltetrazolyl AMPA) *(33)*. The binding sites of iGluR1 and iGluR2, likewise those of iGluR3 and iGluR4, are so close as to be indistinguishable by any conceivable agonist. Subunit specific antagonists are conceivable, because the more open binding site can allow lig-

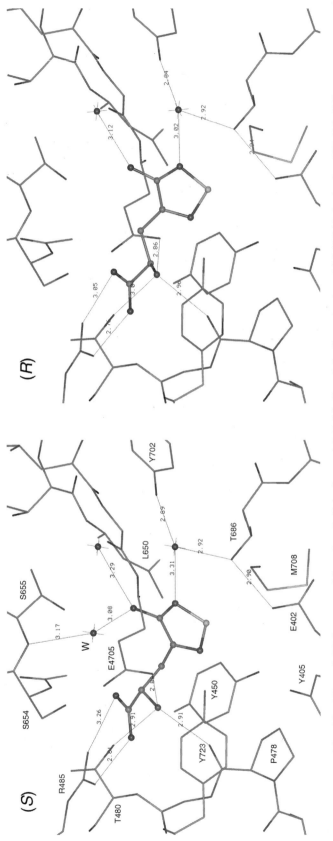

Fig. 12. (S)- and (R)-isomers of 2-amino-3-(3-hydroxy-1,2,5-thiadiazol-4-yl)propionic acid (TDPA) docked to iGluR2 (28).

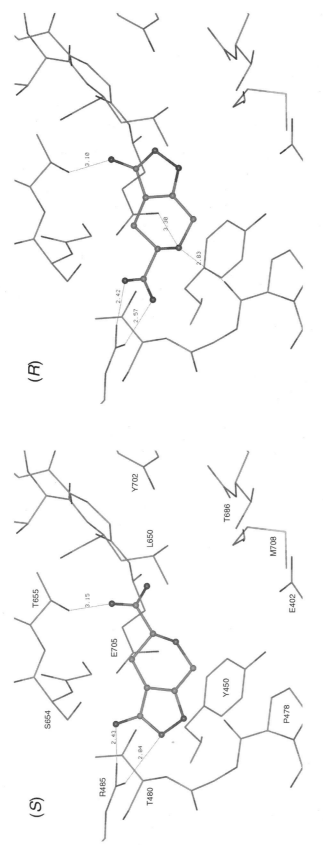

Fig. 13. (*S*)- and (*R*)-5-HPCA docked to iGluR2, according to Glide (*21*).

ands to thread out into less-conserved regions. However, despite the large number of glutamate antagonists known, synthesized in industrial medicinal chemistry programs without the benefit of the new structural data, it has been difficult enough to distinguish between AMPA and kainate receptor antagonism, let alone achieve subunit specificity *(34)*. The existing antagonist scaffolds either cannot or have not been extended in the direction of nonconserved residues. An exception here is Eli Lilly's decahydroquinoline series, where, for example, LY382884 is specific for iGluR5 *(35)*. Clearly there is great opportunity for the structure-based design of subtype-specific AMPA/kainate receptor antagonists, with potential clinical application. To this end, we must first understand the origin of the subtype selectivity displayed by existing ligands. A number of studies have now been published in which models based on the iGluR2 binding site have been used to rationalize activity and selectivity. One of the first such studies was that of Banke et al. *(6)*, which concerned derivatives of the agonist homoibotenic acid (HIBO), in particular Br-HIBO, which had been shown to exhibit binding preference for iGluR1 over iGluR3 *(5)*, as well as different desensitization rates at these receptors. The key finding here, based on homology models and backed up by mutation studies, was that the only significant agonist binding-site difference between the AMPA subunits iGluR1/2 and iGluR3/4 lies in the variability of the tyrosine/phenylalanine at residue 702 (Fig. 4). Moreover, because this residue does not directly interact with any known AMPA ligands, selectivity must depend on the strength of the hydrogen bond network via a water molecule that connects the ligand, Y702, and T686. A similar conclusion was reached by Lampinen et al. *(19)*, who discussed Y702F in the context of an extensive mutation study of binding-site residues.

Banke et al. *(6)* quantified the relationship between iGluR1/2 vs iGluR3/4 selectivity and the Y/F702 switch by measuring the strength of the hydrogen bond between the ligand's distal anionic moiety and the aforementioned water molecule using *ab initio* calculations (density functional theory, DFT), obtaining a good correlation with the experimentally determined selectivity. Though useful, this early model *(see* Fig. 14) was later shown to be imprecise in one respect: although it was clear that the preference of Br-HIBO for iGluR1/2 owed to the strength of the interaction with the interposing receptor water, it was not clear that this water molecule was insufficiently stabilized in iGluR3/4 to be crystallographically observable, as was later shown by the X-ray structure of Br-HIBO bound to (Y702F)iGluR2-S1S2J *(3)*. However, this strengthens rather than negates the role of the hydrogen bond to this water in determining preference for iGluR1/2. Thus, a means of quite accurately predicting the selectivity of agonists for the two classes of AMPA subunits by measuring the relative strength of hydrogen bonding was established, and was soon applied to the willardiine and azawillardiine series *(15,36)* as well as to the design of further HIBO derivatives. 5-Halogenated derivatives of willardiine (1-uracilylalanine) and 6-azawillardiine, the 3,5-(2*H*,4*H*)-dioxo-1,2,4-triazine analog show a remarkable range of potencies and selectivities at iGluR1-5. An almost linear relationship between the size of the 5-halo substituent of the willardiines and the affinity at iGluR5 has been shown, each period increasing the affinity by approximately an order of magnitude *(37)*. This SAR has been discussed with the aid of models *(18,38)*. Here the deciding factor is the size and hydrophobicity of the halogen, which comes into conflict with L650 and T686 in iGluR1-4 as the size increases, but gains favorable van der Waals contacts with the

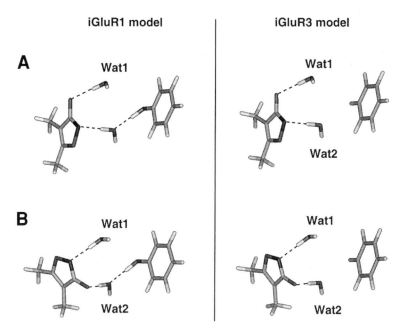

Fig. 14. Simplified perturbative model of AMPA analogs (**A**) and HIBO analogs (**B**) in iGluR1 and iGluR3: effect of swapping orientation of isoxazole ring and mutating tyrosine to phenylalanine, calculated using DFT. $\Delta\Delta E = 1.7$ kcal/mol, i.e., 16-fold relative selectivity.

more open and hydrophobic region around the corresponding valine and serine in iGluR5. Like HIBO derivatives, willardiines also typically show preference for iGluR1/2 over iGluR3/4 via the water network to Y702, as illustrated by the pharmacology and docking of the cyclopentawillardiine CPW-399 *(15)*.

Arriving at a quantitative understanding of the range of observed selectivities of the willardiines between iGluR1/2 and iGluR3/4 is less straightforward; not only does the halogen have an effect, so does subtly changing the electron density of the heterocycle by switching to a triazine. However, a pattern becomes clear if the relative strength of the complex between the heterocycle and the receptor water molecule is calculated using DFT (Fig. 15).

The upshot here is that by controlling the halogen and by altering the azine heterocyclic carboxylate bioisostere among the various diazine- and triazinediones or *N*-oxides, almost any desired combination of subtype specific agonist should be possible, including for iGluR3/4, for which no ligand has yet shown high selectivity.

6. FUTURE PERSPECTIVES

Exploitation of the wealth of information contained in the X-ray structures of ligand–iGluR2 complexes has only recently begun. There is significant scope and opportunity for applying this information to the structure-based design of novel ligands for iGluRs. Such exploration requires interdisciplinary research, including further X-ray crystallography, computational chemistry, medicinal chemistry, and molecular pharmacology. The analysis of the domain-closed agonist complexes discussed previously shows a very narrow binding pocket with limited space to exploit. Thus the design of

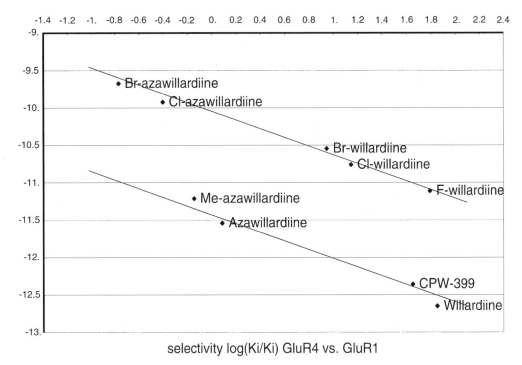

Fig. 15. Correlation between hydrogen bond strength and iGluR1/2 vs 3/4 selectivity among willardiines. Unsubstituted/alkylated willardiines and azawillardiines lie in one series, halogenated derivatives in another. The strength of anion–water complex is calculated as $E[ArO(-)]$ + $E[H_2O]$ – $E[ArO(-)...HOH]$ at B3LYP/6-311+G(d,p).

novel iGluR agonists is a challenging enterprise, because only small modifications with respect to the size and shape of presently known agonists are possible. Nevertheless, as shown earlier, such small modifications may be exploited in the computer-aided design of subunit selective agonists. By contrast, the open *apo* and antagonist states are highly suited to the design of novel iGluR antagonists from widely different structural classes, and the discovery of subunit selective antagonists is within reach.

The determination of the three-dimensional structure of the ternary complex Glu–cyclothiazide–(N754S)iGluR2 and the groundbreaking studies on the mechanism of glutamate receptor densensitization performed by Gouaux and coworkers *(7),* have opened up new and exiting possibilities for the structure-based design of novel allosteric modulators. Together with NMR experiments and molecular dynamics calculations, these structures also represent a step in the direction of understanding the behavior of the full-length iGluR tetramer at the atomic level.

ACKNOWLEDGMENTS

The authors wish to thank Tommy N. Johansen and Darryl S. Pickering, Royal Danish School of Pharmacy, for preliminary data regarding iGluR ligand development and testing; the Lundbeck Foundation for generous financial support, and the Australian Centre for Advanced Computing and Communications (AC3) for donating computing resources.

REFERENCES

1. Armstrong, N., Sun, Y., Chen, G.-Q., and Gouaux, E. (1998) Structure of a glutamate-receptor ligand-binding core in complex with kainate. *Nature* **395,** 913–917.
2. Armstrong, N. and Gouaux, E. (2000) Mechanisms for activation and antagonism of an AMPA-sensitive glutamate receptor: crystal structures of the GluR2 ligand binding core. *Neuron* **28,** 165–181.
3. Hogner, A., Kastrup, J. S., Jin, R., Liljefors, T., Mayer, M. L., Egebjerg, J., et al. (2002) Structural basis for AMPA receptor activation and ligand selectivity: crystal structures of five agonist complexes with the GluR2 binding core. *J. Mol. Biol.* **322,** 93–109.
4. Hogner, A., Greenwood, J. R., Liljefors, T., Lunn, M.-L., Egebjerg, J., Larsen, I. K., et al. (2002) Competitive antagonism of AMPA receptors by ligands of different classes: crystal structures of ATPO bound to the GluR2 ligand-binding core, in comparison with DNQX. *J. Med. Chem.* **46,** 214–221.
5. Coquelle, T., Christensen, J. K., Banke T. G., Madsen, U., Schousboe, A., and Pickering, D. S. (2000) Agonist discrimination between AMPA receptor subtypes. *NeuroReport* **11,** 2643–2648.
6. Banke, T. G., Greenwood, J. R., Christensen, J. K., Liljefors, T., Schousboe, A., and Pickering, D. S. (2001) Identification of amino acid residues in GluR1 responsible for ligand binding and desensitisation. *J. Neurosci.* **21,** 3052–3062.
7. Yu, S., Olson, R., Horning, M., Armstrong, N., Mayer, M., and Gouaux, E. (2002) Mechanism of glutamate receptor desensitization. *Nature* **417,** 245–253.
8. Arinaminpathy, Y., Sansom, M. S. P., and Biggin, B. C. (2002) Molecular Dynamics simulations of the ligand binding domain of the ionotropic glutamate receptor GluR2. *Biophys. J.* **82,** 676–693.
9. McFeeters, R. L. and Oswald, R. E. (2002) Structural mobility of the extracellular ligand-binding core of an ionotropic glutamate receptor. Analysis of NMR relaxation dynamics. *Biochemistry* **41,** 10472–10481.
10. Lunn, M.-L. (2002) From gene to crystal structure. PhD thesis, Royal Danish School of Pharmacy, Copenhagen.
11. Altschul, S. F., Gish, W., Miller, W., Myers, E. W., and Lipman, D. J. (1990) Basic local alignment search tool (BLAST). *J. Mol. Biol.* **215,** 403–410.
12. Schwede T., Diemand A., Guex N., and Peitsch M. C. (2000) Protein structure computing in the genomic era. *Res. Microbiol.* **151,** 107–112. http://www.expasy.ch/swissmod/SWISS-MODEL.html.
13. Brehm, L., Greenwood, J. R., Nielsen, B., Egebjerg, J., Stensbøl, T. B., Bräuner-Osborne, H., et al. (2003) *(S)*-2-Amino-3-(3-hydroxy-7,8-dihydro-6*H*-cyclohept[d]isoxazol-4-yl)propionic acid, a potent and selective agonist at the GluR5 subtype of glutamate receptors. Synthesis, modelling and molecular pharmacology. *J. Med. Chem.* **46,** 1350–1358.
14. Singh, J. and Thornton, J. M. (1992). *Atlas of Protein Side-Chain Interactions,* Vols. 1 and 2. IRL Press, Oxford, UK.
15. Campiani, G., Morelli, E., Nacci, V., Fattorusso, C., Ramunno, A., Novellino, E., et al. (2001) Characterization of the 1*H*-Cyclopentapyrimidine-2,4(1*H,3H*)-dione derivative *(S)*-CPW399 as a novel, potent, and subtype-selective AMPA receptor full agonist with partial desensitization properties. *J. Med. Chem.* **44,** 4501–4504.
16a. Goodford, P. J. (1985) A computational procedure for determining energetically favorable binding sites on biologically important macromolecules. *J. Med. Chem.* **28,** 849–857.
16b. GRID version 20 (2002) Molecular Discovery Ltd., 4 Chandos Street, London, W1A 3AQ.
17. Vriend, G. (1990) A molecular modeling and drug design program. *J. Mol. Graph.* **8,** 52–56.
18. Kizelsztein, P., Eisenstein, M., Strutz, N., Hollmann, M., and Teichberg, V. I. (2000) Mutant cycle analysis of the active and desensitized states of an AMPA receptor induced by willardiines. *Biochemistry* **39,** 12819–12827.

19a.Lampinen, M., Settimo, L., Pentikäinen, O. T., Jouppila, A., Mottershead, D. G., Johnson, M. S., and Keinänen, K. (2002) Discrimination between agonists and antagonists by the α-amino-3-hydroxy-5-methyl-4-isoxazole propionoic acid selective glutamate receptor. A mutation analysis of the ligand-binding domain for GluR-D subunit. *J. Biol. Chem.* **227,** 41,940–41,947.

19b.Pentikäinen, O., Settimo, L., Lampinin, M., Jouppila, A., Keinänen, K., and Johnson, M. S. (2002) Selective agonist binding of AMPA and kainate receptors. P21, Biomolecular Interactions, Molecular Graphics and Modelling Society, Bristol UK, 3–5 April.

20. Jensen, T. V., Jakobsen, K. B., Greenwood, J. R., and Johansen, T. N. (2002) Unpublished data.

21. Glide version 1.8 (2002) Schrödinger, Inc., 1500 S.W. First Avenue, Suite 1180 Portland, OR 97201.

22. Rarey, M., Kramer, B., Lengauer, T., and Klebe G. (1996) A fast flexible docking method using an incremental construction algorithm. *J. Mol. Biol.* **261,** 470–489.

23. Jones, G., Willett, P., and Glen, R. C. (1995) Molecular recognition of receptor sites using a genetic algorithm with a description of desolvation. *J. Mol. Biol.* **245,** 43–53.

24. Morris, G. M., Goodsell, D. S., Halliday, R. S., Huey, R., Hart, W. E., Belew, R. K., and Olsen, A. J. (1998) Automated docking using a Lamarckian genetic algorithm and an empirical binding free energy function. *J. Comput. Chem.* **19,** 1639–1662.

25. Muegge, I. and Rarey, M. (2001) Small molecule docking and scoring in *Rev. Comput. Chem.* Vol. 17, (Lipkowitz, K. and Boyd, D. B., eds.) Wiley-VCH, John Wiley and Sons, Inc., New York, NY, pp. 3–60.

26. Jacobsen, L. (2002) Ligand docking to iGluR2 and PPARγ. Evaluation of the docking program Glide and comparisons with FlexX and GOLD. Master's Thesis, Royal Danish School of Pharmacy, Copenhagen.

27. Stensbøl, T. B., Ulmann, P., Morel, S., Eriksen, B. L., Felding, J., Kromann, H., et al. (2002) Novel 1-hydroxyazole bioisosteres of *(S)*-glutamic acid; synthesis, proteolytic properties and pharmacology. *J. Med. Chem.* **45,** 19–31.

28. Johansen, T. N., Greenwood, J. R., Frydenvang, K., Madsen, U., and Krogsgaard-Larsen, P. (2003). Stereostructure-activity studies on agonists at the AMPA and kainate subtypes of ionotropic glutamate receptors. *Chirality,* **15,** 167–179.

29. Johansen, T. N., Janin, Y. L., Nielsen, B., Frydenvang, K., Bräuner-Osborne, H., Stensbøl, T. B., et al. (2002) 2-Amino-3-(3-hydroxy-1,2,5-thiadiazol-4-yl)propionic acid: resolution, absolute stereochemistry and enantiopharmacology at glutamate receptors. *Bioorg. Med. Chem.* **10,** 2259–2266.

30. Pickering, D. S. (2002) Unpublished data.

31. Ebert, B., Lenz, S. M., Brehm, L., Bregnedal, P., Hansen, J. J., Frederiksen, K., et al. (1994) Resolution, absolute stereochemistry, and pharmacology of the *S*-(+)- and *R*-(–)-isomers of the apparent partial AMPA receptor agonist *(R,S)*-2-amino-3-(3-hydroxy-5-phenylisoxazol-4-yl)propionic acid [*(R,S)*-APPA]. *J. Med. Chem.* **37,** 878–884.

32. Stensbøl, T. B., Jensen, H. S., Nielsen, B., Johansen, T. N., Egebjerg, J., Frydenvang, K., and Krogsgaard-Larsen, P. (2001) Stereochemistry and molecular pharmacology of *(S)*-thio-ATPA, a new potent and selective GluR5 agonist. *Eur. J. Pharmacol.* **411,** 245–253.

33. Vogensen, S. B., Jensen, H. S., Stensbøl, T. B., Frydenvang, K., Bang-Andersen, B., Johansen, T. N., et al. (2000) Resolution, configurational assignment, and enantiopharmacology of 2-amino-3-[3-hydroxy-5-(2-methyl-2*H*-tetrazol-5-yl)isoxazol-4-yl]propionic acid, a potent GluR3- and GluR4-preferring AMPA receptor agonist. *Chirality* **12,** 705–713.

34. Stensbøl, T. B., Madsen, U., and Krogsgaard-Larsen, P. (2002) The AMPA receptor binding site: focus on agonists and competitive antagonists. *Curr. Pharm. Design* **8,** 857–872.

35. Bleakman, R., Schoepp, D. D., Ballyk, B., Bufton, H., Sharpe, E. F., Thomas, K., et al. (1996) Pharmacological discrimination of GluR5 and GluR6 kainate receptor subtypes by (3*S*,4a*R*,6*R*,8a*R*)-6-[2-(1(2)*H*-tetrazol-5-yl)ethyl]decahydroisoquinoline-3-carboxylic acid. *Mol. Pharmacol.* **49,** 581–585.

36. Greenwood, J. R. and T. Liljefors, T. (2001) Using the new X-ray crystal structures of ionotropic glutamate neuroreceptor subunit ligand-protein complexes for understanding subtype selectivity. Molecular Graphics and Modelling Society, Model(l)ing 2001, Erlangen, Germany.

37. Jane, D. E., Hoo, K., Kamboj, R., Deverill, M., Bleakman, D., and Mandelzys, A. Synthesis of willardiine and 6-azawillardiine analogues: pharmacological characterization on cloned homomeric human AMPA and kainate receptor subtypes. *J. Med. Chem.* **40,** 3645–3650.

38. Hogner, A. (2002) AMPA receptor activation and deactivation: a study of protein–ligand interactions of the GluR2 ligand binding core by X-ray crystallography. PhD thesis, Royal Danish School of Pharmacy, Copenhagen.

2

Delineation of the Physiological Role of Kainate Receptors by Use of Subtype Selective Ligands

Sari E. Lauri, Vernon R. J. Clarke, and Graham L. Collingridge

1. INTRODUCTION

Ionotropic glutamate receptors mediate fast excitatory neurotransmission in practically all areas of the central nervous system (CNS). They are also critical for both the induction and expression of synaptic plasticity, and have been implicated in diverse pathological conditions, such as epilepsy, ischemic brain damage, anxiety, and addiction. There are three subtypes of ionotropic glutamate receptors that are named after their high-affinity agonists as α-amino-3-hydroxy-5-methyl-4-isoxazole propionate (AMPA), N-methyl–D-aspartate (NMDA), and kainate (KA) receptors (1).

Whereas the contribution of NMDA and AMPA-type of glutamate receptors to synaptic transmission is well-characterized, much less is known about the functions of kainate receptors. This is mainly owing to significant overlap in the pharmacological profile of the AMPA and KA types of glutamate receptors. Although many antagonists and agonists, including AMPA and KA, have different affinities towards AMPA and KA receptors, few of them are selective enough for pharmacological isolation or activation of one but not the other. Therefore, native receptors have often collectively been referred to as non-NMDA or AMPA/KA receptors, which in most brain areas display properties that are characteristic for AMPA receptors. Elucidation of the physiological functions of native kainate receptors in the CNS has become possible only recently, following the development of selective pharmacological tools as well as genetically engineered mice lacking KA receptor subunits (reviewed in 2,3).

2. MOLECULAR COMPOSITION AND EXPRESSION OF KAINATE RECEPTORS

Molecular cloning techniques have identified five KAR subunits, designated GLU_{K1}, GLU_{K2}, and GLU_{K5-7} (IUPHAR nomenclature [4]; subunits formerly known as KA1, KA2, GluR5-7, respectively [5]). Kainate receptor subunits are comprised of approx 900 amino acids and are thought to share the same transmembrane topology as, and assemble with similar subunit stoichiometry to, other ionotropic glutamate receptors. Thus, each subunit contains four hydrophobic regions within the sequence (designated

From: *Molecular Neuropharmacology: Strategies and Methods*
Edited by: A. Schousboe and H. Bräuner-Osborne © Humana Press Inc., Totowa, NJ

MI-IV), with three believed to form transmembrane domains (MI, III, and IV) with the fourth, MII, forming a reentrant membrane loop. These subunits are believed to assemble as a tetrameric receptor. Additional variation to the primary sequence is created by alternative splicing of GLU_{K5} and GLU_{K7} as well as by RNA editing of the codon that encodes the glutamine/arginine (Q/R) site in GLU_{K5} and GLU_{K6}. Furthermore, the latter can undergo editing at two additional sites, giving rise to further diversity (reviewed in refs. *6–9*).

2.1. Recombinant Receptors

Functional properties of recombinant kainate receptors have been extensively studied in heterologous expression systems, including *Xenopus* oocytes as well as mammalian cells. GLU_{K5}, GLU_{K6}, and GLU_{K7} can form functional homomeric receptors, whereas GLU_{K1} and GLU_{K2} are only expressed as heteromeric combinations with the subunits GLU_{K5-7}. The various glutamate receptor subunits can be distinguished by their agonist affinity as well as desensitization in response to different agonists. For example, GLU_{K5-7} display lower affinity (50–100 nM) to kainate than the high-affinity subunits GLU_{K1} and GLU_{K2} (~ 5 nM), whereas the affinity of the AMPA receptor subunits GLU_{A1-4} to kainate is in the range of 10^{-3}–10^{-5} M. The potency of KA receptors for glutamate is similar to that of AMPA receptors (EC_{50} ~ 0.3 mM), except for the KA receptors containing GLU_{K7} subunits (EC_{50} ~ 6 mM) *(10)*.

Both AMPA and KA receptors desensitize rapidly in the continued presence of glutamate. However, these receptors can be distinguished based on their desensitization in response to kainate. KA receptors typically mediate a rapidly desensitizing current in response to kainate, whereas kainate evokes nondesensitizing agonist responses in AMPA receptors. AMPA has a low activity at homomeric GLU_{K5} but not GLU_{K6} or GLU_{K7}, a phenomenon that appears to be determined by a single amino acid residue. These responses are largely nondesensitizing. In addition, heteromeric $GLU_{K2/K6}$, $GLU_{K1/K7a}$ or $GLU_{K2/K7a}$ (at least) are activated by AMPA *(10)*, suggesting that the introduction of a high-affinity subunit confers sensitivity. In contrast to homomeric GLU_{K5}, AMPA currents in heteromeric $GLU_{K1/K7a}$ and $GLU_{K2/K7a}$ receptors desensitize. Whether the introduction of GLU_{K1} or GLU_{K2} always confers sensitivity to, and desensitization in the presence of, AMPA is not known (for reviews, *see 2,3,6,7*).

2.2. Native Receptors

Kainate receptors are widely expressed in the brain. Early studies using receptor autoradiography identified-high affinity [^{3}H]KA binding sites throughout the nervous system, and in particularly high levels in the stratum lucidum of area CA3 in hippocampus, corresponding to the mossy-fiber nerve terminal region *(11–14)*. Analysis of the mRNA expression of the individual KA receptor subunits by *in situ* hybridization indicated that the expression pattern of the various subunits is distinct but overlapping, and changes during development *(11,15,16)*. In the hippocampus, expression of mRNA for all the subunits has been detected. The GLU_{K1} mRNA occurs mainly in the CA3 field of the hippocampus and dentate gyrus, whereas the GLU_{K2} gene is widely expressed in pyramidal (CA1-CA3) and dentate granule cells. GLU_{K6} is abundantly expressed in pyramidal cells and dentate granule neurons, whereas GLU_{K5} is most prominently expressed in nonpyramidal cells. However, many neurons laying outside the pyramidal cell layer also express GLU_{K6}, and there are some cells within the pyramidal cell layer containing GLU_{K5} transcripts *(15)*. GLU_{K7} appears largely confined to the dentate and scattered interneurones of the stratum oriens. Most patterns of expression are established during late embryonic/early postnatal development. Exceptions include GLU_{K6}, which develops more slowly (postnatal day 12, i.e., *P12*), and a temporally restricted sharp peak in GLU_{K5} expression (P0-5), which declines markedly (by P12) to significant but diminished adult levels *(11)*.

The understanding of the subcellular localization of KA receptor subunits has been hampered by the lack of specific antibodies. By using antibodies recognizing $GLU_{K6/7}$ or $GLU_{K5/6/7}$, it has been shown that kainate receptor immunoreactivity is principally localized at the postsynaptic membranes *(17–19)*. However, in area CA3 unmyelinated axons are also stained *(19)*. Further, using immunogold electron microscopy, expression of $GLU_{K6/7}$ was detected in glutamatergic nerve terminals in the monkey striatum *(20)*.

3. KAINATE RECEPTOR PHARMACOLOGY

Characterization of the physiological roles of kainate receptors in central neurons has been closely linked to the development of selective pharmacological tools for AMPA vs kainate receptor antagonism *(21–23)*. In most central neurons, the response to kainate is dominated by activation of AMPA receptors, and therefore, identification of kainate receptor function is often performed following selective antagonism of the AMPA receptors. In contrast, agonist responses similar to that described for recombinant kainate receptors have been identified in the peripheral nervous system in the dorsal root fibers *(24)* and in the trigeminal ganglia *(25)*. Dorsal root ganglion (DRG) neurons have been widely used as a model for native KA receptors, given that the pharmacological profile of the kainate currents expressed in these neurons closely matches that of homomeric GLU_{K5} *(26)*, and that large amounts of GLU_{K5} mRNA are expressed in neuronal cell bodies *(27)*.

There are several recent reviews that contain detailed descriptions of the development of AMPA- and KA-selective compounds *(2,28–30)*. Here, only an overview of the compounds that have been important in advancing the understanding the physiological functions of the KA receptors is included.

3.1. Competitive Agonists and Antagonists

The "classic" competitive agonists for kainate receptors include kainate and domoate toxins, originally isolated from the red algae *Digenea simplex* and *Chondria armata,* respectively. Most KA receptor-mediated effects show a rank order of agonist potency of domoate > kainate > L-glutamate. The exception within the kainate receptor family appear to be homomeric GLU_{K7} receptors, which display a reduced agonist sensitivity and altered order of potency. Indeed, domoate has no agonist activity at homomeric GLU_{K7} even at 100 μ*M* (*see* **ref. 10**). Kainate and domoate also act on AMPA receptors at higher concentrations. More selective agonists for KA receptors have been developed, and include the AMPA analog (RS)-2-amino-3-(3-hydroxy-5-tert-butylisoxazol-4-yl)propanoic acid (ATPA) *(31),* 5-iodowillardine (5-IW), 5-iodo-6-azawillardine *(32),* and (2S,4R)-4-methylglutamate (SYM2081) *(33)*. Of these, ATPA has been widely used in the hippocampus to study the functional consequences of activating native kainate receptors *(31,34–40)*.

The kainate receptor subunits differ in their responses to a variety of agonists, despite their relatively high primary sequence homology. ATPA and 5-IW are highly active at GLU_{K5} subunits, with practically no binding to GLU_{K6}, and show good selectivity over AMPA receptors *(31,32,41)*. Radioligand studies demonstrate a rank order of binding for ATPA of $GLU_{K5} \gg GLU_{A1-4}$, GLU_{K2} and $GLU_{K7} \gg GLU_{K6}$, GLU_{K6}/GLU_{K2} *(31)*. EC_{50} values obtained from electrophysiological studies where kainate receptor subunits are expressed in isolated cells or primary cultures confirm that ATPA shows good selectivity towards kainate receptors containing, though not necessarily comprising, solely of the GLU_{K5} subunit *(31,15)*. Thus, ATPA acts as a full agonist at homomeric GLU_{K5} ($EC_{50} \sim 2.1\ \mu M$) and in acutely isolated DRG neurones ($EC_{50} \sim 0.6\ \mu M$) *(31)* and as a partial agonist at heteromeric $GLU_{K5/6}$ ($EC_{50} \sim 1.1\ \mu M$) *(15)*. These values were obtained in the presence of concanavalin A, avoiding the complicating issue as to whether this lectin, routinely used to reduce desensitization of kainate receptors *(26,27; see* Subheading 3.2.), increases the affinity of kainate receptors to agonist *(42)* or not *(26,43)*. Although showing no activity at homomeric GLU_{K6} at concentrations of 100 μM–10 m*M* *(15,31,44)*, co-expression of either GLU_{K6}, GLU_{K7}, or GLU_{K2} with GLU_{K5} *(15,44)* does confer sensitivity to ATPA. Perhaps more surprising, ATPA acts as a partial agonist on heteromeric assemblies of GLU_{K2} and GLU_{K6}. The resulting currents are nondesensitizing, differentiating them from those at GLU_{K5}-containing receptors *(15)*. Curiously, this activity is observed despite ATPA having an apparent $K_i > 100\ \mu M$ at these receptors *(31)*.

The quinoxalinedione derivatives 6-cyano-7-nitroquinoxaline (CNQX), 6,7-dinitroquinoxaline-2,3-dione (DNQX), and 2,3-dioxo-6-nitro-1,2,3,4-tetrahydrobenzo[f]quinoxaline-7-sulphonamide (NBQX) are potent competitive antagonists of AMPA and kainate receptors. Of these, NBQX shows the most functional selectivity (approximately threefold) for AMPA over KA receptors *(45),* and has been used in low concentrations (1 μM) to isolate kainate currents in hippocampal interneurons *(46)*.

Recently, decahydroisoquinoxaline derivatives have been identified as glutamate receptor ligands that display selectivity for GLU_{K5} over the other kainate receptor subunits. Of these, LY293558 shows little or no selectivity between homomeric AMPA receptor and GLU_{K5} receptors *(47)*. LY294486 and its active enantiomer LY377770 *(48)* also inhibit AMPA receptors, but show moderate selectivity toward GLU_{K5} *(31)*. Both have been used to study the physiological functions of kainate receptors in the presence of AMPA receptor selective antagonists *(31,34,35)*. The more recently developed LY382884, although less potent on GLU_{K5}-containing receptors expressed in cloned cell lines than LY294486, is the first antagonist that is selective enough to allow the study of the modulation of AMPA-mediated synaptic transmission by GLU_{K5} containing kainate receptors *(39,49,50)*.

3.2. Allosteric Modulators

The first allosteric modulators discovered to act on AMPA/KA receptors were the plant lectins, including concanavalin A (ConA), that block receptor desensitization, probably through binding to *N*-linked oligosaccharides *(51,52)*. It was soon discovered that certain benzothiazides, such as diazoxide and cyclothiazide, also act on the same allosteric site controlling non-NMDA receptor desensitization *(53–55)*. Interestingly, ConA and cyclothiazide show high selectivity for KA and AMPA receptors, respectively *(27)*. This allowed, for the first time, assessment of the contribution made by each receptor type to currents evoked by native receptors (e.g., *26,43,56,57*).

2,3-Benzodiazepines, 1-(4-aminophenyl)-4-methyl-7,8-methylenedioxy-5*H*-2,3-benzodiazepine) (GYKI 52466) and 1-(4-aminophenyl)-3-methylcarbamy-4-methyl-7,8-methylenedioxy-3,4-dihydro-5*H*-2,3-benzodiazepine (GYKI 53655 or LY300168), act as negative allosteric modulators of AMPA receptors *(22,58–60)*. These 2,3-benzodiazepines antagonize kainate receptors only in concentrations well in excess of that required for functional AMPA receptor antagonism, and thus, have enabled pharmacological isolation of currents mediated by KA receptors *(21–23)*. The best selectivity for AMPA over kainate receptor antagonism is provided by GYKI 53655 (and its active isomer LY303070) *(57,61)*.

4. PHARMACOLOGICAL ISOLATION OF KAINATE RECEPTOR-MEDIATED EPSCS

The development of 2,3-benzodiazepines was a key step leading to understanding of the functions of native kainate receptors. GYKI 53655 (10 μM) inhibits currents elicited by AMPA preferring receptors on cortical neurons with at least 200-fold greater potency, compared with those at kainate-preferring receptors on DRG cells *(57)*. This highly selective antagonism of AMPA receptors allowed isolation of native kainate receptor mediated currents in embryonic hippocampal neurons, where application of GYKI 53655 unmasks a small desensitizing response to kainate *(22,62)*.

Synaptic kainate receptor-mediated currents were first described in CA3 pyramidal neurons in response to high frequency stimulation of mossy-fibers *(21,23)*. Application of GYKI53655 in the presence of antagonists to block currents mediated by NMDA, GABA$_A$, and GABA$_B$ receptors inhibited practically all synaptic responses evoked by single pulse stimulation. However, high-frequency mossy-fiber stimulation produced a slow, small amplitude inward current that has a linear I-V relationship and is blocked by AMPA/kainate receptor antagonists NBQX and CNQX, and the GLU$_{K5}$ selective antagonist LY294486 (*see* Fig. 1).

Since their definitive demonstration at the mossy fiber-CA3 synapse *(21,23)*, KA receptor-mediated postsynaptic currents have been described at several synapses in the brain, including the Schaffer collateral-interneuron synapse in the hippocampus *(34,63)*, thalamocortical synapse in the barrel cortex *(64)*, parallel fiber-Golgi cell synapse in the cerebellum *(65)*, external capsule-basolateral amygdala neuron synapse in the amygdala *(66)*, and cone photoreceptors-'Off' bipolar cell synapse in the retina *(67)*. The slow kinetics of the KA receptor-mediated synaptic current differs considerably from the fast onset and rapidly desensitizing response characterized for recombinant kainate receptors. One explanation is that these receptors are located extrasynaptically and their activation depends on glutamate spillover. However, the decay of KA EPSC does not depend on manipulations that modulate glutamate uptake or diffusion in the synaptic cleft at this *(21,23)* or other synapses *(65,68)*. Furthermore, in most synapses described, and also in the mossy fiber-CA3 synapse under certain conditions *(69)*, KA receptor-mediated EPSCs can be evoked by single shock stimulation, arguing against the idea that its kinetic properties are owing to glutamate build-up. In contrast, the functional properties of KA receptors can be altered by their interactions with various cytosolic proteins and by phosphorylation *(70–74)*. Expression of a

A **i) 100 µM PiTX, 50 µM AP5**   **ii) + 50 µM GYKI53655**   **iii) + 10 µM LY294486**

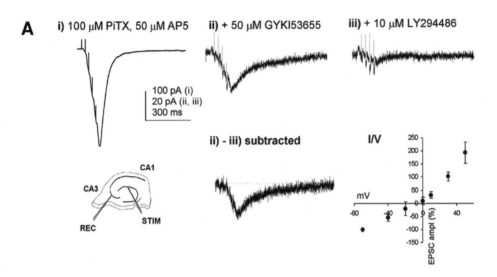

100 pA (i)
20 pA (ii, iii)
300 ms

ii) - iii) subtracted   **I/V**

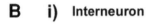

CA1

CA3

REC   STIM

mV

EPSC ampl (%)

B **i) Interneuron**   **ii) CA1 Pyramidal Cell**

GYKI + CNQX   GYKI, GYKI + CNQX

GYKI

control   control

40 pA

50 msec   100 pA

GYKI + CNQX   GYKI, GYKI + CNQX

GYKI

control

8 pA   control   8 pA

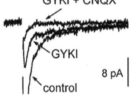

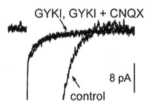
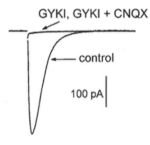

C GYKI + CNQX

GYKI

τ_{slow} = 173.8 ms

Control

τ_{fast} = 3.9 ms   5 pA

50 ms

HP (mV)

60

-80   -40   40

I_{EPSC} (%)

-60

-120

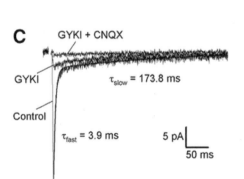
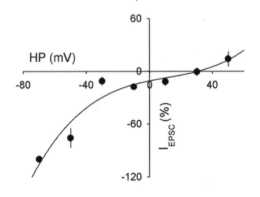

Fig. 1. Pharmacological isolation of kainate receptor-mediated synaptic currents in the central nervous system. **(A)** Synaptic activation of kainate receptors in mossy-fiber synapses in CA3. The traces show EPSCs recorded in the presence of D-AP5 (100 μ*M*) and picrotoxin (100 μ*M*) before **(i)** and after **(ii)** addition of the AMPA-receptor selective antagonist GYKI53655 (50 μ*M*) in response to high-frequency (5 shocks, 100 Hz) stimulation of mossy fibers, and the effect of kainate-receptor selective antagonist LY294486 **(iii).** Although the early phase of the EPSC is mostly blocked by GYKI53655, the late phase is only slightly affected, and is blocked by the AMPA/KA receptor antagonist LY294486. Subtraction of the traces in III–II gives the pure KAR-mediated component of the EPSC which has an approximately linear I–V relationship and a reversal potential close to 0 mV. **(B)** EPSCs in interneurons, but not in pyramidal cells in CA1 have a KA receptor-mediated component. **(i)** Averaged EPSCs recorded from an interneuron, shown at a low (top) and high (bottom) gain, in control conditions, after bath application of 70 μ*M* GYKI 53655, and after addition of 100 μ*M* CNQX. **(ii)** The same experiment done here in pyramidal cells. In these cells, no KAR mediated EPSC could be found. (Reprinted from ref. *(63)* with permission from Nature Publishing Group.) (http://www.nature.com/) **(C)** KAR-mediated synaptic transmission at developing thalamocortical synapses. Superimposed EPSCs in: 'Control' (D-AP5 [100 μ*M*], picrotoxin [50 μ*M*]); 'GYKI' (+GYKI 53655 (25 μ*M* active isomer)); 'GYKI + CNQX' (+ CNQX [100 μ*M*]), and Summary I–V analysis tail of dual-component EPSCs. Continuous lines represent linear regressions. (Reprinted from ref *(64)* with permission from Nature Publishing Group.) (http://www.nature.com/)

different selection of interacting proteins that regulate KA receptor kinetics and targeting might thus provide one explanation for the different functional properties observed in native vs recombinant receptors.

The GLU$_{K5}$ selective antagonists, LY293558 and LY294486, block both postsynaptic currents induced by kainate (200 n*M*) and synaptically released glutamate in CA3 pyramidal neurons *(75)*. The observation that the CA3 region shows a reduced sensitivity to kainate and that synaptically evoked kainate currents are absent in GLU$_{K6}$ knockout mice *(76),* together with the pharmacological profile of these receptors, suggests that they are probably heteromeric assemblies of GLU$_{K5}$/GLU$_{K6}$. However, there are contrasting data that question the contribution of the GLU$_{K5}$ subunit to the EPSCs in CA3. First, expression of GLU$_{K5}$ mRNA in CA3 pyramidal cells is low or is not detected *(11,15)*. Second, ATPA does not depolarize CA3 pyramidal neurons under conditions where large inward currents are induced by nanomolar concentrations of kainate *(35)*. Third, another GLU$_{K5}$ selective antagonist 10 μ*M* LY382884 has little effect on kainate-evoked currents in CA3 *(77)*. However, whereas LY3777770, the active enantiomer of LY294486, and LY382884 show similar potencies at presumed native GLU$_{K5}$-mediated responses in DRG cells (IC$_{50}$ ~ 1 μ*M* *(49,50)*, LY3777770 shows an approx 10-fold greater potency compared with LY382884 at both recombinant homomeric GLU$_{K5}$ and heteromeric GLU$_{K5}$/GLU$_{K6}$ (compare IC$_{50}$s of 0.69 μ*M* and 7.25 μ*M* for LY377770 and LY382884, respectively, at GLU$_{K5}$ and 0.35 μ*M* and 3.61 μ*M* at GLU$_{K5}$/GLU$_{K6}$) *(50)*. Thus, the absence of effect of LY382884 on KA receptor-mediated postsynaptic currents in the CA3 *(77)* is not inconsistent with a role for either homomeric GLU$_{K5}$ or heteromeric GLU$_{K5}$/GLU$_{K6}$. Finally, the action of LY382884, LY293558, and LY294486 on various heteromeric combinations of kainate receptor subunits and on the N-terminal splice variants of GLU$_{K5}$ receptors is not known. This may significantly complicate the interpretation of the pharmacological results in terms of subunit composition of native KA receptors.

On the other hand, the results obtained from the knockout mice are complicated by possible functional compensation between KA receptor subunits. In a recent study *(78)*, the effects of ATPA, LY382884, and LY293558 on kainate currents were studied in dorsal horn neurons of wild-type, GLU_{K5}, and GLU_{K6} knockout mice. LY382884 and LY293558 were effective in antagonizing kainate currents in wild-type and $GLU_{K6-/-}$, but not $GLU_{K5-/-}$, mice *(78)*, thus confirming the selectivity of these compounds on native kainate receptors. Interestingly, the current density in $GLU_{K5-/-}$ mice and wild-types was similar. This shows that GLU_{K5}-dependent functions in wild-type mice can be compensated for in $GLU_{K5-/-}$ mice. Assuming that postsynaptic kainate receptors in CA3 are heteromeric assemblies containing at least GLU_{K5} and GLU_{K6} subunits, in GLU_{K5} knockouts the other kainate receptor subunits would still assemble into functional receptors *(79)*. Moreover, the effects of ATPA and LY382884 in the dorsal horn neurons were greater in $GLU_{K6-/-}$ neurons than in wild-types *(78)*, suggesting that GLU_{K6} deletion increases the GLU_{K5} stoichiometry at the level of individual receptors. This finding further suggests that the stoichiometry of GLU_{K5} influences the antagonism by subunit selective compounds.

5. EVIDENCE FOR PRESYNAPTIC KAINATE RECEPTORS AS MODULATORS OF NEUROTRANSMITTER RELEASE

Presynaptic receptors for neurotransmitters are widespread in the CNS, and can have a crucial role in modulating synaptic transmission *(80)*. Before the synaptic activation of postsynaptic kainate receptors was identified *(21,23)*, a presynaptic locus for the effects of kainate on synaptic transmission in the hippocampus had already been proposed *(81–83)*, supported by biochemical and histological evidence for presynaptic high-affinity KA binding sites *(14)*. It appears that kainate receptors, located presynaptically to the site of recording, modulate transmission at both glutamatergic and GABAergic synapses (reviewed in ref. *84*). Finally, growing evidence supports a role for kainate autoreceptors in the regulation of short- and long-term synaptic plasticity *(49,77,85–87)*.

5.1. Inhibitory Presynaptic Kainate Receptors

Kainate receptors mediate a depression of evoked excitatory synaptic transmission in areas CA1 *(40,88–90)* and CA3 *(35,37,91,92)* of the hippocampus. There is strong evidence that in area CA1 the locus of this effect is presynaptic. Thus, activation of kainate receptors depresses release of L-glutamate from synaptosomes *(88)* and depresses both NMDA and AMPA receptor-mediated components of the evoked EPSC in parallel *(88,90)*. Furthermore, the effects of kainate receptor activation on excitatory synaptic transmission in CA1 are associated with changes in presynaptic Ca^{2+} *(89)*, an increase in paired-pulse facilitation *(35,88,89)*, and a reduction in quantal content, as assessed using $1/CV^2$, but no change in mEPSC amplitude *(90)*.

There is growing evidence to suggest that, within the hippocampus, kainate and ATPA may depress fast excitatory transmission by distinct mechanisms. For example, in area CA3, kainate exerts its effects via an action on presynaptic mossy-fiber excitability *(37,92)*, whereas ATPA is suggested to act indirectly via an increase in interneuronal excitability and the subsequent heterosynaptic action of GABA at presynaptic $GABA_B$ receptors on mossy-fiber terminals *(37)*. Notably, these experiments were done in the presence of elevated extracellular calcium concentration (4 m*M*). At 2 m*M* calcium, $GABA_B$ receptor antagonists have no effect on ATPA-induced depression in area CA3 *(39)*. Similarly, in area CA1, the effects of ATPA are independent of GABA release *(40,88)*. However, in area CA1, depression of glutamatergic transmission in response to kainate and ATPA can be differentiated based on their sensitivity to LY382884 as well as on extracellular calcium concentration *(40)*.

Kainate and ATPA also mediate a depression of evoked GABAergic synaptic transmission in area CA1 *(31,93)* (Fig. 2). The effects of either ligand on GABAergic transmission can be antagonized using the GLU_{K5} receptor antagonist LY294486 *(31)*, implying that, unlike the effects on evoked excitatory transmission in CA1, the effects of either agonist are on the same receptor subtype. The ability of KA receptor agonists to depress both the $GABA_A$ and $GABA_B$ components of the evoked IPSP in parallel and by similar magnitudes suggests that the effect is owing to a presynaptic decrease in GABA release *(31)*. This depression is thought to involve a metabotropic kainate receptor *(36,94)*, an argument lent more weight by the demonstration that kainate receptors can couple to G(i)/G(o) proteins *(95)*.

The simplest explanation for the effect of KA receptor agonists on evoked GABAergic synaptic transmission would be a direct regulation of GABA release probability by kainate receptors located at the axonal presynaptic terminals of interneurones. In support, a modest reduction in mini IPSC frequency in CA1 pyramidal cells has been reported following kainate application *(34,93)*, whereas other studies have observed no effect *(63,65,96)*. The direct depolarization of interneurones via the activation of dendritic or somatic kainate receptors *(34,63)* results in an increase in spontaneous action potential discharge and consequent increase in spontaneous IPSC frequency observed to kainate and ATPA application *(34,36,63,65,83,96)*. Such increases in spontaneous GABA release may underlie alternative indirect mechanisms for the depression of evoked transmission. In support, by mimicking the increase in firing frequency of interneurones seen on kainate application, an analogous depression of the evoked IPSC occurs *(36,46,63)*. Furthermore, raising the threshold for interneurone action potential generation by increasing external divalents abolished the depression of the evoked IPSC by kainate *(63)*. It has been proposed that the increase in GABA concentration as a result of spontaneous interneurone activity leads to a direct reduction of GABA release following the activation of presynaptic $GABA_B$ receptors and a passive shunting of the postsynaptic GABAergic response via the activation of postsynaptic $GABA_A$ receptors *(97)*. However, such an explanation appears insufficient to account for the observed effects of kainate receptor activation on GABAergic transmission in area CA1. Thus, (1) an earlier study observed no effect of kainate on iontophoretically applied GABA *(83)*, arguing against a passive shunt contribution. (2) The magnitude of depression of the evoked $IPSP_B$ was unaltered following pharmacological antagonism of $GABA_A$ receptors *(31)*. (3) The two effects, namely interneurone depolarization and depression of evoked GABAergic transmission, can be dissociated pharmacologically *(34,36)* and couple to separate signaling systems *(36)*. For example, low doses of ATPA (1 μ*M*) have profound effects on interneurone excitability, while having no significant effect on evoked GABAergic responses or postsynaptic properties *(34)*. (4) Previous

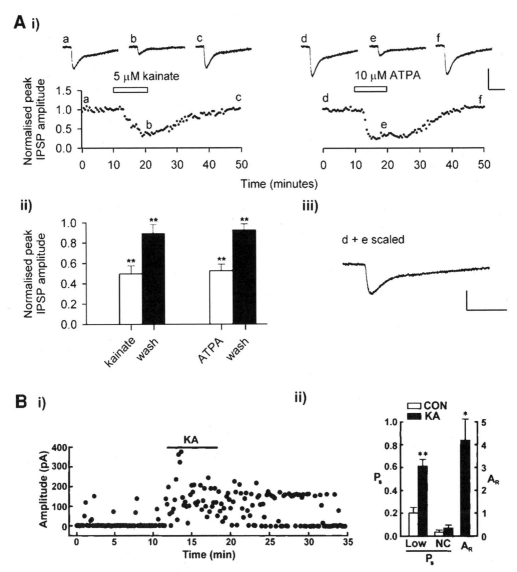

Fig. 2. Effects of presynaptic kainate receptors on GABAergic transmission in CA1. **(A)** Activation of GluR5 depresses inhibitory synaptic transmission. **(i)** Single example and **(ii)** pooled data illustrates the depression of monosynaptic IPSPs, in CA1 pyramidal neurons by kainate (5 μ*M*) and ATPA (10 μ*M*). **(iii)** The maximally depressed response, in the presence of ATPA, is scaled and superimposed with a control response. The lack of change in IPSP shape shows parallel inhibition of GABA$_A$ and GABA$_B$ receptor-mediated IPSPs. Scale bars, 5 mV, 100 ms. (Reprinted from ref. *(31)* with permission from Nature Publishing Group.) (http://www.nature.com/) **(B)** Low concentrations of KAR agonists potentiate GABAergic synapses in CA1. **(i)** Dual patch-clamp recordings from a representative low-release probability pair showing the effect of 300 nM KA on uIPSC amplitude against the time. Note the increase in successful events. **(ii)** Pooled data showing that KA induced a significant increase in the P$_s$ and the average amplitude (A$_R$) of uIPSCs in low-P$_s$ pairs. Coincidental sIPSCs in noncoupled pairs (NC) were not significantly increased by 300 nM KA. (Reprinted from ref. *(86)* with permission from Elsevier Science.)

studies have found no effect of GABA$_B$ receptor antagonists on the kainate-induced depression of GABAergic transmission *(31,63)*.

Regardless of whether the depolarization of interneurones and subsequent increase in IPSC frequency and the depression of evoked transmission are manifestations of the same phenomenon, knockout studies involving mice lacking either GLU$_{K5}$ or GLU$_{K6}$ or both (i.e., double knockouts) suggest that all these effects are absent only in mice lacking both subunits, but are comparable to wild-type control in the absence of either GLU$_{K5-/-}$ or GLU$_{K6-/-}$ mice *(46,98)*. Two separate populations of receptor may exist, with one compensating for the loss of the other, as observed recently in the dorsal horn *(78)*. However, these studies, together with the observations that the effects of ATPA and kainate on interneurone excitability *(34)* and depression of evoked IPSPs *(31)* are antagonized by LY293558 and LY294486, respectively, implies that the native KA receptors responsible for these effects are most likely a heteromeric receptor comprised of both GLU$_{K5}$ and GLU$_{K6}$.

5.2. Facilitatory Presynaptic Kainate Receptors

Facilitatory actions of presynaptic kainate receptors *(82,88)* were long overshadowed by the pronounced inhibitory effects of kainate receptor agonists on both glutamatergic and GABAergic transmission. Recently, facilitation of synaptic transmission via activation of a presynaptic kainate receptors has been described in the GABAergic synapses in area CA1 *(86)*, in the spinal cord *(99)*, and probably most thoroughly, in the mossy-fiber synapse in area CA3 *(39,77,100)*.

Kainate (200 n*M*) renders the mossy-fiber axons more excitable, as evidenced by an increase in the presynaptic fiber volley as well as lowered threshold for antidromic action potentials. At the same time, a kainate-induced suppression of synaptic transmission and depression of presynaptic calcium influx was observed *(92)*. In contrast, application of very low concentrations of kainate (50 n*M*) facilitates synaptic transmission at the mossy fibers *(77,100)*. The kainate-induced facilitation is blocked by LY382884, suggesting a role for GLU$_{K5}$-containing receptors *(77)*. However, only depression of transmission has been observed with ATPA *(35)*. Thus, the facilitatory and depressory effects of kainate could be mediated by different receptor populations that have distinct pharmacological properties.

In contrast to the metabotropic effects described for presynaptic kainate receptors in CA1 *(90,94)*, the effects of kainate in CA3 appear to be mediated by direct depolarization of the presynaptic terminals. The kainate-induced facilitation is not sensitive to antagonists of other receptors (e.g., GABA$_B$), and can be mimicked by elevating the extracellular potassium concentration *(77,100)*. It has been proposed that the facilitation is owing to increased calcium influx that is induced by modest depolarization of the terminals by kainate receptors, whereas a strong depolarization, in response to activation of a larger receptor population, causes the sodium channels to inactivate and thereby depresses transmission *(77,84,88,100–102)*.

GABAergic inhibition between interneurones can also be enhanced by glutamate spillover from neighboring excitatory synapses acting on kainate receptors *(38)*. In CA1 interneurones, an increase in spontaneous action potential discharge and consequent increase in spontaneous IPSC frequency is observed to kainate and ATPA application *(34,36,46,63,83,96,98,103)*. These effects are in part owing to direct depolarization of interneurones via somatic and dendritic KA receptors *(34,63; see* Subheading 5.1.), but also an increase in the efficacy of GABA release is thought to contribute. Activation of presynaptic kainate receptors is proposed to increase the probability of GABA release at interneuron-interneuron synapses, evidenced by kainate receptor mediated increase in the mini IPSC frequency and a decrease in the failure rate of the evoked IPSC in CA1 interneurones *(38,46,* but *see 103).* Although GLU_K5-containing receptors are clearly involved in the postsynaptic activation of interneurones and consequent increase in excitability as demonstrated pharmacologically *(34)* and using knockout mice *(46,98),* increases in efficacy are not mimicked by ATPA and appear absent in GLU_K6–/– mice only. Finally, in a manner analogous to that seen at mossy fibers, kainate has also been shown to decrease the threshold of action potential firing by directly depolarizing the interneuronal axons *(103).* Recently, a biphasic effect of kainate on unitary IPSCs has been described at CA1 pyramidal neurons *(86).* KA receptor agonists increased the success rate of uIPSCs, whereas higher concentrations depressed GABAergic transmission, preferentially on synapses with a high release probability *(86).* Although the mechanism behind these effects is unclear, there is evidence suggesting that the KA receptors increase calcium-dependent GABA release via nonmetabotropic mechanisms *(86,103).*

6. ROLE OF KA RECEPTORS IN SYNAPTIC PLASTICITY

The recently developed GLU_K5 selective compound LY382884 *(47)* is the first antagonist that is selective enough for kainate receptors over AMPA receptors to be used to study the functions of native KA receptors in the presence of intact AMPA receptor-mediated transmission. The use of LY382884 has uncovered a role for kainate receptors in the regulation of short- and long-term synaptic plasticity in the mossy-fiber pathway *(49,77)* as well as at thalamocortical synapses *(87)* (Fig. 3)

Fig. 3. Presynaptic kainate receptors in synaptic plasticity. **(A)** Effects of LY382884 on short- and long-term synaptic plasticity at the mossy fibers. **(i)** LY382884 inhibits facilitation of mossy-fiber EPSCs at frequencies of 25 Hz and higher. Traces from recordings in CA3 pyramidal cells in the presence of D-AP5 (100 μ*M*) and picrotoxin (100 μ*M*) show a response to high-frequency (100 and 50 Hz) 5-pulse stimulation of mossy fibers in before (black) and after washing of the kainate receptor antagonist LY382884 (gray). Right, a pooled data showing the frequency dependent inhibition of 5th EPSC by LY382884. **(ii)** LY382884 specifically blocks the induction of the NMDA-receptor-independent mossy-fiber LTP. (Reprinted from ref. *(49)* with permission from Nature Publishing Group.) (http://www.nature.com/) **(iii)** Mossy-fiber LTP occludes the action of the presynaptic facilitatory kainate receptor. Representative traces showing the responses to five shocks at 50 Hz before and after the induction of LTP, and the lack of effect of LY382884 (10 μ*M*). The lower traces are from a control (i.e., nontetanized) input, showing the typical effect of LY382884 on frequency facilitation. The histogram shows pooled data on the amount of frequency facilitation expressed as a percentage of the frequency facilitation during baseline. (Reprinted from ref. *(77)* with permission from Elsevier Science.) **(B)** The synaptic activation of the presynaptic kainate receptor in developing thalamocortical synapses causes depression during high-frequency transmission. **(i)** EPSC amplitude (percentage of first EPSC in the train) during a 100Hz train and the effect of LY382884 (gray). **(ii)** Developmental regulation of the presynaptic kainate receptor; fifth EPSC amplitude (percentage of the first EPSC in the train) vs age for 100 Hz trains. Inset, examples of responses to trains of stimuli at 100 Hz for P5 and P8. (Reprinted from ref. *(87)* with permission from Elsevier Science.)

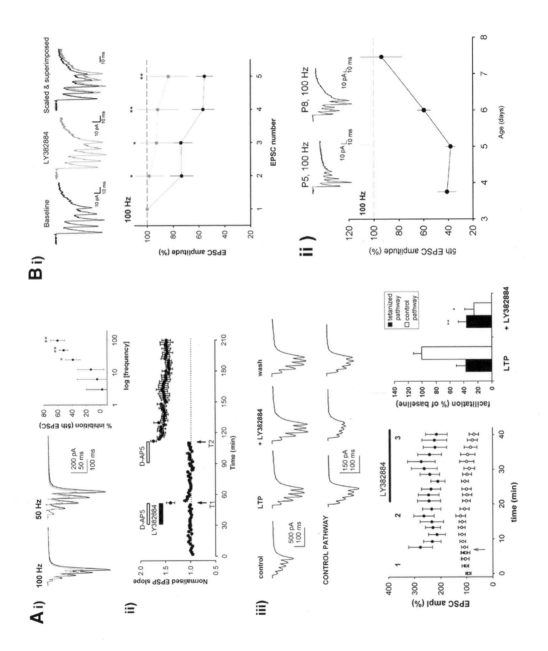

LY382884 at 10 μ*M* has little effect on AMPA receptor-mediated EPSPs, and affected neither monosynaptic GABA$_A$ and GABA$_B$ receptor-mediated synaptic transmission nor passive membrane properties of pyramidal neurons in area CA1 *(49)*. In area CA3, where kainate receptors are expressed in high levels, LY382884 has no effect on low-frequency AMPA-receptor mediated transmission *(39)*, but inhibits pharmacologically isolated kainate EPSCs, induced by brief high-frequency stimulation of mossy fibers by about 40% *(49)*. This effect is not owing to antagonism of the postsynaptic kainate receptor, because LY382884 had little or no effect on the kainate currents in CA3 pyramidal neurons. In contrast, LY382884 significantly inhibited kainate currents at dentate granule cells *(77)*. Together, these data suggests that LY382884 acts as a selective antagonist of presynaptic kainate receptors at mossy fibers *(49,77)*.

Mossy-fiber synaptic transmission is characterized by a large frequency-dependent facilitation, which is mediated by presynaptic mechanisms. The lack of effect of LY382884 on baseline AMPA responses suggested that presynaptic kainate receptors do not contribute to low-frequency synaptic transmission *(49)*. However, LY382884 caused a substantial reduction in the facilitation of AMPA receptor mediated EPSCs at frequencies of 25 Hz or higher *(39,77)*. LY382884 as well as the mixed AMPA/KA antagonist NBQX also blocks facilitation of NMDA EPSCs, induced by a brief 100Hz train and measured as facilitation of a single test pulse 200–1000 ms after the train *(39,100)*. These data suggest that synaptically released glutamate acts on a presynaptic kainate autoreceptor to facilitate mossy-fiber synaptic transmission during high-frequency transmission *(39,77,100)*. This autoreceptor mechanism is thought to act by an ionotropic mechanism, which facilitates presynaptic calcium influx and thereby glutamate release by depolarizing the terminals *(39,77,100)*. In addition, there is evidence suggesting that presynaptic GLU$_K$5-containing kainate receptors on mossy-fiber terminals are directly permeable to Ca^{2+} and linked to calcium release from intracellular stores *(104)*. In support, activation of presynaptic kainate receptors has been shown to amplify presynaptic calcium signals at mossy-fiber terminals *(101)*. Both the inhibition of synaptic facilitation and presynaptic calcium signals by kainate receptor antagonism is evident already on the second stimulus, showing that the kainate autoreceptors can be activated extremely rapidly (< 10 ms) by only a single preceding stimulus *(39,77,101)*. In addition to the homosynaptic facilitatory autoreceptor mechanism, heterosynaptic glutamate release from the associational-commissural fibers has been reported to depress mossy-fiber transmission via activation of a presynaptic kainate receptors *(37)*. These two actions might reflect the concentration-dependent effects (e.g., facilitation vs depression) of glutamate on mossy-fiber transmission.

Kainate receptors have recently been implicated in the induction of LTP in the mossy fibers *(49,85)*. Unlike LTP in the area CA1, induction of mossy-fiber LTP is independent of NMDA-receptor activation and involves presynaptic mechanisms *(105)*. Synaptic activation of the facilitatory presynaptic receptor can account for the role of KA receptors in the induction of mossy-fiber LTP by maintaining a high level of release during high-frequency transmission *(77)*. Furthermore, following induction of LTP, the presynaptic kainate receptor-mediated facilitation of synaptic transmission is lost, suggesting that the mechanism by which presynaptic kainate receptors facilitate transmission is utilized for the expression of mossy-fiber LTP *(77)*. Consistently, LTP also reduces the sensitivity of mossy-fiber transmission to depolarization *(77,106)*.

A role for kainate receptors in the regulation of synaptic plasticity is also emerging in other brain areas *(64,79,86,87,107)*. In developing thalamocortical synapses, the contribution of postsynaptic kainate receptors to transmission decreases during the critical period of development, an effect that can be mimicked by induction of LTP *(64)*. At the same time, kainate autoreceptors mediate short-term depression of synaptic transmission at high frequencies *(87)*. Also this effect is lost after the critical period of experience dependent plasticity and is thought to reflect maturation of the sensory processing network. In the basolateral amygdala, induction of enduring synaptic enhancement by low-frequency stimulation requires activation of GLU_K5 containing KA receptors acting through a calcium-dependent mechanism *(107)*. The induction of this form of heterosynaptic plasticity was blocked by selective antagonists of GLU_K5 kainate receptors and mimicked by the GLU_K5 agonist ATPA. Interestingly, similar to mossy fibers, synaptic plasticity in amygdala is independent of NMDA receptor activation. Whether the requirement of kainate receptors for NMDA-independent forms of synaptic plasticity is a more general principle awaits further studies.

7. CONCLUSIONS

The development of selective pharmacological tools has been critical in advancing the understanding of the physiological functions of kainate receptors. A role for kainate receptors is emerging in both the mediation and modulation of synaptic transmission in several areas of the nervous system. In particular, the recruitment of GLU_K5 containing receptors during high-frequency activity might be important in the control of network excitability, and thus provide an important target for antiepileptic drugs *(50)*.

REFERENCES

1. Watkins, J. C. and Evans, R. H. (1981) Excitatory amino acid transmitters. *Annu. Rev. Pharmacol. Toxicol.* **21,** 165–204.
2. Bleakman, D. and Lodge, D. (1998) Neuropharmacology of AMPA and kainate receptors. *Neuropharmacology* **37,** 1187–1204.
3. Lerma, J., Paternain, A. V., Rodriguez-Moreno, A., and Lopez-Garcia, J. C. (2001) Molecular physiology of kainate receptors. *Physiol. Rev.* **81,** 971–998.
4. Lodge, D. and Dingledine, R. (2000) *The IUPHAR Compendium of Receptor Characterization and Classification,* 2nd ed., IUPHAR Media, London, pp. 189–194.
5. Hollmann, M. and Heinemann, S. (1994) Cloned glutamate receptors. *Annu. Rev. Neurosci.* **17,** 31–108.
6. Bettler, B. and Mulle, C. (1995) Review: neurotransmitter receptors. II. AMPA and kainate receptors. *Neuropharmacology* **34,** 123–139.
7. Chittajallu, R., Braithwaite, S. P., Clarke, V. R. J., and Henley, J. M. (1999) Kainate receptors: subunits, synaptic localization and function. *Trends Pharmacol. Sci.* **20,** 26–35.
8. Dingledine, R., Borges, K., Bowie, D., and Traynelis, S. F. (1999) The glutamate receptor ion channels. *Pharmacol. Rev.* **51,** 7–61.
9. Ozawa, S., Kamiya, H., and Tsuzuki, K. (1998) Glutamate receptors in the mammalian central nervous system. *Prog. Neurobiol.* **54,** 581–618.
10. Schiffer, H. H., Swanson, G. T., and Heinemann, S. F. (1997) Rat GluR7 and a carboxy-terminal splice variant, GluR7b, are functional kainate receptor subunits with a low sensitivity to glutamate. *Neuron* **19,** 1141–1146.

11. Bahn, S., Volk, B., and Wisden, W. (1994) Kainate receptor gene expression in the developing rat brain. *J. Neurosci.* **14,** 5525–5547.

12. Foster, A. C., Mena, E. E., Monaghan, D. T., and Cotman, C. W. (1981) Synaptic localization of kainic acid binding sites. *Nature* **289,** 73–75.

13. Monaghan, D. T. and Cotman, C. W. (1982) The distribution of [3H]kainic acid binding sites in rat CNS as determined by autoradiography. *Brain Res.* **252,** 91–100.

14. Represa, A., Tremblay, E., and Ben Ari, Y. (1987) Kainate binding sites in the hippocampal mossy fibers: localization and plasticity. *Neuroscience* **20,** 739–748.

15. Paternain, A. V., Herrera, M. T., Nieto, M. A., and Lerma, J. (2000) GluR5 and GluR6 kainate receptor subunits coexist in hippocampal neurons and coassemble to form functional receptors. *J. Neurosci.* **20,** 196–205.

16. Wisden, W. and Seeburg, P. H. (1993) A complex mosaic of high-affinity kainate receptors in rat brain. *J. Neurosci.* **13,** 3582–3598.

17. Huntley, G. W., Rogers, S. W., Moran, T., Janssen, W., Archin, N., Vickers, J. C., et al. (1993) Selective distribution of kainate receptor subunit immunoreactivity in monkey neocortex revealed by a monoclonal antibody that recognizes glutamate receptor subunits GluR5/6/7. *J. Neurosci.* **13,** 2965–2981.

18. Petralia, R. S., Wang, Y. X., and Wenthold, R. J. (1994) Histological and ultrastructural localization of the kainate receptor subunits, KA2 and GluR6/7, in the rat nervous system using selective antipeptide antibodies. *J. Comp. Neurol.* **349,** 85–110.

19. Siegel, S. J., Janssen, W. G., Tullai, J. W., Rogers, S. W., Moran, T., Heinemann, S. F., and Morrison, J. H. (1995) Distribution of the excitatory amino acid receptor subunits GluR2(4) in monkey hippocampus and colocalization with subunits GluR5-7 and NMDAR1. *J. Neurosci.* **15,** 2707–2719.

20. Charara, A., Blankstein, E., and Smith, Y. (1999) Presynaptic kainate receptors in the monkey striatum. *Neuroscience* **91,** 1195–1200.

21. Castillo, P. E., Malenka, R. C., and Nicoll, R. A. (1997) Kainate receptors mediate a slow postsynaptic current in hippocampal CA3 neurons. *Nature* **388,** 182–186.

22. Paternain, A. V., Morales, M., and Lerma, J. (1995) Selective antagonism of AMPA receptors unmasks kainate receptor-mediated responses in hippocampal neurons. *Neuron* **14,** 185–189.

23. Vignes, M. and Collingridge, G. L. (1997) The synaptic activation of kainate receptors. *Nature* **388,** 179–182.

24. Agrawal, S. G. and Evans, R. H. (1986) The primary afferent depolarizing action of kainate in the rat. *Br. J. Pharmacol.* **87,** 345–355.

25. Sahara, Y., Noro, N., Iida, Y., Soma, K., and Nakamura, Y. (1997) Glutamate receptor subunits GluR5 and KA-2 are coexpressed in rat trigeminal ganglion neurons. *J. Neurosci.* **17,** 6611–6620.

26. Huettner, J. E. (1990) Glutamate receptor channels in rat DRG neurons: activation by kainate and quisqualate and blockade of desensitization by Con A. *Neuron* **5,** 255–266.

27. Partin, K. M., Patneau, D. K., Winters, C. A., Mayer, M. L., and Buonanno, A. (1993) Selective modulation of desensitization at AMPA versus kainate receptors by cyclothiazide and concanavalin A. *Neuron* **11,** 1069–1082.

28. Bleakman, D., Gates, M. R., Ogden, A. M., and Mackowiak, M. (2002) Kainate receptor agonists, antagonists and allosteric modulators. *Curr. Pharm. Des.* **8,** 873–885.

29. Fletcher, E. J. and Lodge, D. (1996) New developments in the molecular pharmacology of alpha-amino-3-hydroxy-5-methyl-4-isoxazole propionate and kainate receptors. *Pharmacol. Ther.* **70,** 65–89.

30. Lees, G. J. (2000) Pharmacology of AMPA/kainate receptor ligands and their therapeutic potential in neurological and psychiatric disorders. *Drugs* **59,** 33–78.

31. Clarke, V. R. J., Ballyk, B. A., Hoo, K. H., Mandelzys, A., Pellizzari, A., Bath, C. P., et al. (1997) A hippocampal GluR5 kainate receptor regulating inhibitory synaptic transmission. *Nature* **389,** 599–603.

32. Jane, D. E., Hoo, K., Kamboj, R., Deverill, M., Bleakman, D., and Mandelzys, A. (1997) Synthesis of willardiine and 6-azawillardiine analogs: pharmacological characterization on cloned homomeric human AMPA and kainate receptor subtypes. *J. Med. Chem.* **40,** 3645–3650.

33. Gu, Y. P. and Huang, L. Y. (1991) Block of kainate receptor channels by Ca2+ in isolated spinal trigeminal neurons of rat. *Neuron* **6,** 777–784

34. Cossart, R., Esclapez, M., Hirsch, J. C., Bernard, C., and Ben Ari, Y. (1998) GluR5 kainate receptor activation in interneurons increases tonic inhibition of pyramidal cells. *Nat. Neurosci.* **1,** 470–478.

35. Vignes, M., Clarke, V. R. J., Parry, M. J., Bleakman, D., Lodge, D., Ornstein, P. L., and Collingridge, G. L.(1998) The GluR5 subtype of kainate receptor regulates excitatory synaptic transmission in areas CA1 and CA3 of the rat hippocampus. *Neuropharmacology* **37,** 1269–1277.

36. Rodriguez-Moreno, A., Lopez-Garcia, J. C., and Lerma, J. (2000) Two populations of kainate receptors with separate signaling mechanisms in hippocampal interneurons. *Proc. Natl. Acad. Sci. USA* **97,** 1293–1298.

37. Schmitz, D., Frerking, M., and Nicoll, R. A. (2000) Synaptic activation of presynaptic kainate receptors on hippocampal mossy fiber synapses. *Neuron* **27,** 327–338.

38. Cossart, R., Tyzio, R., Dinocourt, C., Esclapez, M., Hirsch, J. C., Ben Ari, Y., and Bernard, C. (2001) Presynaptic kainate receptors that enhance the release of GABA on CA1 hippocampal interneurons. *Neuron* **29,** 497–508.

39. Lauri, S. E., Delany, C., Clarke, V.R J., Bortolotto, Z. A., Ornstein, P. L., Isaac, J. T. R., and Collingridge, G. L. (2001a) Synaptic activation of a presynaptic kainate receptor facilitates AMPA receptor-mediated synaptic transmission at hippocampal mossy fibre synapses. *Neuropharmacology* **41,** 907–915.

40. Clarke, V. R. J. and Collingridge, G. L. (2002) Characterisation of the effects of ATPA, a GLU_K5 receptor selective agonist, on excitatory synaptic transmission in area CA1 of rat hippocampal slices. *Neuropharmacology* **42,** 889–902.

41. Swanson, G. T., Green, T., and Heinemann, S. F. (1998) Kainate receptors exhibit differential sensitivities to (S)-5-iodowillardiine. *Mol. Pharmacol.* **53,** 942–949.

42. Paternain, A. V., Rodriguez-Moreno, A., Villarroel, A., and Lerma, J. (1998) Activation and desensitization properties of native and recombinant kainate receptors. *Neuropharmacology* **37,** 1249–1259.

43. Wong, L. A. and Mayer, M. L. (1993) Differential modulation by cyclothiazide and concanavalin A of desensitization at native alpha-amino-3-hydroxy-5-methyl-4-isoxazolepropionic. *Mol. Pharmacol.* **44,** 504–510.

44. Cui, C. and Mayer, M. L. (1999) Heteromeric kainate receptors formed by the coassembly of GluR5, GluR6, and GluR7. *J. Neurosci.* **19,** 8281–8291.

45. Wilding, T. J. and Huettner, J. E. (1996) Antagonist pharmacology of kainate- and alpha-amino-3-hydroxy-5-methyl-4-isoxazolepropionic acid-preferring receptors. *Mol. Pharmacol.* **49,** 540–546.

46. Mulle, C., Sailer, A., Swanson, G. T., Brana, C., O'Gorman, S., Bettler, B., and Heinemann, S. F. (2000) Subunit composition of kainate receptors in hippocampal interneurons. *Neuron* **28,** 475–484.

47. Bleakman, R., Schoepp, D. D., Ballyk, B., Bufton, H., Sharpe, E. F., Thomas, K., et al. (1996) Pharmacological discrimination of GluR5 and GluR6 kainate receptor subtypes by (3S,4aR,6R,8aR)-6-[2-(1(2)H-tetrazole-5-yl)ethyl]decahyd roisdoquinoline-3 carboxylic-acid. *Mol. Pharmacol.* **49,** 581–585.

48. O'Neill, M. J., Bond, A., Ornstein, P. L., Ward, M. A., Hicks, C. A., Hoo, K., et al. (1998) Decahydroisoquinolines: novel competitive AMPA/kainate antagonists with neuroprotective effects in global cerebral ischaemia. *Neuropharmacology* **37,** 1211–1222.

49. Bortolotto, Z. A., Clarke, V. R., Delany, C. M., Parry, M. C., Smolders, I., Vignes, M., et al. (1999) Kainate receptors are involved in synaptic plasticity. *Nature* **402,** 297–301.

50. Smolders, I., Bortolotto, Z. A., Clarke, V. R. J., Warre, R., Khan, G. M., O'Neill, M. J., et al. (2002) Antagonists of GLU(K5)-containing kainate receptors prevent pilocarpine-induced limbic seizures. *Nat. Neurosci.* **5,** 796–804.

51. Mayer, M. L. and Vyklicky, L., Jr. (1989) Concanavalin A selectively reduces desensitization of mammalian neuronal quisqualate receptors. *Proc. Natl. Acad. Sci. USA* **86,** 1411–1415.

52. Thalhammer, A., Everts, I., and Hollmann, M. (2002) Inhibition by lectins of glutamate receptor desensitization is determined by the lectin's sugar specificity at kainate but not AMPA receptors. *Mol. Cell. Neurosci.* **21,** 521–33.

53. Yamada, K. A. and Rothman, S. M. (1992) Diazoxide blocks glutamate desensitization and prolongs excitatory postsynaptic currents in rat hippocampal neurons. *J. Physiol.* **458,** 409–423.

54. Palmer, A. J. and Lodge, D. (1993) Cyclothiazide reverses AMPA receptor antagonism of the 2,3-benzodiazepine, GYKI 53655. *Eur. J. Pharmacol.* **244,** 193–194.

55. Patneau, D. K., Vyklicky, L., Jr., and Mayer, M. L. (1993) Hippocampal neurons exhibit cyclothiazide-sensitive rapidly desensitizing responses to kainate. *J. Neurosci.* **13,** 3496–3509.

56. Pook, P., Brugger, F., Hawkins, N. S., Clark, K. C., Watkins, J. C., and Evans, R. H. (1993) A comparison of the actions of agonists and antagonists at non-NMDA receptors of C fibres and motoneurones of the immature rat spinal cord in vitro. *Br. J. Pharmacol.* **108,** 179–184.

57. Wilding, T. J. and Huettner, J. E. (1995) Differential antagonism of alpha-amino-3-hydroxy-5-methyl-4- isoxazolepropionic acid-preferring and kainate-preferring receptors by 2,3-benzodiazepines. *Mol. Pharmacol.* **47,** 582–587.

58. Tarnawa, I., Farkas, S., Berzsenyi, P., Pataki, A., and Andrasi, F. (1989) Electrophysiological studies with a 2,3-benzodiazepine muscle relaxant: GYKI 52466. *Eur. J. Pharmacol.* **167,** 193–199.

59. Donevan, S. D. and Rogawski, M. A. (1993) GYKI 52466, a 2,3-benzodiazepine, is a highly selective, noncompetitive antagonist of AMPA/kainate receptor responses. *Neuron* **10**(1), 51–59.

60. Zorumski, C. F., Yamada, K. A., Price, M. T., and Olney, J. W. (1993) A benzodiazepine recognition site associated with the non-NMDA glutamate receptor. *Neuron* **10,** 61–67.

61. Bleakman, D., Ballyk, B. A., Schoepp, D. D., Palmer, A. J., Bath, C. P., Sharpe, E. F., et al. (1996) Activity of 2,3-benzodiazepines at native rat and recombinant human glutamate receptors in vitro: stereospecificity and selectivity profiles. *Neuropharmacology* **35,** 1689–1702.

62. Wilding, T. J. and Huettner, J. E. (1997) Activation and desensitization of hippocampal kainate receptors. *J. Neurosci.* **17,** 2713–2721.

63. Frerking, M., Malenka, R. C., and Nicoll, R. A. (1998) Synaptic activation of kainate receptors on hippocampal interneurons. *Nat. Neurosci.* **1,** 479–486.

64. Kidd, F. L. and Isaac, J. T. R. (1999) Developmental and activity-dependent regulation of kainate receptors at thalamocortical synapses. *Nature* **400,** 569–573.

65. Bureau, I., Dieudonne, S., Coussen, F., and Mulle, C. (2000) Kainate receptor-mediated synaptic currents in cerebellar Golgi cells are not shaped by diffusion of glutamate. *Proc. Natl. Acad. Sci. USA* **97,** 6838–6843.

66. Li, H. and Rogawski, M. A. (1998) GluR5 kainate receptor mediated synaptic transmission in rat basolateral amygdala in vitro. *Neuropharmacology* **37,** 1279–1286.

67. DeVries, S. H. and Schwartz, E. A. (1999) Kainate receptors mediate synaptic transmission between cones and 'Off' bipolar cells in a mammalian retina. *Nature* **397,** 157–160.

68. Kidd, F. L. and Isaac, J. T. (2001) Kinetics and activation of postsynaptic kainate receptors at thalamocortical synapses: role of glutamate clearance. *J. Neurophysiol.* **86,** 1139–1148.

69. Cossart, R., Epsztein, J., Tyzio, R., Becq, H., Hirsch, J., Ben Ari, Y., and Crepel, V. (2002) Quantal release of glutamate generates pure kainate and mixed AMPA/kainate EPSCs in hippocampal neurons. *Neuron* **35,** 147–159.

70. Raymond, L. A., Blackstone, C. D., and Huganir, R. L. (1993) Phosphorylation and modulation of recombinant GluR6 glutamate receptors by cAMP-dependent protein kinase. *Nature* **361,** 337–341.

71. Wang, L. Y., Taverna, F. A., Huang, X. P., MacDonald, J. F., and Hampson, D. R. (1993) Phosphorylation and modulation of a kainate receptor (GluR6) by cAMP-dependent protein kinase. *Science* **259,** 1173–1175.

72. Traynelis, S. F. and Wahl, P. (1997) Control of rat GluR6 glutamate receptor open probability by protein kinase A and calcineurin. *J. Physiol.* **503,** 513–531.

73. Garcia, E. P., Mehta, S., Blair, L. A., Wells, D. G., Shang, J., Fukushima, T., et al. (1998) SAP90 binds and clusters kainate receptors causing incomplete desensitization. *Neuron* **21,** 727–739.

74. Hirbec, H., Francis, J. C., Lauri, S. E., et al. (2003) Rapid and differential regulation of AMPA and Kainate receptors at hippocampal mossy fibre synapses by PICK1 and GRIP. *Neuron* **37,** 625–638.

75. Vignes, M., Bleakman, D., Lodge, D., and Collingridge, G. L. (1997) The synaptic activation of the GluR5 subtype of kainate receptor in area CA3 of the rat hippocampus. *Neuropharmacology* **36,** 1477–1481.

76. Mulle, C., Sailer, A., Perez-Otano, I., Dickinson-Anson, H., Castillo, P. E., Bureau, I., et al. (1998) Altered synaptic physiology and reduced susceptibility to kainate-induced seizures in GluR6-deficient mice. *Nature* **392,** 601–605.

77. Lauri, S. E., Bortolotto, Z. A., Bleakman, D., Ornstein, P. L., Lodge, D., Isaac, J. T., and Collingridge, G. L. (2001b) A critical role of a facilitatory presynaptic kainate receptor in mossy fiber LTP. *Neuron* **32,** 697–709.

78. Kerchner, G. A., Wilding, T. J., Huettner, J. E., and Zhuo, M. (2002) Kainate receptor subunits underlying presynaptic regulation of neurotransmitter release in the dorsal horn. *J. Neurosci.* **22,** 8010–8017.

79. Huettner, J. E. (2001) Kainate receptors: knocking out plasticity. *Trends Neurosci.* **24,** 365–366.

80. McGehee, D. S. and Role, L. W. (1996) Presynaptic ionotropic receptors. *Curr. Opin. Neurobiol.* **6,** 342–349.

81. Coyle, J. T. (1983) Neurotoxic action of kainic acid. *J. Neurochem.* **41,** 1–11.

82. Kehl, S. J., McLennan, H., and Collingridge, G. L. (1984) Effects of folic and kainac acids on synaptic responses of hippocampal neurones. *Neuroscience* **11(1),** 111–124.

83. Fisher, R. S., and Alger, B. E. (1984) Electrophysiological mechanisms of kainic acid-induced epileptiform activity in the rat hippocampal slice. *J. Neurosci.* **4(5),** 1312–1323

84. Kullmann, D. M. (2001) Presynaptic kainate receptors in the hippocampus: slowly emerging from obscurity. *Neuron* **32,** 561–564.

85. Contractor, A., Swanson, G., and Heinemann, S. F. (2001) Kainate receptors are involved in short- and long-term plasticity at mossy fiber synapses in the hippocampus. *Neuron* **29,** 209–216.

86. Jiang, L., Xu J., Nedergaard, M., and Kang, J. (2001) A kainate receptor increases the efficacy of GABAergic synapses. *Neuron* **30,** 503–513.

87. Kidd, F. L., Coumis, U., Collingridge, G. L., Crabtree, J. W., and Isaac, J. T. (2002) A presynaptic kainate receptor is involved in regulating the dynamic properties of thalamo-cortical synapses during development. *Neuron* **34,** 635–646.

88. Chittajallu, R., Vignes, M., Dev, K. K., Barnes, J. M., Collingridge, G. L., and Henley, J. M. (1996) Regulation of glutamate release by presynaptic kainate receptors in the hippocampus. *Nature* **379,** 78–81.

89. Kamiya, H. and Ozawa, S. (1998) Kainate receptor-mediated inhibition of presynaptic Ca2+ influx and EPSP in area CA1 of the rat hippocampus. *J. Physiol.* **509 (Pt 3),** 833–845.

90. Frerking, M., Schmitz, D., Zhou, Q., Johansen, J., and Nicoll, R. A. (2001) Kainate receptors depress excitatory synaptic transmission at CA3-CA1 synapses in the hippocampus via a direct presynaptic action. *J. Neurosci.* **21,** 2958–2966.

91. Contractor, A., Swanson, G. T., Sailer, A., O'Gorman, S., and Heinemann, S. F. (2000) Identification of the kainate receptor subunits underlying modulation of excitatory synaptic transmission in the CA3 region of the hippocampus. *J. Neurosci.* **20,** 8269–8278.

92. Kamiya, H. and Ozawa, S. (2000) Kainate receptor-mediated presynaptic inhibition at the mouse hippocampal mossy fibre synapse. *J. Physiol.* **523,** 653–665.

93. Rodríguez-Moreno, A., Herreras, O., and Lerma, J. (1997) Kainate receptors presynaptically downregulate GABAergic inhibition in the rat hippocampus. *Neuron* **19,** 893–901

94. Rodríguez-Moreno, A. and Lerma, J. (1998) Kainate receptor modulation of GABA release involves a metabotropic function. *Neuron* **20,** 1211–1218.

95. Cunha, R. A., Malva, J. O., and Ribeiro, J. A. (1999) Kainate receptors coupled to G(i)/G(o) proteins in the rat hippocampus. *Mol. Pharmacol.* **56,** 429–433.

96. Vignes, M. (2001) Regulation of spontaneous inhibitory synaptic transmission by endogenous glutamate via non-NMDA receptors in cultured rat hippocampal neurons. *Neuropharmacology* **40,** 737–748.

97. Frerking, M., Petersen, C. C., and Nicoll, R. A. (1999) Mechanisms underlying kainate receptor-mediated disinhibition in the hippocampus. *Proc. Natl. Acad. Sci. USA* **96,** 12917–12922.

98. Bureau, I., Bischoff, S., Heinemann, S. F., and Mulle, C. (1999) Kainate receptor-mediated responses in the CA1 field of wild-type and GluR6-deficient mice. *J. Neurosci.* **19,** 653–663

99. Kerchner, G. A., Wang, G. D., Qiu, C. S., Huettner, J. E., and Zhuo, M. (2001) Direct presynaptic regulation of GABA/glycine release by kainate receptors in the dorsal horn: an ionotropic mechanism. *Neuron* **32,** 477–488.

100. Schmitz, D., Mellor, J., and Nicoll, R. A. (2001) Presynaptic kainate receptor mediation of frequency facilitation at hippocampal mossy fiber synapses. *Science* **291,** 1972–1976.

101. Kamiya, H., Ozawa, S., and Manabe, T. (2002) Kainate receptor-dependent short-term plasticity of presynaptic Ca2 influx at the hippocampal mossy fiber synapses. *J. Neurosci.* **22(21),** 9237–9243.

102. Kamiya, H. (2002) Kainate receptor-dependent presynaptic modulation and plasticity. *Neurosci. Res.* **42,** 1–6.

103. Semyanov, A. and Kullmann, D. M. (2001) Kainate receptor-dependent axonal depolarization and action potential initiation in interneurons. *Nat. Neurosci.* **4,** 718–723.

104. Lauri, S. E., Bortolotto, Z. A., Bleakman D., et al. (2003) A role for Ca^{2+} stores in kainate-dependent synaptic facilitation and LTP at mossy fibre synapses in the hippocampus. *Neuron* **39,** 327–341.

105. Nicoll, R. A. and Malenka, R. C. (1995) Contrasting properties of two forms of long-term potentiation in the hippocampus. *Nature* **377,** 115–118.

106. Mellor, J., Nicoll, R. A., and Schmitz, D. (2002) Mediation of hippocampal mossy fiber long-term potentiation by presynaptic Ih channels. *Science* **295,** 143–147.

107. Li, H., Chen, A., Xing, G., Wei, M. L., and Rogawski, M. A. (2001) Kainate receptor-mediated heterosynaptic facilitation in the amygdala. *Nat. Neurosci.* **4,** 612–620.

3

Molecular Pharmacology of the Metabotropic Glutamate Receptors

Anders A. Jensen

1. INTRODUCTION

(S)-Glutamate (Glu) is the major excitatory neurotransmitter in the central nervous system (CNS), where it plays a key role in a wide range of brain functions, such as neural plasticity, memory formation, and neural development *(1)*. On the other hand, Glu can also act as a neurotoxin under certain conditions, especially when energy supply is reduced. Excessive glutamatergic signaling has been implicated in acute neurotoxic insults such as ischemia, stroke, and epilepsy, and in multiple chronic neurodegenerative states like Parkinson's disease, amyotrophic lateral sclerosis (ALS), Huntington's chorea, and dementia. Furthermore, glutamatergic mechanisms have been proposed to contribute to psychiatric disorders like schizophrenia and anxiety, and modulation of glutamatergic transmission has been shown to be beneficial on certain forms of pain *(2–4)*.

Glu mediates its synaptic actions through two distinct types of receptors: ionotropic and metabotropic glutamate receptors (iGluRs and mGluRs, respectively) *(3–6)*. The iGluRs are tetrameric ligand-gated ion channels, which directly mediate electrical signaling of nerve cells and the fast excitatory transmission of Glu. The 16 iGluR subunits cloned to date have been divided into three heterogeneous classes based on their respective selective agonists *N*-methyl-D-aspartate (NMDA), 2-amino-3-(3-hydroxy-5-methyl-4-isoxazolyl)-propionate (AMPA), and kainate *(4,5)*. In contrast, the mGluRs are seven transmembrane, G protein coupled receptors (GPCRs), which modulate the electrical signal more indirectly through second-messenger systems *(3,6)*.

The present review will outline the molecular pharmacology of the mGluRs. Particular emphasis will be paid to the molecular mechanisms underlying mGluR signaling and the regulation of it mediated by intracellular proteins.

2. THE MGLUR FAMILY: PHARMACOLOGY, TRANSDUCTION, AND SYNAPTIC LOCALIZATION

The mGluRs belong to family C of the GPCR superfamily, which also includes a calcium-sensing receptor (CaR), two γ-aminobutyric acid type B (GABA$_B$) receptors,

From: *Molecular Neuropharmacology: Strategies and Methods*
Edited by: A. Schousboe and H. Bräuner-Osborne © Humana Press Inc., Totowa, NJ

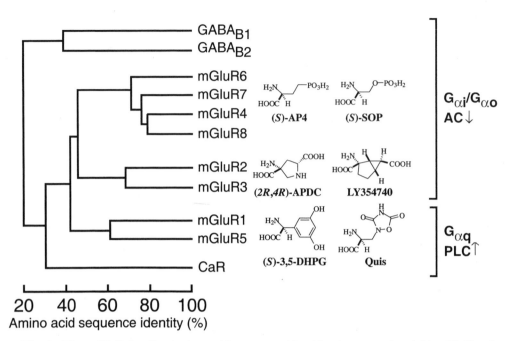

Fig. 1. The mGluR family. Amino acid sequence identities between the eight mGluR subtypes and other family C GPCRs are depicted together with prototypic agonists for and G protein coupling preferences of Group I, II, and III mGluRs.

families of pheromone and taste receptors, and four orphan receptors *(2,3,6–9)*. The eight mGluR subtypes cloned to date are classified into three subgroups based on their amino acid sequence identities, agonist pharmacologies, and signal-transduction properties (Fig. 1). Group I consists of mGluR1 and mGluR5; Group II of mGluR2 and mGluR3; and Group III of mGluR4, mGluR6, mGluR7, and mGluR8. The amino acid sequence identity between mGluRs from the same subgroup is 60–80%, whereas mGluRs from different subgroups only exhibit approx 40% identity (Fig. 1). The presence of splicing sites in the carboxy termini of Group I and III mGluRs give rise to multiple splice variants of these receptors, whereas Group II mGluRs are not subjected to splice variation (Fig. 2).

Because recent comprehensive reviews have described the plethora of mGluR ligands presently available, the medicinal chemistry aspects of mGluRs will not be addressed in depth here *(2,10,11)*. When it comes to the basic agonist pharmacology, Group I mGluRs are selectively activated by *(S)*-3,5-dihydrophenylglycine [*(S)*-3,5-DHPG] and *(S)*-quisqualate (Quis), although Quis also is a potent AMPA receptor agonist. The Group II mGluRs are selectively activated by structurally constrained Glu analogs such as *(2R,4R)*-4-aminopyrrolidine-2,4-dicarboxylate [*(2R,4R)*-APDC] and *(1S,2S,5R,6S)*-2-aminobicyclo[3.1.0]hexane-2,6-dicaboxylate (LY354740) and the Group III mGluRs by *(S)*-4-phosphono-2-aminobutyrate [*(S)*-AP4] and *(S)*-serine-*O*-phosphate [*(S)*-SOP] (Fig. 1). In contrast to these subgroup-selective agonists, very few subtype-selective ligands derived from amino acids have been reported *(2,10,11)*.

The Group I mGluRs are coupled to $G_{\alpha q}$ proteins and stimulation of phospholipase C (PLC). Stimulation of phosphoinositide hydrolysis increases the formation of the

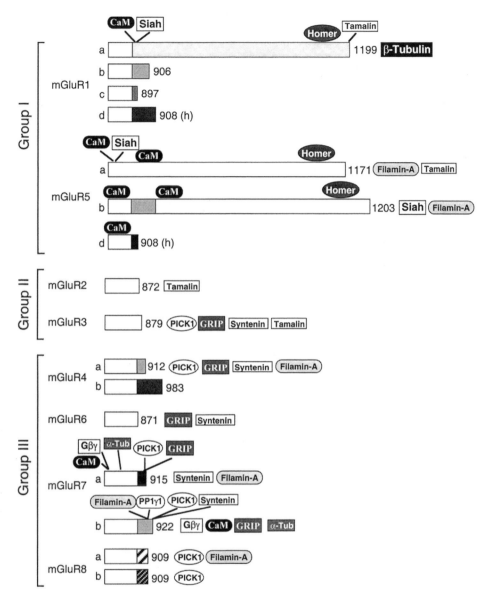

Fig. 2. Splice variations in the carboxy terminal of the mGluR. The carboxy termini of rat mGluR splice variants (or human, indicated with "h") and the intracellular proteins binding to them. The white regions of the termini represent the identical sequences in the splice variants of a certain mGluR subtype. The total numbers of amino acid residues in the splice variants are given at the end of the carboxy termini next to the proteins for which the binding sites in the mGluRs have not been identified. Siah, Seven in absentia homolog-1A; α-Tub, α-tubulin; PP1γ1, catalytic γ1-subunit of protein phosphatase 1C; PICK1, Protein Interacting with C Kinase; GRIP, Glutamate Receptor Interacting Protein; CaM, calmodulin; $G_{\beta\gamma}$, βγ-subunits of G protein.

second messengers inositol-1,4,5-triphosphate (IP_3) and diacylglycerol, which in turn gives rise to numerous metabolic events such as mobilization of Ca^{2+} from intracellular stores and activation of protein kinase C (PKC). Group II and III mGluRs are coupled to $G_{\alpha i}/G_{\alpha o}$ proteins and reduction of adenylyl cyclase (AC) catalyzed cAMP formation,

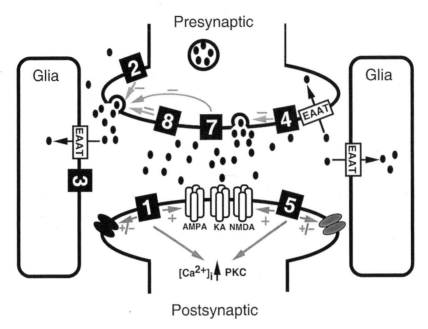

Fig. 3. Synaptic localization of the mGluRs. The predominant localizations of the seven mGluR subtypes expressed in the CNS. The typical localizations of iGluRs and excitatory amino acid transporters (EAATs) are given, and the regulation of Glu release, iGluR signaling and ion channel activities mediated by the mGluRs is shown.

which subsequently affects the activities of a wide range of kinases, phosphatases, and lipases. In addition to these signaling pathways, activation of mGluRs has been demonstrated to stimulate AC, activate cGMP phosphodiesterase, increase arachidonic acid release, and activate phospholipase D through other pathways *(3)*. Furthermore, mGluRs modulate the functions of a wide range of ion channels through G_α as well as $G_{\beta\gamma}$ signaling *(3,12)*. Finally, a recent demonstration of mGluR1 coupling to activation of Src-family tyrosine kinases in the mossy fiber-CA3 synapse in hippocampus underlines the ability of the mGluR to signal through G protein-independent pathways *(13)*.

In the CNS, the mGluRs are localized at pre- and postsynaptic sites of glutamatergic and GABAergic neurons *(3)*. Group I mGluRs are predominantly localized in the peripheral parts of postsynaptic densities, where they increase neuronal excitability through stimulation of PLC, potentiation of AMPA and NMDA receptor signaling, and modulation of other ion channels (Fig. 3) *(12,14–17)*. mGluR2 is localized to preterminal axons of glutamatergic neurons, where it functions as a negative feedback mechanism to attenuate the release of Glu *(15,18,19)*. In contrast, mGluR3 is predominantly localized in glial cells, and its contribution to the regulation of glutamatergic transmission is still poorly understood *(15,18,19)*. Because of their remote localization from the synaptic cleft, it has been suggested that Group II mGluRs only are activated during periods of high synaptic activity *(19)*. In contrast, the Group III receptors mGluR4, mGluR7, and mGluR8 are located in or near presynaptic active zones and function as autoreceptors regulating the release of Glu or other neurotransmitters (Fig. 3) *(15,18,19)*. The localization of mGluR7 at the active zone of the presynaptic terminal and its function as an autoreceptor are consistent with its low sensitivity towards Glu compared with other mGluRs *(2,10,20)*. Finally, mGluR6 is found exclusively in ON bipolar cells in the retina where it is localized postsynaptically, and plays a key role in the transmission between photoreceptor cells and the ON bipolar cells in the visual system *(21)*.

3. SIGNAL TRANSDUCTION THROUGH THE MGLUR

The cloning of the mGluRs and CaR in the early 1990s introduced a unique GPCR type that shared no amino acid sequence identity with any other known protein *(3,6,7)*. The mGluR consists of an unusually large extracellular amino terminal domain (ATD) of ~550 residues; a cysteine rich region (CRR) of 60 residues, including nine highly conserved cysteines; and a seven transmembrane moiety (7TM) constituted by seven membrane-spanning α-helical segments (TMs) connected by alternating intracellular (i1, i2, i3) and extracellular loops (e1, e2, e3), and an intracellular carboxy terminal (Fig. 4).

All family C GPCRs exist as constitutive dimeric (or oligomeric) complexes in the cell membrane, and the dimerization has been demonstrated to crucial for their function. The mGluRs and CaR form homodimers *(22–24)*, whereas the GABA$_B$ and taste receptors undergo heterodimerization *(8,9)*. In contrast to the GABA$_B$ receptor, where the GABA$_B$1 and GABA$_B$2 subunits assemble through a coil-coil interaction between their carboxy termini *(8)*, the mGluR and CaR homodimers are predominantly based on interactions between the two ATDs *(25–28)*. In the following sections on mGluR signaling, studies of CaR and GABA$_B$ receptors will be included when appropriate.

3.1. The Amino Terminal Domain (ATD)

The era of molecular pharmacology studies of mGluRs was initiated in 1993 with two studies by the groups of Nakanishi and O'Hara *(29,30)*. Pharmacological characterization of a series of mGluR1/mGluR2 chimeras revealed that agonist binding to the family C GPCR takes place exclusively to its ATD *(29)*, an observation that subsequently has been supported by studies of mGluR4/mGluR1, CaR/mGluR1, mGluR1/CaR, and GABA$_B$1/mGluR1 chimeras *(31–34)*. At the same time O'Hara and coworkers discovered a weak but significant amino acid sequence identity between the mGluR ATD and a family of bacterial periplasmic binding proteins (PBPs) *(30,35)*. Using the crystal structure of one of these PBPs, the leucine/isoleucine/valine binding protein (LIVBP), as a template, a molecular model of the mGluR1 ATD was constructed, and based on the model residues involved in agonist binding to the receptor were identified *(30)*. Subsequently, crystal structures of LIVBP and other PBPs were used as templates for molecular models of the ATDs of other mGluRs, CaR and GABA$_B$1, as well as of the agonist binding S1S2-domain of the iGluR *(26,36–38)*.

Although these models have been very useful in studies of family C GPCRs, the high-resolution X-ray structures of the mGluR1 ATD homodimer recently published by the groups of Jingami and Morikawa still constitute a milestone in the field *(39,40)*. Leading up to the publication of the crystal structures, it was shown that the ATDs of mGluR1, mGluR4, and mGluR8 could be expressed as soluble homodimeric proteins that were secreted from transfected cells and displayed ligand binding characteristics similar to those of the wild-type receptors *(41–43)*. The crystal structures of the soluble

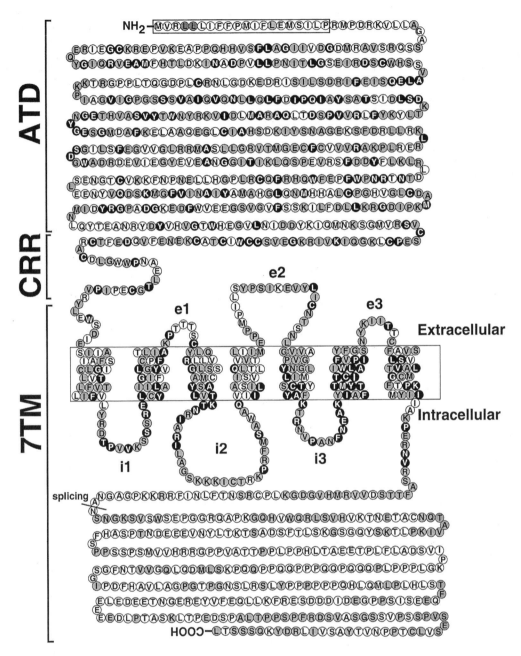

Fig. 4. The topographical structure of the rat mGluR1a in the plasma membrane. The signal peptide, amino terminal domain (ATD), cysteine-rich region (CRR), and seven transmembrane moiety (7TM) are shown. The amino acid residues of mGluR1a identical in all eight mGluR subtypes are given as solid circles, and the additional amino acid residues identical in mGluR1a and mGluR5a are given as gray circles.

mGluR1 ATD protein (the S33-S522 segment of the receptor) have verified the bilobial, clam-shelf structure of the ATD proposed by O'Hara et al. *(30),* where Lobe I is constituted by the S33-V205 and S344-G477 regions and Lobe II by the P206-Q343

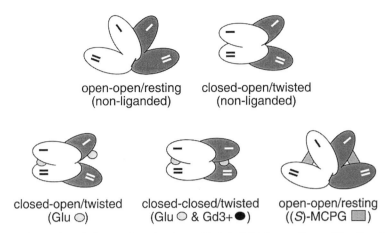

open-open/resting
(non-liganded)

closed-open/twisted
(non-liganded)

closed-open/twisted
(Glu ○)

closed-closed/twisted
(Glu ○ & Gd3+ ●)

open-open/resting
((*S*)-MCPG ▢)

Fig. 5. Conformational fluctuations in the mGluR1 ATD homodimer. The five forms of the mGluR1 ATD homodimer solved as high-resolution X-ray structures. The mGluR1 ATD homodimer is either nonliganded or complexed with Glu, with Glu and Gd^{3+}, or with the competitive antagonist *(S)*-methyl-4-carboxyphenylglycine [*(S)*-MCPG]. The identities of the two ATD lobes (Lobe I and II) are indicated in each of the conformations.

and R478-S522 regions. Furthermore, the structures have refined the information about the agonist binding site (or *orthosteric* site) and have underlined the essential role of homodimerization for mGluR signaling.

3.1.1. mGluR1 ATD Homodimer Conformations

The mGluR1 ATD homodimer has been crystallized in two nonliganded forms and in forms complexed with Glu, with Glu and Gd^{3+}, and with the competitive antagonist *(S)*-methyl-4-carboxyphenylglycine [*(S)*-MCPG] (Fig. 5) *(39,40)*. Analogously to the PBP, each of the two protomers in the mGluR1 ATD homodimer oscillates between an "open" and a "closed" conformation in the absence of agonist (Fig. 5) *(35,39)*. Glu binds in the crevice between the two lobes in the ATD and stabilizes the closed conformation through interactions with residues located on both lobes. Because it is physically impossible for Glu to access its binding site in the closed ATD conformation, the agonist has to bind to the open or an intermediate conformation, after which the ATD closes around it. This speculation is supported by one of the crystal structures, where Glu is bound to an open ATD via interactions with Lobe I residues (Fig. 5) *(39)*.

The equilibrium between the open and the closed ATD bears a resemblance to the widely accepted "two state model" for family A GPCR function *(44)*. Interestingly, this model can also be applied to describe the "intermolecular" conformational changes in the ATD homodimer elicited by the intramolecular events *(39,45)*. The ATD homodimer fluctuates between a "resting" and a "twisted" conformation, and the relative orientation of the ATDs in the two states differs by 70° (Fig. 5). All crystal structures with one or both ATDs closed exist in the "twisted" conformation, so ATD closure seems to trigger the intermolecular rearrangement in the homodimer. Because this twist ultimately is stabilized by Glu binding, the "closed-open/twisted" and/or "closed-closed/twisted" conformations are likely to be the active conformation(s). However, the molecular mechanisms responsible for the translation of this "activation twist" into 7TM signaling are poorly understood (Subheading 3.4.).

A. Agonists

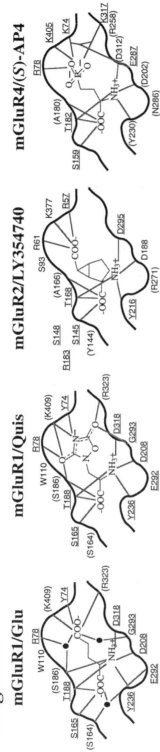

mGluR1/Glu

mGluR1/Quis

mGluR2/LY354740

mGluR4/(S)-AP4

B. Competitive antagonists

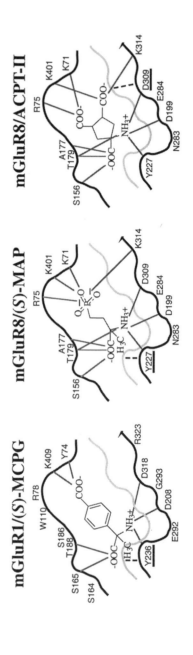

mGluR1/(S)-MCPG

mGluR8/(S)-MAP

mGluR8/ACPT-II

Fig. 6. Orthosteric ligand recognition by the mGluR. Schematic representations of the orthosteric sites of selected mGluRs with agonists (**A**) or competitive antagonists (**B**) bound. The interactions of the compounds with residues in Lobe I and Lobe II (the top and the bottom of the binding pocket, respectively) are depicted. For reasons of clarity, the side chains of the amino acid residues participating in the binding are not depicted. (A) Agonists. The orthosteric sites in the crystal structure of the mGluR1 ATD homodimer complexed with Glu and in molecular models of mGluR1, mGluR2, and mGluR4 ATDs with subgroup-selective agonists Quis, LY354740, and *(S)*-AP4 docked, respectively *(39,46,49,50)*. Residues demonstrated to be crucial for agonist binding affinity and/or potency in mutagenesis studies are underlined, whereas residues for which mutations have little or no impact on agonist binding are given in parentheses. (B) Competitive antagonists. The orthosteric sites in the mGluR1 ATD homodimer complexed with *(S)*-MCPG and in a molecular model of the mGluR8 ATD with competitive antagonists *(S)*-MAP4 and ACPT-II docked *(40,51)*. The positions of the Lobe II residues in the closed ATD conformation are given in gray, and the residues presenting the steric or ionic hindrance to proper ATD closure around the antagonists are underlined.

3.1.2. The Orthosteric Site of the mGluR

According to the crystal structures, the orthosteric site of mGluR1 is constituted by 14 amino acid residues distributed across both sites of the crevice formed by the two ATD lobes (Fig. 6A) *(39)*. The residues on the Lobe I site predominantly form hydrogen bonds to the functional groups of Glu, directly or through water molecules, whereas the Lobe II residues interact with Glu via salt bonds, hydrogen bonds via water molecules, or cation-π interactions. Glu binding is also stabilized by van der Waals interactions with the side chain of W110. Finally, the closed ATD conformation is further stabilized by several interactions between residues on both sides of the crevice (for example, W110-E292 and S166-N235) *(46)*.

In mutagenesis studies, T188, S165, R78, and Y74 in Lobe I and Y236, E292, G293, D208, and D318 in Lobe II have been demonstrated to be crucial residues for [³H]Quis and/or [³H]Glu binding to mGluR1, whereas mutations of S164, S186, K409, and R323 have little or no impact on radioligand binding to the receptor *(30,47,48)*. In functional assays, alanine mutations of T188, Y236, D208, and D318 render mGluR1 virtually nonresponsive to both Glu and Quis, and mutations of S165 and E292 impair receptor signaling significantly *(30,48)*. Interestingly, mutations of Y74, R78, and G293 impair the response elicited by Glu much more severely than that of Quis *(47,48)*. The different contributions of the 3 residues to the potencies of Glu and Quis strongly suggest that the 1,2,4-oxadiazol-3,5-dione group of Quis binds to additional residues in the distal part of the orthosteric site or at least binds in a different fashion than the γ-carboxylate group of Glu. Based on molecular modeling, Bertrand et al. have proposed that the glycine moiety of Quis binds similarly to the receptor as that of Glu, whereas its bulky 1,2,4-oxadiazol-3,5-dione group is able to bind directly to S186, R78, and G293 and not via water molecules (Fig. 6A) *(46)*. Thus, binding to residues on both side of the crevice seems to be optimized with Quis, and the fact that Quis fits better in the closed ATD conformation than Glu would explain its higher affinity and potency at mGluR1 than the endogenous agonist *(46)*.

Like the PBP structures before them, the crystal structures of the mGluR1 ATD homodimer have been applied as templates for molecular models of Group II and III mGluR ATDs *(46,49,50)*. The glycine moieties of agonists for mGluR2 and mGluR4 appear to coordinate to residues corresponding to those involved in Glu binding to mGluR1 *(46,49,50)*. Hence, the distal acidic function of the agonist seems to be the major determinant of its affinity, potency, and selectivity. In a model of the mGluR2 ATD, the distal groups of three different Group II mGluR agonists all coordinate with R57, R61, and K377 in Lobe I via hydrogen bonds and ionic interactions, but not with any residues in Lobe II (Fig. 6A) *(46)*. Mutations of S145, T168, R57, Y216, and D295 in mGluR2 (corresponding to S165, T188, Y74, Y236, and D318 in mGluR1) have been shown to eliminate [3H]LY354740 binding or reduce its affinity significantly, and π-cation interactions between the side chain of Y144 and the agonist has been proposed to contribute to the binding as well *(49)*. Mutations of residues S148 and R183 located at some distance from the orthosteric site were also shown to reduce the [3H]LY354740 affinity of mGluR2, and these residues were proposed to stabilize the 3D-structure of the agonist binding region or to promote ATD closure *(49)*.

In mGluR4 ATD models, the distal acidic groups of agonists *(S)*-AP4 and *(S)*-SOP bind to numerous residues at both lobes (Fig. 6A) *(36,46)*. Mutations of S159, T182, R78, K405, and E287 (corresponding to S165, T188, R78, K409, and G293 in mGluR1) have been shown to abolish [3H]-*(S)*-AP4 binding completely *(36,50)*. Whereas single mutations of K74 and K317 (corresponding to Y74 and R323 in mGluR1) had no impact on binding, a K74A/K317A mGluR4 mutant was shown to be unable to bind the radioligand. Thus, it was proposed that either or both of the lysines can form ion pair bonds to the phosphonate group of *(S)*-AP4, and that the carboxylate group of E287 stabilizes spatial orientations of the lysines favorable for these interactions (Fig. 6A) *(50)*.

Although not all residues in mGluR2 and mGluR4 corresponding to the 14 residues in the orthosteric site of mGluR1 have been investigated in mutagenesis studies, the binding modes of Quis, LY354740, and *(S)*-AP4 to their respective mGluRs clearly are significantly different. For example, Y230 in mGluR4 does not appear to contribute significantly to [3H]-*(S)*-AP4 binding, whereas mutations of the corresponding tyrosine in mGluR1 and mGluR2 have detrimental effects on agonist binding to the receptors *(48–50)*. Furthermore, D208 and D318 in mGluR1 are crucial for [3H]Quis binding to the receptor, whereas mutations of D202 and D312 in mGluR4 do not impair [3H]-*(S)*-AP4 binding. In contrast, K405 is crucial for agonist binding to and activation of mGluR4, whereas a K409A mGluR1 mutant displays wild-type agonist pharmacology *(48,50)*. Differences between the overall structures of the various mGluR binding pockets and in the distribution of water molecules in them may account for some of these differences. Mutel and coworkers have claimed that the spatial dimensions of the orthosteric site in mGluR2 could be significantly different from those in mGluR1, and that this opens up for numerous new contacts between agonist and receptor *(49)*.

All in all, affinity and potency of an agonist at a particular mGluR subtype seem to correlate with the nature of its interactions with Lobe I and II and to what extent it is capable of stabilizing the closed ATD. Selectivity, on the other hand, arises from sterical hindrance or ionic repulsion between the distal acidic group of the agonist and side chains of "distal" residues present in some mGluR subtypes but not in others. The Y74/R57/K74 residue in mGluR1/2/4 appears to be a particular important determinant of selectivity. The side chains of arginine and lysine residues are characterized by higher degrees of flexibilities than that of the phenol group of tyrosine, and this seems to permit the binding of longer and wider agonist molecules to mGluR2 and mGluR4 than to mGluR1. Furthermore, the Y74/R57/K74 residue has been proposed to influence the spatial orientation of the distal acidic group in the binding pocket, as the distal group appears to be projected more towards Lobe I in mGluR2 and mGluR4 than in mGluR1 *(46)*.

Analogously to the mGluR agonist, a competitive antagonist binds to the orthosteric site in the open or an intermediate ATD conformation. However, in contrast to the agonist, it possesses a structural element that hampers the ATD contraction process. In the mGluR1 ATD homodimer structure complexed with *(S)*-MCPG, the antagonist binds to many of the same residues as Glu (Fig. 6B) *(40)*. However, the phenol group of Y236 collides with the amino acid back bone of *(S)*-MCPG, and this steric hindrance prevents proper ATD closure (Fig. 6B). Recently, the information gained from the crystal structure has been supplemented by an elegant study, in which the efficacies of two competitive antagonists at mGluR8 have been altered radically by single point mutations in the orthosteric site of the receptor *(51)*. In the study, ACPT-II was shown to be a full agonist at a D309A mGluR8 mutant, whereas *(S)*-MAP4, the α-methylated analog of *(S)*-AP4, displayed partial and full agonism at Y227F and Y227A mGluR8 mutants, respectively. The molecular basis proposed for the conversions of the antagonists into agonists are depicted in Fig. 6B. Removal of ionic (D309A) or steric (Y227A) hindrances to the ATD closure allows the contraction process to run its course in spite of the ligand present in the binding pocket. In the Y227F mGluR8 mutant, the ATD is unable to close up completely around *(S)*-MAP, and thus the ligand displays partial agonism *(51)*. Hence, the efficacy of a mGluR ligand depends on the degree of ATD closure it allows, a correlation that resembles the one reported for iGluRs, where decreasing degrees of domain closure have been observed in crystal structures of the GluR2-S1S2 construct complexed with full agonists Glu and AMPA, partial agonist kainate, and antagonist CNQX *(52,53)*.

3.1.3. The ATD Homodimer Interfaces

The mGluR ATD homodimer possesses two interfaces: a permanent one constituted by regions in the two Lobe Is, and another constituted by regions in the two Lobe IIs, which is found exclusively in the "twisted" ATD homodimer (Fig. 7) *(39,40,45)*.

The ATD homodimer is assembled via noncovalent interactions and a disulfide bond mediated by a conserved cysteine (C140 in mGluR1) between the two Lobe Is *(25–28,39,54)*. The importance of the covalent interaction is not clear, because mutation of the conserved cysteine has been reported to have different impact on the dimerization states of mGluR1, mGluR5, and CaR in different studies *(25–28,54)*. On the other hand, the noncovalent interactions between residues on both sides of the "Lobe I interface" are not only essential for homodimer assembly, they also appear to be key determinants of the activation state of the mGluR or CaR. Both in the "resting" and the "twisted" ATD homodimer, the Lobe I interface is predominantly composed of the so-called B- and C-helices of the two ATDs (S112-I123 and S165-F178 in mGluR1) *(39)*. Residues in the two helices form intermolecular bonds to residues in the opposite ATD but very few of these interactions are conserved in both conformations (Fig. 7). Furthermore, the interface in the "resting" ATD homodimer includes an additional region,

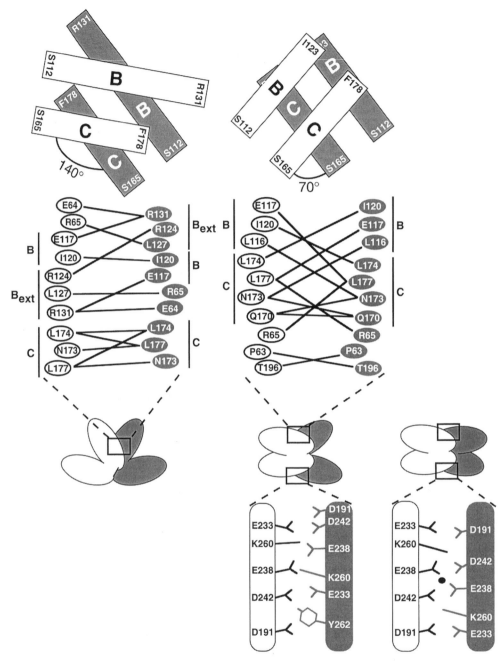

Fig. 7. Interfaces in the mGluR1 ATD homodimer. The two interfaces in the "open-open/resting," "closed-open/twisted," and "closed-closed/twisted" mGluR1 ATD homodimer are depicted. The Lobe I interface: The spatial rearrangement of the Lobe I interface upon the 70° activation twist is illustrated, and the intermolecular bonds in the "resting" and "twisted" homodimer conformations are given. The interactions are predominantly formed by residues from the two pairs of B- and C-helices and, in the "resting" homodimer, from the extension region of helix B (B_{ext}) *(39)*. The Lobe I interface is almost identical in the two "twisted" conformations, and therefore it is only depicted for the "closed-open/twisted" homodimer. The Lobe II interface: Clusters of acidic residues on the two Lobe IIs line this interface of the "twisted" ATD homodimer. The basic K260 residue alleviates the repulsion between the negatively charged residues, hereby allowing the "twisted" conformations to exist. The binding site of Gd^{3+} in the "closed-closed/twisted" conformation is indicated *(40)*.

the segment immediately following the B-helix (R124-R131 in mGluR1). This region forms a helical extension of the B-helix, and residues in it interact with residues in the opposite Lobe I. In the "twisted" ATD homodimer, however, the region has become disordered, and residues in the region no longer form intermolecular bonds (Fig. 7).

Alanine mutations of 5 residues (I116, I120, L174, L177, and F178) located in the B- and C-helices in the central core of the interface of both "resting" and "twisted" conformations have been shown to inhibit mGluR1 signaling significantly *(48)*. Mutation of I120 in particular had a detrimental effect on mGluR1 signaling, which was ascribed to the close vicinity of the residue to the rotation axis of the ATD homodimer *(48)*. On the other hand, random saturation mutagenesis of the A116-P136 region in CaR (corresponding to R124-K155 in mGluR1) have been shown to give rise to mutants with increased Ca^{2+} potencies and constitutive activities, and deletion of the Q117-S137 region in the receptor causes hypersensitivity to Ca^{2+} *(55,56)*. The functional importance of the A116-L123 region in CaR (R124-R131 in mGluR1) is further supported by the considerable number of "activating" somatic mutations localized in this region of the receptor *(7,55)*. Thus, the extension segment of helix B seems to exert a structural constraint on the "resting" ATD homodimer, hereby regulating the excitability of the receptor. Removal of the intermolecular interactions mediated by the region reduces the energy barrier between the "resting" and "twisted" ATD homodimer conformations and increases the fraction of receptors existing in the "twisted" activated state.

The activation twist in the ATD homodimer creates a new interface between residues in the two Lobe IIs. In mGluR1, the interface is constituted by four negatively charged residues (D191, E233, E238, and D242) located on each side of the cleft between the two lobes (Fig. 7). Tsuchiya et al. have proposed that the "Lobe II interface" in the theoretical "open-open/twisted" ATD homodimer is so electrostatically unfavorable that this conformation is virtually nonexisting *(40)*. In the "closed-open/twisted" conformation, on the other hand, the amino group of K260 in the closed ATD is located close enough to form intermolecular salt bridges with the carboxylate groups of E238 and D242 in the open ATD, and K260 in the open ATD can form an ionic interaction with E238 in the closed ATD (Fig. 7). These interactions alleviate the repulsive forces between the negatively charged residues in the interface and permit the "closed-open/twisted" conformation to exist. The arrangement of the four acidic residues and K260 is different in the "closed-closed/twisted" conformation. In the crystal structure of the mGluR1 ATD homodimer complexed with Glu and the allosteric mGluR1 potentiator Gd^{3+}, the metal ion binds to E238 and D242 on both sides of the interface and thus reduces the intermolecular repulsion and stabilizes the conformation (Fig. 7) *(40)*. However, according to Tsuchiya et al., the orientations of the two K260 residues allow the "closed-closed/twisted" conformation to exist in the absence of the metal ion as well *(40)*. Interestingly, Ca^{2+} has also been reported to be an allosteric potentiator of mGluR1 signaling *(57)*, and because Gd^{3+} and Ca^{2+} have similar ionic radii, Ca^{2+} could be speculated to act through binding to the Lobe II interface. However, the allosteric potentiation of mGluR1 displayed by Ca^{2+} has been disputed in a recent study *(58)*.

The importance of the Lobe II interface for the activation state of the mGluR is supported by several observations. Mutation of E238 in mGluR1 yields an elevated basal signaling of the receptor *(45,59)*. Furthermore, Zn^{2+} acts as a noncompetitive antagonist via binding to an engineered metal ion site in a K260H mGluR1 mutant, presumably by interfering with the activation twist or by stabilizing an inactive ATD homodimer conformation *(45,59)*. Finally, several somatic mutations giving rise to decreased Ca^{2+} potencies are localized in the D215-P221 region of CaR, which corresponds to E233-S239 in mGluR1 *(7)*.

Although the two interfaces in the mGluR ATD homodimer serve different purposes for receptor function, both appear to be crucial for the activation state of the receptor. Thus, these regions are highly attractive targets for allosteric modulation of mGluR signaling. Numerous allosteric modulators of mGluR function have been reported, but so far all of them seem to act through the 7TMs of the receptors (Subheading 3.3.1.).

3.2. The Cysteine-Rich Region (CRR)

The CRR is an enigmatic region of the mGluR. The region is also found in CaR and in the taste and pheromone receptors of family C, but, interestingly, it is absent in both $GABA_B$ receptor subunits *(7–9)*.

It is generally accepted that the CRR is important for the signal transduction through the mGluR and CaR homodimers, but whether the region functions as a scaffold maintaining the correct relative orientation of ATDs and 7TMs for signal transference, or whether it participates actively in the process, has not been addressed. The crystal structure of the mGluR1 ATD homodimer offers no clues to the roles of the CRR for receptor function, because the crystallized S33-S522 region does not include the region. However, cysteines in the CRR of CaR have been demonstrated to form disulfide bonds important for proper receptor function *(60)*. Serine substitutions of all nine conserved cysteines in the CRR region of CaR have been shown to reduce dramatically cell-surface expression levels of the receptor *(61)*. In contrast, deletion of the entire CRR has no effect on the cell-surface expression of CaR, although it abolishes receptor signaling *(60)*. There do not appear to be disulfide bonds between the ATD and the CRR of CaR, but noncovalent interactions could exist between the two domains *(62)*.

3.3. The 7-Transmembrane Moiety (7TM)

In contrast to the detailed insight into ATD structure–function aspects gained from the GluR1 ATD homodimer crystal structures, very little is known about the structural arrangement of TMs, loops, and carboxy terminal in the mGluR 7TM. The presence of 7 TMs is generally assumed based on the presence of seven hydrophobic regions in hydrophobicity plots of the receptor sequences and from their ability to couple to G proteins. The mGluR 7TM shares no amino acid sequence identity with the classical rhodopsin-like family A GPCR. Furthermore, with the exception of two cysteines in e1 and e2 conserved throughout the GPCR superfamily, none of the fingerprint motifs of the family A GPCR can be found in the mGluR 7TM *(63)*. Instead, other motifs are conserved in the mGluR family (Fig. 4).

Considering the lack of amino acid sequence identity, the recently published high-resolution X-ray structure of the rhodopsin receptor is not suited as a template for a molecular model of the mGluR 7TM *(64)*. However, studies of the G protein coupling of the mGluR and delineation of the binding sites for a new generation of allosteric modulators at the receptors have helped to shed light on the structural organization of the mGluR 7TM and constitute the first steps towards a 3D model of this moiety.

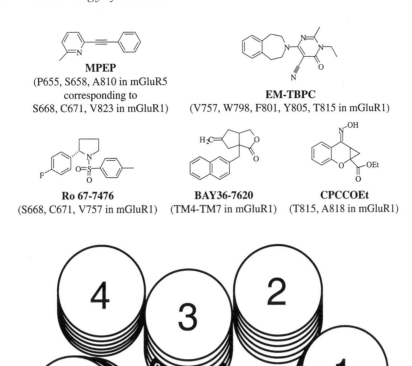

Fig. 8. Allosteric sites in the mGluR 7TM. The chemical structures of the allosteric inhibitors MPEP, EM-TBPC, BAY36-7620, and CPCCOEt, and the allosteric potentiator Ro 67-7476 are given. The crucial residues or regions for the subtype selectivity of each of the compounds are stated and shown in a schematic illustration of the mGluR1 7TM viewed from the extracellular side of the cell membrane. The residues in mGluR1 corresponding to those involved in MPEP binding to mGluR5 are also shown in the illustration.

3.3.1. Allosteric Sites in the 7TM of the MGluR

In recent years numerous allosteric modulators of mGluR function have been published, many of which have been shown to act exclusively at the 7TMs of the receptors (Fig. 8). All of the modulators are characterized by high degrees of mGluR subtype-specificities, which probably can be ascribed to the lower degree of amino acid sequence identities in the 7TMs compared with the orthosteric sites of the mGluRs. Specific residues and regions in mGluR1 or mGluR5 have been identified as determinants of the subtype-selectivities of allosteric inhibitors CPCCOEt, MPEP, BAY36-

7620, and EM-TBPC and allosteric potentiator Ro 67-7476 (Fig. 8). The mGluR1 specificity of BAY36-7620 has been shown to arise from its interactions with residues in TM4-TM7 *(65)*. The binding pockets of CPCCOEt and MPEP, antagonists of mGluR1 and mGluR5, respectively, have been demonstrated to overlap each other *(66,67)*. The mGluR1 specificity of allosteric activator Ro 67-7476 has been shown to arise from interactions with S668 and C671 in TM3 and V757 in TM5, and the two TM3 residues correspond to the two TM3 residues in mGluR5 critical for MPEP antagonism *(67,68)*. Finally, residues in TM5, TM6, and TM7 of mGluR1 have been demonstrated to involved in the binding of the antagonist EM-TBPC *(65,69)*. Hence, the allosteric inhibitors and activators seem to act through a common allosteric site formed by TM3, TM5, TM6, and TM7, or at least through overlapping sites in this crevice (Fig. 8). This speculation is supported by recent studies, where allosteric inhibitors of Group I mGluRs have displayed weak potencies as allosteric potentiators of mGluR4 signaling *(70,71)*.

Analogously to the common binding pocket for allosteric modulators in mGluRs, numerous allosteric potentiators and inhibitors of muscarinic acetylcholine receptors have been proposed to act through a common site located in the extracellular regions of the receptors *(72)*. Furthermore, it is interesting that the common allosteric site in the mGluR is composed of the same TMs forming the orthosteric site in the monoamine GPCRs *(63)*. However, in spite of their overlapping binding sites in the Group I mGluR 7TM, MPEP, BAY36-7620, and CPCCOEt appear to antagonize receptor function in distinct ways. BAY36-7620 and MPEP have been shown to be inverse agonists at mGluR1a and mGluR5a, respectively, whereas CPCCOEt is a neutral antagonist of mGluR1a *(65–67)*.

3.3.2. G protein Coupling of the mGluR

In the family A GPCR, the α-helical N- and C-termini of i3 and the N-terminal of i2 with its highly conserved DRY tripeptide have been identified as critical determinants of G protein coupling specificity and efficiency, respectively *(63,73)*. The G protein coupling of the mGluR is bound to be significantly different, however, as the sizes and amino acid sequences of the intracellular loops in the mGluR are completely different from those in the family A GPCR (Fig. 4). The coupling process has been investigated in studies of chimeric or point mutated mGluRs expressed alone or in combination with chimeric G proteins *(74–78)*. All four intracellular regions of the receptor appear to contribute to the coupling *(74,75)*. The i2 has been proposed to play the role of i3 in the family A GPCR, because its N- and C-termini appear to fold into amphipathic α-helices, and because the C-terminal of i2 in mGluR1 has been shown to be crucial for its $G_{\alpha q}$-coupling *(74–76)*. The highly conserved i3, on the other hand, is important for the coupling efficacy of the mGluR, much like the i2 in the family A GPCR *(74–76)*. In mGluR1a, specific residues in i2 have been proposed as crucial determinants of its $G_{\alpha q}$ (K690, T695, K697, and S702) and $G_{\alpha s}$ (K690, C694/T695, and P698) coupling, and R775 and F781 in i3 have been shown to be crucial for its coupling efficacy to both pathways *(76)*.

The involvement of the carboxy terminal in the G protein coupling of the mGluR is perhaps best illustrated by the significant differences in agonist potencies, signaling kinetics, and basal activities between the "long-tailed" and "short-tailed" Group I mGluR splice variants. The differences contrast the pharmacological profiles of Group III mGluRs, where no significant differences have been reported between the splice variants of mGluR4, mGluR7, or mGluR8. In *Xenopus* oocytes, activation of the "long-tailed" Group I mGluR splice variants (mGluR1a, mGluR5a, and mGluR5b) elicit rapid and transient Ca^{2+} signals, whereas the "short-tailed" mGluR1 splice variants generate much slower responses *(77,79,80)*. Furthermore, the long-tailed Group I mGluRs display higher agonist potencies than do the short-tailed splice variants, and a significant degree of constitutive activity reflecting spontaneous coupling to G proteins *(77,79–82)*. Finally, mGluR1a couples to $G_{\alpha s}$ and stimulation of AC, whereas the "short-tailed" mGluR1 splice variants do not *(77,79,80)*.

Most of these pharmacological differences indicate that the presence of a long carboxy terminal yields a more efficient G protein coupling to the Group I mGluR. A basic tetrapeptide (R877-R878-K879-K880) in the proximal part of the carboxy termini of all mGluR1 splice variants has originally been claimed to obstruct G protein coupling to the short-tailed mGluR1, an obstruction that supposedly should be masked in mGluR1a by the long carboxy terminal of the receptor *(77)*. Subsequently, however, the RRKK motif has been shown to be an ER retention signal regulating mGluR1b transport to the plasma membrane *(83)*. Furthermore, the RRK tripeptide has been demonstrated to target mGluR1b to axons in neurons and to the apical compartment in epithelial cells, whereas mGluR1a was localized predominantly in the dendritic and the basolateral compartment *(84)*. The differential trafficking and targeting of the mGluR1 splice variants observed in these studies were also ascribed to masking of the RRKK motif by the long carboxy terminal of mGluR1a. Furthermore, it was suggested the different expression patterns of the long- and short-tailed receptors would account for the subtle differences in their signaling kinetics *(84)*.

3.4. mGluR Signaling: From Agonist Binding to G protein Coupling

When it comes to the molecular mechanisms responsible for the translation of agonist binding to the ATD into 7TM signaling, two general schemes have been proposed: a "mechanical transfer" model and a "direct interaction" model (Fig. 9A). Both of these models originate from the 70° "activation twist" in the mGluR1 ATD homodimer. Based on the signal-transduction mechanism of certain tyrosine kinase receptors, Kunishima et al. have proposed that the structural rearrangement of the ATD dimer could cause the two 7TMs in the homodimer to contract, hereby creating a structural motif recognizable to the G protein (Fig. 9A) *(39)*. The overall organization of the two 7TMs in the inactive homodimer could be conserved in the active state (a contact dimer), or the 14 TMs could be "reshuffled," analogous to the "domain swapping" proposed to occur in family A GPCR dimerization (Fig. 9A) *(85)*. In the direct interaction model, the activation twist in the ATD homodimer is speculated to bring ATD segments into direct contact with extracellular loops and/or TMs, hereby stabilizing an active 7TM conformation (Fig. 9A). In this model, ATD segments act as a "built-in" agonist in the mGluR, a principle already known from trypsin and glycoprotein hormone GPCRs *(63)*.

Both of the signal-transduction models have their strengths and weaknesses. As mentioned previously, CaR/mGluR1 and mGluR1/CaR chimeras display pharmacologies similar to wild-type CaR and mGluR1, respectively *(32,33)*. Considering the weak amino acid sequence identity between CaR and mGluR1 (Fig. 1), this observation is

A

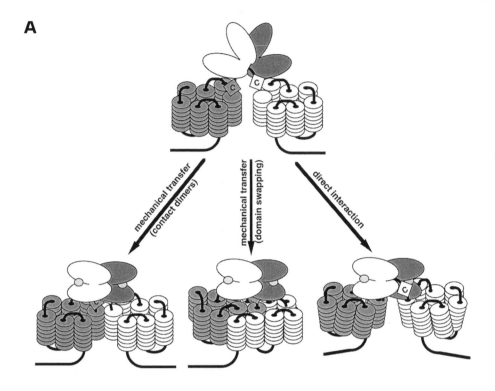

B

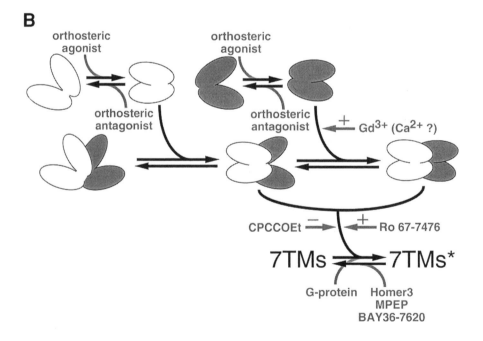

Fig. 9. Signal transduction through the mGluR homodimer. (**A**) Two models for the translation of agonist binding into 7TM signaling of the mGluR homdimer. The "C"-domains in the figure represent the CRRs in the mGluR homodimer. In the "mechanical transfer" model, the activation twist in the ATD homodimer causes a contraction of the two 7TMs, thereby creating one or several intracellular motifs capable of binding the G protein. In the active homodimer, each of the 7TMs may form an integrated unit that touches the other (a contact dimer). Alternatively, the TMs of the two 7TMs may be "reshuffled" into two "new" units (domain swapping). In the "direct interaction model", the activation twist brings ATD regions into contact with extracellular loops or TMs, thereby stabilizing the active conformation(s) of the 7TM(s). (**B**) The multiple equilibriums of mGluR function. The actions of agonists, competitive antagonists, allosteric modulators, and intracellular proteins on the different equilibriums are shown.

certainly more convincingly explained by the mechanical model than by a direct interaction between ATD and 7TM regions. On the other hand, Gd^{3+} has been shown to be an agonist at CaR acting exclusively at its 7TM, and mutations in the extracellular loops of the receptor have been shown to give rise to constitutive activity or increased sensitivity to Ca^{2+} *(7,33,86,87)*. It seems that ligand binding to or mutations in 7TM regions are more likely to mimic or facilitate the binding of an "ATD probe" to the region than to induce a dramatical rearrangement of 7TM regions.

In a recent study, insertion of peptide linkers between the ATD and the 7TM in $GABA_B1$ and $GABA_B2$ was demonstrated not to affect $GABA_B$ receptor signaling *(88)*. This observation seems to advocate for a direct interaction between ATDs and 7TMs in the $GABA_B$ heterodimer. However, considering the differences between the $GABA_B$ heterodimer and the mGluR homodimer, and in particular the lack of CRRs in the $GABA_B$ subunits, it is questionable whether conclusions about mGluR signaling can be extrapolated from this study.

Regardless of the signal-transduction mechanism of the mGluR, the G protein coupling of the receptor should be re-evaluated in light of the essential role of the dimerization. The two 7TMs in the mGluR homodimer could each bind a heterotrimeric G protein, but the coupling of one 7TM to a G protein could also be speculated to obstruct the G protein association of the other. Alternatively, the G protein coupling region could be composed of intracellular regions from both 7TMs, or G_α and $G_{\beta\gamma}$ could bind to different 7TMs in the homodimer. Hence, instead of the classical 1:1 stoichiometry for GPCR:G protein interactions, the mGluR:G protein coupling clearly has to be described with 2:2 or 2:1. Bai and colleagues have shown that co-expression of two nonfunctional CaR mutants with mutations in the ATD and 7TM, respectively, leads to partial recovery of function *(89)*. Thus, the presence of one wild-type ATD and one wild-type 7TM in the homodimer appears to be sufficient for signaling. On the other hand, it does not exclude that intracellular regions in the "mutated 7TM" can contribute to the G protein coupling of the homodimer.

Analogously to the ATD dimer, the 7TM level of the mGluR homodimer oscillates between inactive and active conformations *(90)*. The 7TMs-7TMs* equilibrium is regulated by the equilibrium between the inactive and active ATD homodimer conformations, which in turn is regulated by the equilibrium between the open and the closed ATD (Fig. 9B). Both the "closed-open/twisted" and "closed-closed/twisted" ATD homodimer conformations could be capable of stabilizing an active 7TM conformation, and the fact that Gd^{3+}, an allosteric potentiator of mGluR1 function, has been complexed with the "closed-closed/twisted" conformation strongly indicates that at least this conformation is active (Fig. 5 and Fig. 7) *(40)*. In addition to the input from the ATD homodimer, the 7TMs-7TMs* equilibrium is also modulated by the interactions of the 7TMs with intracellular signaling components such as G proteins *(78)* and Homer proteins (Subheading 4.2.2.) *(91)*.

The conformational oscillations at the 7TM level of the mGluR homodimer are reflected in the constitutive activity of the "long-tailed" Group I mGluR. Competitive antagonists are not able to suppress the basal signaling of mGluR1a and mGluR5a *(67,82)*. In contrast, binding of BAY36-7620 and MPEP to their respective "long-tailed" Group I mGluRs stabilize an inactive 7TM conformation, and thus the allosteric inhibitors display inverse agonism (Fig. 9B) *(65,67)*. CPCCOEt, on the other hand, is a neutral antagonist unable to inhibit the basal signaling of mGluR1a *(66)*. It is tempting to speculate that CPCCOEt binding to the extracellular part of the mGluR1 7TM interferes with the "cross-talk" taking place between ATDs and 7TMs according to the "direct interaction" model (Fig. 8 and Fig. 9A). Conversely, binding of allosteric potentiator Ro 67-7476 to the mGluR1 7TM facilitates the stabilization of the active conformation exerted by the ATD homodimer, whereas the compound has no effect on the 7TM-7TM* equilibrium in the absence of orthosteric agonist (Fig. 9B) *(68)*. So far, no 7TM-binding mGluR agonist have been identified, but the activation of CaR by Gd^{3+} suggests that stabilization of 7TMs* by direct ligand binding is feasible *(33)*.

4. REGULATION OF MGLUR SIGNALING

Like other GPCRs, the mGluRs are subjected to desensitization events such as phosphorylation, internalization (endocytosis), and downregulation, and both homologous and heterologous regulation mechanisms appear to be involved in the process (Subheading 4.1.). Furthermore, in recent years it has become clear that the expression, signaling, and desensitization properties of mGluRs are regulated by their interactions with intracellular adaptor and scaffolding proteins (Subheading 4.2.).

4.1. Desensitization of the mGluR

The principal component of heterologous (agonist-independent) desensitization of GPCRs is the rapid phosphorylation catalyzed by PKC and cAMP-dependent protein kinase (PKA). In studies using primary cultures and recombinant systems, the role of kinase-mediated phosphorylation of intracellular serines or threonines for the desensitization of the mGluR has been investigated *(92–101)*.

PKC has been demonstrated to desensitize the PLC/Ca^{2+} signaling of mGluR1a via phosphorylation of T695 in i2 of the receptor *(101)*. In accordance with the crucial role of T695 for the $G_{\alpha q}$-coupling of mGluR1a *(76)*, PKC was shown to attenuate this effector pathway selectively, having no significant impact on the $G_{\alpha s}$-mediated stimulation of AC by the receptor (Fig. 10) *(101)*. Desensitization of Group I mGluR-mediated PLC/Ca^{2+} signaling has been shown to involve a switch from facilitation to inhibition of excitatory transmission in hippocampus *(102)*, and PKC-mediated phosphorylation of T695 in mGluR1 could be an important component of these events. Interestingly, the mechanisms underlying the PKC regulation of internalization of mGluR1a appear to differ from those responsible for the desensitization. The distal part of the carboxy terminal in mGluR1a (S894-L1194) has been shown to be important for the PKC-mediated internalization of the receptor *(103)*.

PKC-mediated phosphorylation of mGluR5 has a profound impact on the signaling of the receptor. Not less than six intracellular serine and threonine residues in the receptor are substrates for PKC phosphorylation, and single alanine mutations of T606 and S613 in i1, T665 in i2, and S881 and S890 in the carboxy terminal have been shown to reduce the desensitization of the receptor significantly *(99)*. In addition to the uncoupling of mGluR5 from $G_{\alpha q}$, phosphorylation of S881 and S890 also competitively antagonizes the Ca^{2+}-dependent binding of calmodulin (CaM) to two binding sites in the carboxy terminal of the receptor (Fig. 2 and Fig. 10) *(104)*. Conversely, CaM binding inhibits phosphorylation of the two residues. Although the activities of some ion channels can be modulated by CaM binding, the functional implications of the displacement of CaM from mGluR5 have not been investigated *(104)*. Interestingly, it has been shown that mGluR5 can be desphosphorylated by a Ca^{2+}-dependent protein phosphatase activated by NMDA receptor signaling in native and recombinant systems (Fig. 10) *(105)*. Hence, in addition to the potentiation of NMDA receptor signaling exerted by Group I mGluRs, the NMDA receptor is able to reverse the desensitization of mGluR5. This is a highly interesting reciprocal positive feedback mechanism in amplification and induction of NMDA-dependent processes; for example, in connection with synaptic plasticity and neurotoxic insults.

Although the T840 residue in i3 of mGluR5a also is phosphorylated by PKC, this event does not contribute to the desensitization of the receptor. Instead, phosphorylation of the threonine gives rise to the characteristic $[Ca^{2+}]_i$ oscillations elicited by activation of mGluR5 *(106)*. It has been proposed that phosphorylation/dephosphorylation cycles switch the mGluR5-mediated $[Ca^{2+}]_i$ oscillations on and off. The corresponding residue in mGluR1a, D854, is not a substrate for PKC phosphorylation, and transient application of agonist induces single-peaked intracellular Ca^{2+} mobilization in mGluR1a-transfected cells *(106)*.

Activation of PKC and PKA also leads to phosphorylation and functional uncoupling of Group II and III mGluRs *(92,93,107–109)*. PKA targets a serine residue conserved in the carboxy terminal of the Group II and III mGluRs (S843 in mGluR2) *(109)*. Furthermore, PKC-mediated phosphorylation plays an essential role in a complex array of intracellular events regulating the signaling through mGluR7 and possible other Group III mGluRs (Subheading 4.2.1.).

In contrast to the second messenger-dependent protein kinases, GPCR kinases (GRKs) specifically phosphorylate the agonist-occupied GPCR, and this creates a motif in the intracellular receptor regions recognizable to β-arrestin. Binding of β-arrestin to the GPCR displaces the G protein and initiates receptor sequestration. Homologous (agonist-dependent) desensitization of mGluRs has so far only been investigated for mGluR1a and not in great detail. Expression of various GRKs in HEK293 cells has been shown to lead to phosphorylation and desensitization of mGluR1a, although there are some discrepancies as to which of the GRKs that are able to phosphorylate the receptor *(110,111)*. Upon sustained agonist exposure, mGluR1a internalizes rapidly in a GRK/β-arrestin-dependent fashion *(94,103,111,112)*. The S869-V893 region in the proximal part of the carboxy terminal of the receptor has been demonstrated to be important for this process *(103)*.

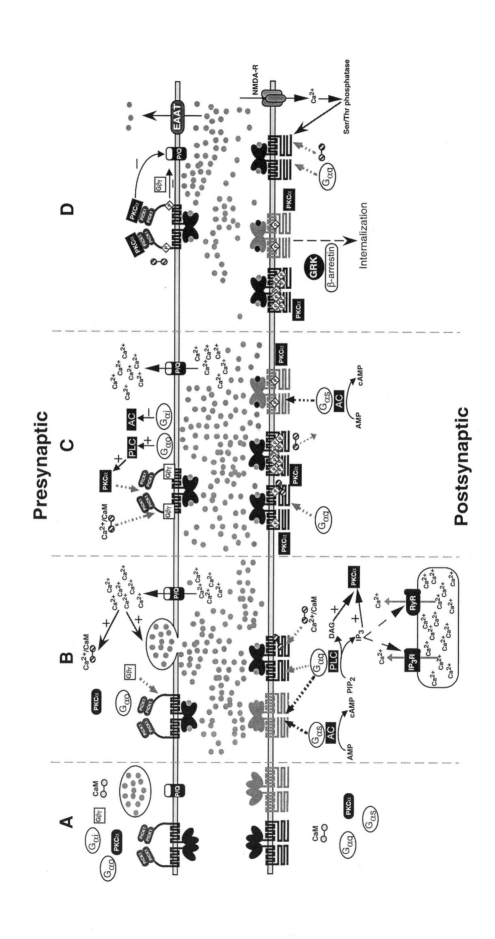

Presynaptic

Postsynaptic

A B C D

CaM
Gβγ
G$_{\alpha i}$
PKCα
G$_{\alpha q}$
PICK1
PICK1
P/Q

Ca^{2+}/CaM
Ca^{2+} Ca^{2+} Ca^{2+} Ca^{2+} Ca^{2+} Ca^{2+} Ca^{2+}

CaM
PKCα
G$_{\alpha q}$
G$_{\alpha s}$

AMP cAMP PIP$_2$
AC
G$_{\alpha s}$
G$_{\alpha q}$
PLC
DAG
IP$_3$
PKCα
IP$_3$R
RyR
Ca^{2+}

Ca^{2+}/CaM
Gβγ
PKCα
PICK1
PICK1
PLC +
AC −
G$_{\alpha q}$
G$_{\alpha i}$

Ca^{2+} Ca^{2+} Ca^{2+} Ca^{2+} Ca^{2+} Ca^{2+} Ca^{2+} Ca^{2+}
P/Q

PKCα
AMP cAMP
AC
G$_{\alpha s}$

PKCα
PICK1
PICK1
Gβγ −
P/Q −

EAAT

NMDA-R
Ca^{2+}
Ser/Thr phosphatase

GRK
β-arrestin
Internalization

PKCα
G$_{\alpha q}$

Fig. 10. Synaptic regulation of mGluR signaling. Events in the regulation of presynaptic mGluR7 signaling and in the rapid desensitization of postsynaptic mGluR1 (gray) and mGluR5 (black) function. The presynaptic and postsynaptic events depicted in **(A), (B), (C).** and **(D)** are not necessarily simultaneous. The actions of Homer proteins (Fig. 11) and GRKs are not included in this figure. The mGluR7 part of the figure is closely based on a similar figure in ref. *(148)*. (A) The resting neuron. Presynaptic: mGluR7a is localized in clusters at presynaptic active zones thanks to its association with PICK1. The receptor, PKCα and CaM are inactive. Postsynaptic: The Group I mGluRs, PKCα and CaM, are inactive. (B) Depolarization, Glu release, and mGluR activation. Presynaptic: Depolarization of the neuron triggers Ca^{2+} entry through voltage-gated Ca^{2+} channels (P/Q). The increase in $[Ca^{2+}]_i$ activates CaM and causes vesicular release of Glu. The released Glu activates mGluR7, which couples to the heterotrimeric G protein. Postsynaptic: Glu activates mGluR1 and mGluR5, which couple to their respective G proteins. PKCα is activated by second messengers IP_3 and diacylglycerol (DAG), and CaM is activated by the Ca^{2+} mobilized from intracellular stores and binds to the carboxy terminal of mGluR5 (and mGluR1). (C) PKC-mediated phosphorylation of the mGluRs. Presynaptic: Coupling of mGluR7 to $G_{\alpha i}$ leads to inhibition of adenylyl cyclase (AC), whereas coupling to $G_{\alpha o}$ stimulates phospholipase C (PLC), which in turn activates PKCα. Activated PKCα binds to PICK1 at mGluR7a and initiates phosphorylation of the receptor. Alternatively, activated CaM binds to the carboxy terminal. Postsynaptic: PKCα-mediated phosphorylation of mGluR1 and mGluR5 uncouples the receptors from $G_{\alpha q}$ and displaces Ca^{2+}/CaM from the receptors. Phosphorylation of T840 in mGluR5 leads to $[Ca^{2+}]_i$ oscillations. (D) Termination of Glu release and desensitized mGluRs. Presynaptic: PKCα-mediated phosphorylation of Ser^{862} and/or binding of Ca^{2+}/CaM to the carboxy terminal of mGluR7a displace $G_{\beta\gamma}$. The activity of the P/Q channel is inhibited directly by PKCα or by the released $G_{\beta\gamma}$. The synaptic release of Glu is terminated, and already released Glu is transported back into the presynaptic neuron by excitatory amino acid transporters (EAATs). Postsynaptic: mGluR1 and mGluR5 are desensitized and internalize. NMDA receptor activity may lead to dephosphorylation of mGluR5 and a reversal of its desensitized state.

4.2. The mGluR As a Complex Partner

Besides its roles in signaling and the regulation of it, the carboxy terminal of the mGluR has been shown to be essential for trafficking and targeting of the receptor to specific cellular compartments. Specific regions important for trafficking and targeting have been identified in the carboxy termini of some mGluRs, but in most cases the molecular mechanisms are poorly understood *(84,113,114)*.

Several intracellular proteins have been demonstrated to bind to the carboxy termini of mGluRs (Fig. 2). Although the physiological significance of most of these interactions remains to be explored, two meticulously studied cases of interactions between mGluRs and cytoplasmic proteins will be described here.

4.2.1. mGluR7 and Its Intracellular Partners

Activation of the presynaptic Group III mGluR leads to a dual signal transduction. In addition to the slow kinetics of its $G_{\alpha i}/G_{\alpha o}$-mediated inhibition of adenylyl cyclase activity, the receptor also controls neurotransmission at a much faster pace through its $G_{\beta\gamma}$-subunits. In the case of mGluR7, receptor activation has been shown to cause opening of K^+ inward rectifying channels (GIRKs) and to inhibit the activities of voltage-gated Ca^{2+} channels *(115,116)*. mGluR7 signaling inhibits presynaptic P/Q-type Ca^{2+} channel activity through $G_{\alpha o}$- and $G_{\beta\gamma}$-mediated coupling to PLC, intracellular Ca^{2+} mobilization and PKC activation *(117)*. In pull-down experiments, $G_{\beta\gamma}$, CaM, and PICK1 have been demonstrated to bind directly to the carboxy terminal of mGluR7 (Fig. 2). Together with PKC-mediated phosphorylation of mGluR7, these interactions have been demonstrated to be crucial for the targeting and signaling of the receptor (Fig. 10).

Analogously to its association with mGluR5 (Subheading 4.1.), CaM binds in a Ca^{2+}-dependent manner to a lysine-rich segment in the proximal region of the carboxy terminal of mGluR7 (Fig. 2). Because the binding site for $G_{\beta\gamma}$ is partially overlapping this segment, CaM and $G_{\beta\gamma}$ binding to mGluR7 are mutually exclusive *(115,118,119)*. Hence, prebound $G_{\beta\gamma}$ can be displaced from mGluR7 by CaM activated by the Ca^{2+} influx upon presynaptic depolarization. Because $G_{\beta\gamma}$ signaling in turn inhibits the Ca^{2+} channel activity, the system represents a very elegant example of negative feedback (Fig. 10).

When activated mGluR7 is phosphorylated by PKC, PKA, or cGMP-dependent protein kinase (PKG) *(118,120)*. The phosphorylation takes place exclusively at the S862 residue located in the CaM binding segment of the carboxy termini and abolishes binding of CaM to the receptor *(118,119,121)*. Hence, PKC, PKA, and PKG have been proposed to regulate the interaction between mGluR7 and CaM and ultimately the function of mGluR7 *(118,121)*. However, in a recent study, a S862E mGluR7 mutant mimicking the phosphorylated mGluR7 displayed a wild type-like coupling to GIRK channel coupling, and PKC has been shown to be able to attenuate the functional response of a S862A mutant unable to become phosphorylated *(120)*. Hence, there seems to be other determinants of $G_{\beta\gamma}$ signaling of mGluR7 than CaM binding to its carboxy terminal.

Protein Interacting with C Kinase (PICK1) is a 46.5-kDa PSD-95/Dlg1/ZO-1 (PDZ) domain-containing protein that interacts with and is phosphorylated by the active form of PKC *(122)*. Using yeast two-hybrid screens, three different groups have demonstrated that PICK1 also binds to mGluR7a, and that the last three residues in the carboxy terminal of mGluR7a (L913-V914-I915) and the PDZ domain of PICK1 are crucial for the interaction (Fig. 2) *(123–125)*. The ability of PICK1 to dimerize enables it to act as a scaffolding protein, and it has been demonstrated to be essential for the formation of mGluR7 clusters at the presynaptic terminal *(124,126)*. Furthermore, trimeric mGluR7a/PICK1/PKC complexes have been demonstrated in brain homogenates *(124,125)*. There are conflicting reports on the importance of the formation of these trimeric complexes for the PKCα-mediated phosphorylation of mGluR7 *(123,125)*. However, binding of PICK1 to mGluR7a has been shown to be crucial for the inhibition of P/Q-type Ca^{2+} channels mediated by the receptor, indicating that PICK1 exerts an important regulation of the synaptic transmission through the mGluR7a complex *(127)*. The molecular mechanism underlying the inhibition has not been delineated but the role of PICK1 resembles that of Homer proteins on the coupling of Group I mGluR coupling to Ca^{2+} and K^+ channels (Subheading 4.2.2.).

PICK1 is not the only PDZ domain-containing protein interacting with mGluRs *(128,129)*. Using yeast two-hybrid and GST pull-down assays, Dev and colleagues have recently demonstrated that Glutamate Receptor Interacting Protein (GRIP) and syntenin bind to several mGluRs (Fig. 2) *(128)*. GRIP was shown to bind to the same region of the carboxy terminal of mGluR7a as PICK1. Furthermore, mGluR7a and

GRIP were shown to be co-localized in hippocampal neurons, so GRIP could supplement the actions of PICK1 on mGluR7 trafficking in vivo *(128)*. Analogously, syntenin has been shown to bind to the same site in mGluR7b as PICK1 *(129)*. Because PICK1, GRIP, and syntenin also regulate AMPA receptor expression and clustering, these proteins appear to play key roles in the regulation of the overall glutamatergic transmission *(128,130)*.

All Group III mGluRs except mGluR4b and mGluR6 are subjected to PKC-mediated phosphorylation, presumably targeted at a conserved serine residue corresponding to the S862 residue in mGluR7a *(121)*. Furthermore, these splice variants have also been claimed to bind PICK1 *(123)*. However, the binding affinities of PICK1 to mGluR4a, mGluR8a, and mGluR8b are significantly lower than its binding to mGluR7a and mGluR7b, so the complex formations between PICK1 and mGluR4a/8a/8b need to be reproduced in other systems than the yeast two-hybrid system before any final conclusions can be drawn. Finally, all Group III mGluRs except mGluR4b appear to bind the $G_{\beta\gamma}$ subunits, whereas controversy exists whether CaM binds to other Group III mGluRs than mGluR7 *(118,119)*. Thus, it remains to be seen whether the intracellular protein network regulating mGluR7 function also regulates the targeting and signaling of other Group III mGluRs.

4.2.2. Group I mGluRs and Homer

With their discovery in 1997 of a novel gene being upregulated by synaptic activity in hippocampus and cortex, Worley and coworkers initiated an exciting area of investigations into synaptic regulation of trafficking, clustering, and signaling of the "long-tailed" Group I mGluRs *(131,132)*. The Homer (alternatively termed "vesl" for "VASP/Ena-related gene upregulated during seizure and long-term potentiation") family consists of seven proteins encoded by the genes *homer1-homer3* (Fig. 11) *(133)*. All family members possess a highly conserved EVH1 domain (homologous to Ena/VASP proteins) as their amino-terminal region, and all except Homer1a have a carboxy terminal region with a predicted coil-coil secondary structure, which mediates specific Homer dimerization (Fig. 11). The EVH1 domain, on the other hand, binds to PPXXFR motifs present in the distal parts of the carboxy termini of mGluR1a, mGluR5a, and mGluR5b; in intracellular IP$_3$ and ryanodine receptors; and in the scaffolding protein Shank *(133–137)*. Hence, the dimerization of Homer1b, -1c, -1d, -2a, -2b, and -3 enable them to link the mGluR with various transmembrane and intracellular proteins (Fig. 11). In contrast to the dimeric Homer proteins that are constitutively expressed in the postsynaptic terminal, Homer1a is upregulated rapidly during synaptic activity *(133)*. Being unable to dimerize, Homer1a thus functions as a physiological "dominant-negative," which antagonizes the complex formation between the Group I mGluR and dimeric Homers during states of seizures and long-term potentiation.

The roles of Homer proteins for "long-tailed" Group I mGluR signaling are depicted in Fig. 11. The dimeric Homer mediates a direct coupling of the mGluR to IP$_3$/ryanodine receptors gating the intracellular Ca^{2+} pools. The close proximity of the IP$_3$ generated by PLC stimulation to its intracellular receptors results in an optimized Ca^{2+} mobilization upon mGluR activation *(138)*. Conversely, binding of the dimeric Homer protein to the mGluR reduces the inhibition of K^+ and Ca^{2+} channels exerted by the receptor through $G_{\alpha q/11}$ and $G_{\beta\gamma}$ pathways, respectively (Fig. 11) *(139)*. Hence, the

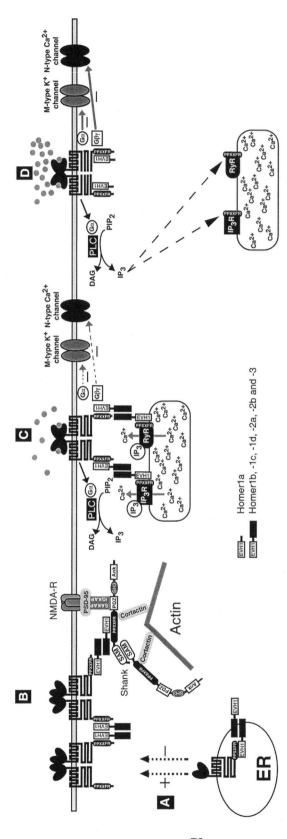

Fig. 11. Homer: a multitasking adaptor protein for "long-tailed" Group I mGluRs. The actions of Homer proteins on "long-tailed" Group I mGluR trafficking (**A**), clustering (**B**), and signaling (**C,D**). (A) Homer proteins inhibit and/or facilitate mGluR trafficking from endoplasmic reticulum (ER) to the plasma membrane. (B) Binding of a dimeric Homer to the mGluR enables the formation of receptor clusters, either directly with other mGluRs or with the NMDA-R/PSD-95/GKAP complex through Shank proteins. (C) The dimeric Homer bridges the mGluR to intracellular IP_3 and ryanodine receptors, and the proximity of receptor and intracellular Ca^{2+} pools optimizes signaling through the PLC/Ca^{2+} pathway. Furthermore, binding of the dimeric Homer to the mGluR reduces its G_α- and $G_{\beta\gamma}$-mediated inhibition of ion channels. (D) Upon long-term potentiation, Homer1a is rapidly upregulated and antagonizes all the actions of dimeric Homer proteins on mGluR trafficking, clustering, and signaling competitively.

upregulation of Homer1a during long-term potentiation switches Group I mGluR signaling from PLC stimulation to attenuation of ion channel activity. Finally, a recent study of native mGluR1a and overexpressed mGluR5 in cultured neurons has demonstrated an alternate function for Homer proteins *(91)*. In the study, binding of Homer3 to mGluR1a and mGluR5 was shown to "silence" the constitutive activity of the receptors. Mutations in the PPXXFR motif of mGluR5 or upregulation of Homer1a in the neurons disrupted the mGluR5-Homer3 interaction and unmasked the constitutive activity *(91)*.

The role of Homer as a "chaperone" regulating the trafficking and targeting of "long-tailed" Group I mGluRs to specific neuronal regions appears to be complex. Homer1b has been shown to inhibit cell-surface expression of mGluR5 by retaining the receptor in endoplasmic reticulum in both mammalian cells and neurons *(140,141)*. In contrast, Homer1a and Homer1c have been reported to increase cell-surface expression of mGluR1a in some cell lines and not to affect it in others *(142,143)*. In neurons, Group I mGluRs are localized perisynaptically adjacent to iGluRs *(14)*, and Homer proteins seem to play a key role in this targeting. Co-expression of Group I mGluRs with Homer1b or Homer1c have been shown to lead to a predominant dendritic localization, whereas Homer1a promotes both dendritic and axonal targeting of the receptors *(142,144)*.

Binding of the dimeric Homer to the Group I mGluR also facilitates its incorporation into clusters with other Group I mGluRs *(133,139,143,144)* or with other receptors through the actions of scaffolding proteins (Fig. 11). A highly interesting example of the latter is the bridging of the mGluR/Homer and NMDA-R/PSD-95/GKAP complexes mediated by Shank proteins, a family of postsynaptic density proteins *(136,137)*. The Shank protein consists of five domains: ankyrin repeats, a Src homology domain 3 (SH3) domain, a PDZ domain, a proline-rich domain, and a sterile α motif (SAM). The PDZ and proline-rich domains mediate the association with the NMDA-R/PSD-95/GKAP and mGluR/Homer complexes, respectively *(136,137)*. Shank proteins form dimers, and they have been shown to associate with other proteins, such as somatostatin receptors and the actin-binding proteins cortactin and fodrin. Hence, the dimeric Homer not only facilitates the functional cross-talk between "long-tailed" Group I mGluRs and NMDA receptors, but also links the receptors to the cytoskeleton and/or other proteins. AMPA and kainate receptors also interact with PDZ domain-containing proteins, but so far scaffolding proteins enabling clustering of these with the Group I mGluR/Homer complex have not been identified *(130,145)*.

4.2.3. Complexes Between mGluRs and Other GPCRs

In addition to the key importance of the homodimerization for mGluR function, two cases of heterodimerization (or heterooligomerization) between mGluRs and other GPCRs have been reported recently.

The mGluR1a and the adenosine A_1 receptor are co-localized in several CNS regions, and in a recent study the receptors were shown to form oligomeric complexes in mammalian cell lines and cerebellar synaptosomes *(146)*. In cells co-expressing mGluR1a and A_1R, glutamate/adenosine synergism was observed, as the signal elicited by Quis was potentiated by preincubation with adenosine receptor agonist R-PIA and conversely preincubation with Quis enhanced the signal through A_1R. The carboxy terminal of the mGluR1a was found to be essential for the heterodimerization, because the "short-tailed" splice variant mGluR1b did not interact with A_1R *(146)*. The authors proposed that the mGluR1a/A_1R interaction either is based in "pure" heterooligomerization, or that it arises from co-localization of two receptors in signaling complexes owing to adaptor/scaffolding proteins.

Heterodimerization between a Group I mGluR (mGluR1a or mGluR5) and another family C GPCR, CaR, has been demonstrated in hippocampal and cerebellar neurons and in mammalian cell lines *(147)*. The functional implications of the heterodimerization were not investigated in detail, but the mGluR1a/CaR heterodimer was shown to internalize upon sustained exposure to Ca^{2+} and Glu. Gama et al. proposed that the interface of the mGluR1a/CaR heterodimer could be made up by the same covalent and noncovalent interactions as in the mGluR and CaR homodimers (Subheading 3.1.3.). This is quite interesting considering that the sequence identity between these regions in mGluR1a and CaR is lower than between two different mGluR subtypes. To date, the existence of mGluR/mGluR heterodimers has not been reported.

5. FUTURE OUTLOOK

Being an evolutionary fusion product between a procaryotic binding protein and an origin 7TM, the mGluR is bound to have a unique signal mechanism. The crystal structures of the mGluR1 ATD homodimer have shed light on the first phase of mGluR signaling, whereas structure–function aspects of the CRR and 7TM still are poorly understood. Hence, one of the major challenges in the future exploration of mGluRs and other family C GPCRs will be to merge the fragmentized knowledge of the different receptor domains into a greater understanding of the signal transduction through the receptors.

Six years after the discovery of the Homer family, the contributions of these proteins to Group I mGluRs expression, trafficking, and signaling have already become a textbook example of the crucial impact intracellular proteins can have on GPCR function. Future studies will undoubtedly refine and add to our current insight into the regulation of mGluR function exerted by the numerous adaptor and scaffolding proteins interacting with them. Considering the extensive use of yeast two-hybrid screen systems, additional intracellular mGluR partners undoubtedly will be reported in the future. Finally, the studies of mGluR desensitization reported so far have only scratched the surface of the phenomenon. Thus, the future of mGluR research promises to be just as exciting as the last decade.

ACKNOWLEDGMENTS

The author wishes to thank the Lundbeck Foundation for financial support, and Professor Hans Bräuner-Osborne for his critical reading of the manuscript.

REFERENCES

1. Dingledine, R. and McBain, C. J. (1999) Glutamate and aspartate, in *Basic Neurochemistry* (Siegel, G. J., Agranoff, B. W., Albers, R. W., Fisher, S. K., and Uhler, M. D., eds.), Lippencott-Raven, Philadelphia, pp. 315–333.
2. Bräuner-Osborne, H., Egebjerg, J., Nielsen, E. Ø., Madsen, U., and Krogsgaard-Larsen, P. (2000) Ligands for glutamate receptors: design and therapeutic prospects. *J. Med. Chem.* **43,** 2609–2645.

3. Conn, P. J. and Pin, J.-P. (1997) Pharmacology and functions of metabotropic glutamate receptors. *Annu. Rev. Pharmacol. Toxicol.* **37,** 205–237.

4. Dingledine, R., Borges, K., Bowie, D., and Traynelis, S. F. (1999) The glutamate receptor ion channels. *Pharmacol. Rev.* **51,** 7–61.

5. Hollmann, M. and Heinemann, S. (1994) Cloned glutamate receptors. *Annu. Rev. Neurosci.* **17,** 31–108.

6. Nakanishi, S. (1994) Metabotropic glutamate receptors: synaptic transmission, modulation, and plasticity. *Neuron* **13,** 1031–1037.

7. Brown, E. M. (1999) Physiology and patophysiology of the extracellular calcium sensing receptor. *Am. J. Med.* **106,** 238–253.

8. Möhler, H. and Fritschy, J.-M. (1999) GABA$_B$ receptors make it to the top—as dimers. *Trends Pharmacol. Sci.* **20,** 87–89.

9. Nelson, G., Hoon, M. A., Chandrashekar, J., Zhang, Y., Ryba, N. J., and Zuker, C. S. (2001) Mammalian sweet taste receptors. *Cell* **106,** 381–390.

10. Schoepp, D. D., Jane, D. E., and Monn, J. A. (1999) Pharmacological agents acting at subtypes of metabotropic glutamate receptors. *Neuropharmacology* **38,** 1431–1476.

11. Pin, J.-P., de Colle, C., Bessis, A. S., and Acher, F. (1999) New perspectives for the development of selective metabotropic glutamate receptor ligands. *Eur. J. Pharmacol.* **375,** 277–294.

12. Anwyl, R. (1999) Metabotropic glutamate receptors: electrophysiological properties and role in plasticity. *Brain Res. Brain Res. Rev.* **29,** 83–120.

13. Heuss, C., Scanziani, M., Gahwiler, B. H., and Gerber, U. (1999) G protein-independent signaling mediated by metabotropic glutamate receptors. *Nat. Neurosci.* **2,** 1070–1077.

14. Lujan, R., Nusser, Z., Roberts, J. D., Shigemoto, R., and Somogyi, P. (1996) Perisynaptic location of metabotropic glutamate receptors mGluR1 and mGluR5 on dendrites and dendritic spines in the rat hippocampus. *Eur. J. Neurosci.* **8,** 1488–1500.

15. Shigemoto, R., Kinoshita, A., Wada, E., et al. (1997) Differential presynaptic localization of metabotropic glutamate receptor subtypes in the rat hippocampus. *J. Neurosci.* **17,** 7503–7522.

16. Bordi, F. and Ugolini, A. (1999) Group I metabotropic glutamate receptors: implications for brain diseases. *Prog. Neurobiol.* **59,** 55–79.

17. Hermans, E. and Challiss, R. A. (2001) Structural, signalling and regulatory properties of the group I metabotropic glutamate receptors: prototypic family C G protein-coupled receptors. *Biochem. J.* **359,** 465–484.

18. Schoepp, D. D. (2001) Unveiling the functions of presynaptic metabotropic glutamate receptors in the central nervous system. *J. Pharmacol. Exp. Ther.* **299,** 12–20.

19. Cartmell, J. and Schoepp, D. D. (2000) Regulation of neurotransmitter release by metabotropic glutamate receptors. *J. Neurochem.* **75,** 889–907.

20. Shigemoto, R., Kulik, A., Roberts, J. D., Ohishi, H., Nusser, Z., Kaneko, T., and Somogyi, P. (1996) Target-cell-specific concentration of a metabotropic glutamate receptor in the presynaptic active zone. *Nature* **381,** 523–525.

21. Nakanishi, S., Nakajima, Y., Masu, M., et al. (1998) Glutamate receptors: brain function and signal transduction. *Brain Res. Rev.* **26,** 230–235.

22. Romano, C., Yang, W.-L., and O'Malley, K. L. (1996) Metabotropic glutamate receptor 5 is a disulfide-linked dimer. *J. Biol. Chem.* **271,** 28612–28616.

23. Bai, M., Trivedi, S., and Brown, E. M. (1998) Dimerization of the extracellular calcium-sensing receptor (CaR) on the cell surface of CaR-transfected HEK293. *J. Biol. Chem.* **273,** 23605–23610.

24. Pace, A. J., Gama, L., and Breitwieser, G. E. (1999) Dimerization of the calcium-sensing receptor occurs within the extracellular domain and is eliminated by Cys→Ser mutations at Cys[101] and Cys[236]. *J. Biol. Chem.* **274,** 11629–11634.

25. Tsuji, Y., Shimada, Y., Takeshita, T., et al. (2000) Cryptic dimer interface and domain organization of the extracellular region of metabotropic glutamate receptor subtype 1. *J. Biol. Chem.* **275,** 28144–28151.

26. Ray, K., Hauschild, B. C., Steinbach, P. J., Goldsmith, P. K., Hauache, O., and Spiegel, A. M. (1999) Identification of the cysteine residues in the amino-terminal extracellular domain of the human Ca^{2+} receptor critical for dimerization. *J. Biol. Chem.* **274,** 27642–27650.

27. Ray, K. and Hauschild, B. C. (2000) Cys-140 is critical for metabotropic glutamate receptor-1 (mGluR-1) dimerization. *J. Biol. Chem.* **275,** 34245–34251.

28. Romano, C., Miller, J. K., Hyrc, K., et al. (2000) Covalent and noncovalent interactions mediate metabotropic glutamate teceptor mGlu$_5$ dimerization. *Mol. Pharmacol.* **59,** 46–53.

29. Takahashi, K., Tsuchida, K., Tanabe, Y., Masayuki, M., and Nakanishi, S. (1993) Role of the large extracellular domain of metabotropic glutamate receptors in agonist selectivity determination. *J. Biol. Chem.* **268,** 19341–19345.

30. O'Hara, P. J., Sheppard, P. O., Thøgersen, H., et al. (1993) The ligand-binding domain in metabotropic glutamate receptor is related to bacterial periplasmic binding proteins. *Neuron* **11,** 41–52.

31. Tones, M. A., Bendali, N., Flor, P. J., Knöpfel, T., and Kuhn, R. (1995) The agonist selectivity of a class III metabotropic glutamate receptor, human mGluR4a, is determined by the N-terminal extracellular domain. *NeuroReport* **7,** 117–120.

32. Bräuner-Osborne, H., Jensen, A. A., Sheppard, P. O., O'Hara, P., and Krogsgaard-Larsen, P. (1999) The agonist-binding domain of the calcium-sensing receptor is located at the amino-terminal domain. *J. Biol. Chem.* **274,** 18382–18386.

33. Hammerland, L. G., Krapcho, K. J., Garrett, J. E., Alasti, N., Hung, B. C. P., Simin, R. T., et al. (1999) Domains determing ligand specificity for Ca^{2+} receptors. *Mol. Pharmacol.* **55,** 642–648.

34. Malitschek, B., Schweizer, C., Keir, M., et al. (1999) The N-terminal domain of γ-aminobutyric acid$_B$ receptors is sufficient to specify agonist and antagonist binding. *Mol. Pharmacol.* **56,** 448–454.

35. Quiocho, F. A. and Ledvina, P. S. (1996) Atomic structure and specificity of bacterial periplasmic receptors for active transport and chemotaxis: variation of common themes. *Mol. Microbiol.* **20,** 17–25.

36. Hampson, D. R., Huang, X.-P., Pekhletski, R., Peltekova, V., Hornby, G., Thomsen, C., and Thøgersen, H. (1999) Probing the ligand-binding domain of the mGluR4 subtype of metabotropic glutamate receptor. *J. Biol. Chem.* **274,** 33488–33495.

37. Galvez, T., Parmentier, M.-L., Joly, C., et al. (1999) Mutagenesis and modeling of the GABA$_B$ receptor extracellular domain support a Venus Flytrap mechanism for ligand binding. *J. Biol. Chem.* **274,** 13362–13369.

38. Stern-Bach, Y., Bettler, B., Hartley, M., Sheppard, P. O., O'Hara, P. J., and Heinemann, S. F. (1994) Agonist selectivity of glutamate receptors is specified by two domains structurally related to bacterial amino acid-binding proteins. *Neuron* **13,** 1345–1357.

39. Kunishima, N., Shimada, Y., Tsuji, Y., et al. (2000) Structural basis of glutamate recognition by a dimeric metabotropic glutamate receptor. *Nature* **407,** 971–976.

40. Tsuchiya, D., Kunishima, N., Kamiya, N., Jingami, H., and Morikawa, K. (2002) Structural views of the ligand-binding cores of a metabotropic glutamate receptor complexed with an antagonist and both glutamate and Gd^{3+}. *Proc. Natl. Acad. Sci. USA* **99,** 2660–2665.

41. Okamoto, T., Sekiyama, N., Otsu, M., Shimada, Y., Sato, A., Nakanishi, S., and Jingami, H. (1998) Expression and purification of the extracellular ligand binding region of metabotropic glutamate receptor subtype 1. *J. Biol. Chem.* **273,** 13089–13096.

42. Han, G. and Hampson, D. R. (1999) Ligand binding of the amino-terminal domain of the mGluR4 subtype of metabotropic glutamate receptor. *J. Biol. Chem.* **274,** 10008–10013.

43. Peltekova, V., Han, G., Soleymanlou, N., and Hampson, D. R. (2000) Constraints on proper folding of the amino terminal domains of group III metabotropic glutamate receptors. *Mol. Brain. Res.* **76,** 180–190.

44. Lefkowitz, R. J., Cotecchia, S., Samama, P., and Costa, T. (1993) Constitutive activity of receptors coupled to guanine nucleotide regulatory proteins. *Trends Pharmacol. Sci.* **14,** 303–307.

45. Jensen, A. A., Greenwood, J. R., and Bräuner-Osborne, H. (2002) The dance of the clams: twists and turns in the family C GPCR homodimer. *Trends Pharmacol Sci.* **23,** 491–493.

46. Bertrand, H.-O., Bessis, A.-S., Pin, J.-P., and Acher, F. C. (2002) Common and selective molecular determinants involved in metabotopic glutamate receptor agonist activity. *J. Med. Chem.* **45,** 3171–3183.

47. Jensen, A. A., Sheppard, P. O., O'Hara, P. J., Krogsgaard-Larsen, P., and Bräuner-Osborne, H. (2000) The role of Arg[78] in the metabotropic glutamate receptor mGlu[1] for agonist binding and selectivity. *Eur. J. Pharmacol.* **397,** 247–253.

48. Sato, T., Shimada, Y., Nagasawa, N., Nakanishi, S., and Jingami, H. (2003) Amino acid mutagenesis of the ligand binding site and the dimer interface of the metabotropic glutamate receptor 1: Identification of crucial residues for setting the activated state. *J. Biol. Chem.* **278,** 4314–4321.

49. Malherbe, P., Knoflach, F., Broger, C., et al. (2001) Identification of essential residues involved in the glutamate binding pocket of the group II metabotropic glutamate receptor. *Mol. Pharmacol.* **60,** 944–954.

50. Rosemond, E., Peltekova, V., Naples, M., Thøgersen, H., and Hampson, D. R. (2002) Molecular determinants of high affinity binding to group III metabotropic glutamate receptors. *J. Biol. Chem.* **277,** 7333–7340.

51. Bessis, A. S., Rondard, P., Gaven, F., et al. (2002) Closure of the Venus flytrap module of mGlu8 receptor and the activation process: insights from mutations converting antagonists into agonists. *Proc. Natl. Acad. Sci. USA* **99,** 11097–11102.

52. Armstrong, N., Sun, Y., Chen, G.-Q., and Goaux, E. (1998) Structure of a glutamate-receptor ligand-binding core in complex with kainate. *Nature* **395,** 913–917.

53. Armstrong, N. and Gouaux, E. (2000) Mechanisms for activation and antagonism of an AMPA-sensitive glutamate receptor: crystal structures of the GluR2 ligand binding core. *Neuron* **28,** 165–181.

54. Zhang, Z., Sun, S., Quinn, S. J., Brown, E. M., and Bai, M. (2001) The extracellular calcium-sensing receptor dimerizes through multiple types of intermolecular interactions. *J. Biol. Chem.* **276,** 5316–5322.

55. Jensen, A. A., Spalding, T. A., Burstein, E. S., et al. (2000) Functional importance of the Ala[116]-Pro[136] region in the calcium-sensing receptor: constitutive activity and inverse agonism in a family C G protein coupled receptor. *J. Biol. Chem.* **275,** 29547–29555.

56. Reyes-Cruz, G., Hu, J., Goldsmith, P. K., Steinbach, P. J., and Spiegel, A. M. (2001) Human Ca^{2+} receptor extracellular domain. Analysis of function of lobe I loop deletion mutants. *J. Biol. Chem.* **276,** 32145–32151.

57. Kubo, Y., Miyashita, T., and Murata, Y. (1998) Structural basis for a Ca^{2+}-sensing function of the metabotropic glutamate receptors. *Science* **279,** 1722–1725.

58. Nash, M. S., Saunders, R., Young, K. W., Challiss, R. A., and Nahorski, S. R. (2001) Reassessment of the Ca^{2+} sensing property of a type I metabotropic glutamate receptor by simultaneous measurement of inositol 1,4,5-trisphosphate and Ca^{2+} in single cells. *J. Biol. Chem.* **276,** 19286–19293.

59. Jensen, A. A., Sheppard, P. O., Jensen, L. B., O'Hara, P. J., and Bräuner-Osborne, H. (2001) Construction of a high affinity zinc binding site in the metabotropic glutamate receptor mGluR1. Noncompetitive antagonism from the amino-terminal domain of a family C G protein-coupled receptor. *J. Biol. Chem.* **276,** 10110–10118.

60. Hu, J., Hauache, O., and Spiegel, A. M. (2000) Human Ca^{2+} receptor cysteine rich-domain. *J. Biol. Chem.* **275,** 16382–16389.

61. Fan, G.-F., Ray, K., Zhao, X., Goldsmith, P. K., and Spiegel, A. M. (1998) Mutational analysis of the cysteines in the extracellular domian of the human Ca^{2+} receptor: effects on cell surface expression, dimerization and signal transduction. *FEBS Lett.* **436,** 353–356.

62. Hu, J., Reyes-Cruz, G., Goldsmith, P. K., and Spiegel, A. M. (2001) The Venus's-flytrap and cystein-rich domains of the human Ca^{2+} receptor are not linked by disulfide bonds. *J. Biol. Chem.* **276,** 6901–6904.

63. Gether, U. (2000) Uncovering molecular mechanisms involved in activation of G protein-coupled receptors. *Endocr. Rev.* **21,** 90–113.

64. Palczewski, K., Kumasaka, T., Hori, T., et al. (2000) Crystal structure of rhodopsin: a G protein-coupled receptor. *Science* **289,** 739–745.

65. Carroll, F. Y., Stolle, A., Beart, P. M., et al. (2001) BAY36–7620: a potent non-competitive mGlu1 receptor antagonist with inverse agonist activity. *Mol. Pharmacol.* **59,** 965–973.

66. Litschig, S., Gasparini, F., Rueegg, D., et al. (1999) CPCCOEt, a noncompetitive metabotropic glutamate receptor 1 antagonist, inhibits receptor signaling without affecting glutamate binding. *Mol. Pharmacol.* **55,** 453–461.

67. Pagano, A., Rüegg, D., Litschig, S., et al. (2000) The non-competitive antagonists MPEP and CPCCOEt interact with overlapping binding pockets in the transmembrane region of group-I glutamate receptors. *J. Biol. Chem.* **275,** 35750–35758.

68. Knoflach, F., Mutel, V., Jolidon, S., et al. (2001) Positive allosteric modulators of metabotropic glutamate 1 receptor: characterization, mechanism of action, and binding site. *Proc. Natl. Acad. Sci. USA* **98,** 13402–13407.

69. Malherbe, P. N. K., Knoflach, F., Zenner, M. T., et al. (2003) Mutational analysis and molecular modeling of the allosteric binding site of a novel, selective, non-competitive antagonist of the metabotropic glutamate 1 receptor. *J. Biol. Chem.* **278,** 8340–8347.

70. Flor, P. J., Maj, M., Dragic, Z., et al. (2002) Positive allosteric modulators of metabotropic glutamate receptor subtype 4: pharmacological and molecular characterization. *Neuropharmacology* **43,** 286.

71. Mathiesen, J. M., Svendsen, N., Bräuner-Osborne, H., Thomsen, C., and Ramirez, M. T. (2003) Positive allosteric modulation of human metabotropic glutamate receptor 4 (hmGluR4) by SIB-1893 and MPEP. *Br. J. Pharmacol.* **138,** 1026–1030.

72. Tucek, S. and Proska, J. (1995) Allosteric modulation of muscarinic acetylcholine receptors. *Trends Pharmacol. Sci.* **16,** 205–212.

73. Wess, J. (1998) Molecular basis of receptor/G protein-coupling selectivity. *Pharmacol. Ther.* **80,** 231–264.

74. Pin, J.-P., Joly, C., Heinemann, S. F., and Bockaert, J. (1994) Domains involved in the specificity of G protein activation in phospholipase C-coupled metabotropic glutamate receptors. *EMBO J.* **13,** 342–348.

75. Gomeza, J., Joly, C., Kuhn, R., Knöpfel, T., Bockaert, J., and Pin, J.-P. (1996) The second intracellular loop of metabotropic glutamate receptor 1 cooperates with the other intracellular domains to control coupling to G proteins. *J. Biol. Chem.* **271,** 2199–2205.

76. Francesconi, A. and Duvoisin, R. M. (1998) Role of the second and third intracellular loops of metabotropic glutamate receptors in mediating dual signal transduction activation. *J. Biol. Chem.* **273,** 5615–5624.

77. Mary, S., Gomeza, J., Prézeau, L., Bockaert, J., and Pin, J.-P. (1998) A cluster of basic residues in the carboxyl-terminal tail of the short metabotropic glutamate receptor 1 variants impairs their coupling to phospholipase C. *J. Biol. Chem.* **273,** 425–432.

78. Parmentier, M. L., Joly, C., Restituito, S., Bockaert, J., Grau, Y., and Pin, J.-P. (1998) The G protein-coupling profile of metabotropic glutamate receptors, as determined with exogenous G proteins, is independent of their ligand recognition domain. *Mol. Pharmacol.* **53,** 778–786.

79. Joly, C., Gomeza, J., Brabet, I., Curry, K., Bockaert, J., and Pin, J.-P. (1995) Molecular, functional, and pharmacological characterization of the metabotropic glutamate receptor type 5 splice variants: comparison with mGluR1. *J. Neurosci.* **15,** 3970–3981.

80. Pin, J.-P., Waeber, C., Prezeau, L., Bockaert, J., and Heinemann, S. F. (1992) Alternative splicing generates metabotropic glutamate receptors inducing different patterns of calcium release in Xenopus oocytes. *Proc. Natl. Acad. Sci. USA* **89,** 10331–10335.

81. Flor, P. J., Gomeza, J., Tones, M. A., Kuhn, R., Pin, J.-P., and Knöpfel, T. (1996) The C-terminal domain of the mGluR1 metabotropic glutamate receptor affects sensitivity to agonists. *J. Neurochem.* **67,** 58–63.

82. Prézeau, L., Gomeza, J., Ahern, S., Mary, S., Galvez, T., Bockaert, J., and Pin, J.-P. (1996) Changes in the carboxyl-terminal of metabotropic glutamate receptor 1 by alternative splicing generate receptors with differing agonist-independent activity. *Mol. Pharmacol.* **49,** 422–429.

83. Chan, W. Y., Soloviev, M. M., Ciruela, F., and McIlhenney, R. A. J. (2001) Molecular determinants of metabotropic glutamate 1b trafficking. *Mol. Cell. Neurosci.* **17,** 577–588.

84. Francesconi, A. and Duvoisin, R. (2002) Alternative splicing unmasks dendritic and axonal targeting signals in metabotropic glutamate receptor 1. *J. Neurosci.* **22,** 2196–2205.

85. Gouldson, P. R., Higgs, C., Smith, R. E., Dean, M. K., Gkoutos, G. V., and Reynolds, C. A. (2000) Dimerization and domain swapping in G protein-coupled receptors. A computational study. *Neuropsychopharmacology* **23,** S60-S77.

86. Zhao, X., Hauache, O., Goldsmith, P. K., Collins, R., and Spiegel, A. M. (1999) A missense mutation in the seventh transmembrane domain constitutively activates the human Ca^{2+} receptor. *FEBS Lett.* **448,** 180–184.

87. Hu, J., Reyes-Cruz, G., Chen, W., Jacobson, K. A., and Spiegel, A. M. (2002) Identification of acidic residues in the extracellular loops of the seven-transmembrane domain of the human Ca^{2+} receptor critical for response to Ca^{2+} and a positive allosteric modulator. *J. Biol. Chem.* **277,** 46622–46631.

88. Margeta-Mitrovic, M., Jan, Y. N., and Jan, L. Y. (2001) Ligand-induced signal transduction within heterodimeric $GABA_B$ receptor. *Proc. Natl. Acad. Sci. USA* **98,** 14643–14648.

89. Bai, M., Trivedi, S., Kifor, O., Quinn, S. J., and Brown, E. M. (1999) Intermolecular interactions between dimeric calcium-sensing receptor monomers are important for its normal function. *Proc. Natl. Acad. Sci. USA* **96,** 2834–2839.

90. Parmentier, M. L., Prezeau, L., Bockaert, J., and Pin, J. P. (2002) A model for the functioning of family 3 GPCRs. *Trends Pharmacol. Sci.* **23,** 268–274.

91. Ango, F., Prezeau, L., Muller, T., et al. (2001) Agonist-independent activation of metabotropic glutamate receptors by the intracellular protein Homer. *Nature* **411,** 962–965.

92. Alagarsamy, S., Sorensen, S. D., and Conn, P. J. (2001) Coordinate regulation of metabotropic glutamate receptors. *Curr. Opin. Neurobiol.* **11,** 357–362.

93. De Blasi, A., Conn, P. J., Pin, J.-P., and Nicoletti, F. (2001) Molecular determinants of metabotropic glutamate receptor signaling. *Trends Pharmacol. Sci.* **22,** 114–120.

94. Dale, L. B., Babwah, A. V., and Ferguson, S. S. (2001) Mechanisms of metabotropic glutamate receptor desensitization: role in the patterning of effector enzyme activation. *Neurochem. Int.* **41,** 319–326.

95. Catania, M. V., Aronica, E., Sortino, M. A., Canonico, P. L., and Nicoletti, F. (1991) Desensitization of metabotropic glutamate receptors in neuronal cultures. *J. Neurochem.* **56,** 1329–1335.

96. Schoepp, D. D. and Johnson, B. G. (1988) Selective inhibition of excitatory amino acid-stimulated phosphoinositide hydrolysis in the rat hippocampus by activation of protein kinase C. *Biochem. Pharmacol.* **37,** 4299–4305.

97. Balazs, R., Miller, S., Romano, C., de Vries, A., Chun, Y., and Cotman, C. W. (1997) Metabotropic glutamate receptor mGluR5 in astrocytes: pharmacological properties and agonist regulation. *J. Neurochem.* **69,** 151–163.

98. Alaluf, S., Mulvihill, E. R., and McIlhinney, R. A. J. (1995) Rapid agonist mediated phosphorylation of the metabotropic glutamate receptor 1α by protein kinase C in permanently transfected BHK cells. *FEBS Lett.* **367,** 301–305.

99. Gereau, R. W. and Heinemann, S. F. (1998) Role of protein kinase C phosphorylation in rapid desensitization of metabotropic glutamate receptor 5. *Neuron* **20,** 143–151.

100. Ciruela, F., Giacometti, A., and McIlhinney, R. A. (1999) Functional regulation of metabotropic glutamate receptor type 1c: a role for phosphorylation in the desensitization of the receptor. *FEBS Lett.* **462,** 278–282.

101. Francesconi, A. and Duvoisin, R. (2000) Opposing effects of protein kinase C and protein kinase A on metabotropic glutamate receptor signaling: selective desensitization of the inositol trisphosphate/Ca^{2+} pathway by phosphorylation of the receptor-G protein-coupling domain. *Proc. Natl. Acad. Sci. USA* **97,** 6185–6190.

102. Rodriguez-Moreno, A., Sistiaga, A., Lerma, J., and Sanchez-Prieto, J. (1998) Switch from facilitation to inhibition of excitatory synaptic transmission by group I mGluR desensitization. *Neuron* **21,** 1477–1486.

103. Mundell, S. J., Pula, G., Carswell, K., Roberts, P. J., and Kelly, E. (2003) Agonist-induced internalization of metabotropic glutamate receptor 1A: structural determinants for protein kinase C- and G protein-coupled receptor kinase-mediated internalization. *J. Neurochem.* **84,** 294–304.

104. Minakami, R., Jinnai, N., and Sugiyama, H. (1997) Phosphorylation and calmodulin binding of the metabotropic glutamate receptor subtype 5 (mGluR5) are antagonistic in vitro. *J. Biol. Chem.* **272,** 20291–20298.

105. Alagarsamy, S., Marino, M. J., Rouse, S. T., Gereau IV, R. W., Heinemann, S. F., and Conn, P. J. (1999) Activation of NMDA receptors reverses desensitization of mGluR5 in native and recombinant systems. *Nat. Neurosci.* **2,** 234–240.

106. Kawabata, S., Tsutsumi, R., Kohara, A., Yamaguchi, T., Nakanishi, S., and Okada, M. (1996) Control of calcium oscillations by phosphorylation of metabotropic glutamate receptors. *Nature* **383,** 89–92.

107. Swartz, K. J., Merritt, A., Bean, B. P., and Lovinger, D. M. (1993) Protein kinase C modulates glutamate receptor inhibition of Ca^{2+} channels and synaptic transmission. *Nature* **361,** 165–168.

108. Macek, T. A., Schaffhauser, H., and Conn, P. J. (1998) Protein kinase C and A3 adenosine receptor activation inhibit presynaptic metabotropic glutamate receptor (mGluR) function and uncouple mGluRs from GTP-binding proteins. *J. Neurosci.* **18,** 6138–6146.

109. Schaffhauser, H., Cai, Z., Hubalek, F., Macek, T. A., Pohl, J., Murphy, T. J., and Conn, P. J. (2000) cAMP-dependent protein kinase inhibits mGluR2 coupling to G proteins by direct receptor phosphorylation. *J. Neurosci.* **20,** 5663–5670.

110. Dale, L. B., Bhattacharya, M., Anborgh, P. H., Murdoch, B., Bhatia, M., Nakanishi, S., and Ferguson, S. S. (2000) G protein-coupled receptor kinase-mediated desensitization of metabotropic glutamate receptor 1A protects against cell death. *J. Biol. Chem.* **275,** 38213–38220.

111. Sallese, M., Salvatore, L., D'Urbano, E., et al. (2000) The G protein-coupled receptor kinase GRK4 mediates homologous desensitization of metabotropic glutamate receptor 1. *FASEB J.* **14,** 2569–2580.

112. Doherty, A. J., Coutinho, V., Collingridge, G. L., and Henley, J. M. (1999) Rapid internalization and surface expression of a functional, fluorescently tagged G protein coupled glutamate receptor. *Biochem. J.* **341,** 415–422.

113. Stowell, J. N. and Craig, A. M. (1999) Axon/dendrite targeting of metabotropic glutamate receptors by their cytoplasmic carboxy-terminal domains. *Neuron* **22,** 525–536.

114. McCarthy, J. B., Lim, S. T., Elkind, N. B., Trimmer, J. S., Duvoisin, R. M., Rodriguez-Boulan, E., and Caplan, M. J. (2001) The C-terminal tail of the metabotropic glutamate receptor subtype 7 is necessary but not sufficient for cell surface delivery and polarized targeting in neurons and epithelia. *J. Biol. Chem.* **276,** 9133–9140.

115. O'Connor, V., El Far, O., Bofill-Cardona, E., et al. (1999) Calmodulin dependence of presynaptic metabotropic glutamate receptor signaling. *Science* **286,** 1180–1184.

116. Fagni, L., Chavis, P., Ango, F., and Bockaert, J. (2000) Complex interactions between mGluRs, intracellular Ca^{2+} stores and ion channels in neurons. *Trends Neurosci.* **23,** 80–88.

117. Perroy, J., Prezeau, L. M. D. W., Shigemoto, R., Bockaert, J., and Fagni, L. (2000) Selective blockade of P/Q-type calcium channels by the metabotropic glutamate receptor type 7 involves a phospholipase C pathway in neurons. *J. Neurosci.* **20,** 7896–7904.

118. Nakajima, Y., Yamamoto, T., Nakayama, T., and Nakanishi, S. (1999) A relationship between protein kinase C phosphorylation and calmodulin binding to the metabotropic glutamate receptor subtype 7. *J. Biol. Chem.* **274,** 27573–27577.

119. El Far, O., Bofill-Cardona, E., Airas, J. M., O'Connor, V., Boehm, S., Freissmuth, M., et al. (2001) Mapping of calmodulin and Gβγ binding domains within the C-terminal region of the metabotropic glutamate receptor 7A. *J. Biol. Chem.* **276,** 30662–30669.

120. Sorensen, S. D., Macek, T. A., Cai, Z., Saugstad, J. A., and Conn, P. J. (2002) Dissociation of protein kinase-mediated regulation of metabotropic glutamate receptor 7 (mGluR7) interactions with calmodulin and regulation of mGluR7 function. *Mol. Pharmacol.* **61,** 1303–1312.

121. Airas, J. M., Betz, H., and El Far, O. (2001) PKC phosphorylation of a conserved serine residue in the C-terminus of group III metabotropic glutamate receptors inhibits calmodulin binding. *FEBS Lett.* **494,** 60–63.

122. Staudinger, J., Zhou, J., Burgess, R., Elledge, S. J., and Olson, E. N. (1995) PICK1: a perinuclear binding protein and substrate for protein kinase C isolated by the yeast two-hybrid system. *J. Cell Biol.* **128,** 263–271.

123. Dev, K. K., Nakajima, Y., Kitano, J., Braithwaite, S. P., Henley, J. M., and Nakanishi, S. (2000) PICK1 interacts with and regulates PKC phosphorylation of mGLUR7. *J. Neurosci.* **20,** 7252–7257.

124. Boudin, H., Doan, A., Xia, J., Shigemoto, R., Huganir, R. L., Worley, P., and Craig, A. M. (2000) Presynaptic clustering of mGluR7a requires the PICK1 PDZ domain binding site. *Neuron* **28,** 485–497.

125. El Far, O., Airas, J., Wischmeyer, E., Nehring, R. B., Karschin, A., and Betz, H. (2000) Interaction of the C-terminal tail region of the metabotropic glutamate receptor 7 with the protein kinase C substrate PICK1. *Eur. J. Neurosci.* **12,** 4215–4221.

126. Boudin, H. and Craig, A. M. (2001) Molecular determinants for PICK1 synaptic aggregation and mGluR7a receptor coclustering: role of the PDZ, coiled-coil, and acidic domains. *J. Biol. Chem.* **276,** 30270–30276.

127. Perroy, J., El Far, O., Bertaso, F., Pin, J., Betz, H., Bockaert, J., and Fagni, L. (2002) PICK1 is required for the control of synaptic transmission by the metabotropic glutamate receptor 7. *EMBO J.* **21,** 2990–2999.

128. Hirbec, H., Perestenko, O., Nishimune, A., Meyer, G., Nakanishi, S., Henley, J. M., and Dev, K. K. (2002) The PDZ proteins PICK1, GRIP, and syntenin bind multiple glutamate receptor subtypes. Analysis of PDZ binding motifs. *J. Biol. Chem.* **277,** 15221–15224.

129. Enz, R. and Croci, C. (2003) Different binding motifs in the metabotropic glutamate receptor type 7b for filamin-A, PP1C, PICK1 and syntenin allow the formation of multimeric protein complexes. *Biochem. J.* **372,** 183–191.

130. Dong, H., O'Brien, R. J., Fung, E. T., Lanahan, A. A., Worley, P. F., and Huganir, R. L. (1997) GRIP: a synaptic PDZ domain-containing protein that interacts with AMPA receptors. *Nature* **386,** 279–284.

131. Brakeman, P. R., Lanahan, A. A., O'Brien, R., Roche, K., Barnes, C. A., Huganir, R. L., and Worley, P. F. (1997) Homer: a protein that selectively binds metabotropic glutamate receptors. *Nature* **386,** 279–284.

132. Kato, A., Ozawa, F., Saitoh, Y., Hirai, K., and Inokuchi, K. (1997) vesl, a gene encoding VASP/Ena family related protein, is upregulated during seizure, long-term potentiation and synaptogenesis. *FEBS Lett.* **412,** 183–189.

133. Xiao, B., Tu, J. C., Petralia, R. S., et al. (1998) Homer regulates the association of group I metabotropic glutamate receptors with multivalent complexes of homer-related, synaptic proteins. *Neuron* **21,** 707–716.
134. Kato, A., Ozawa, F., Saitoh, Y., Fukazawa, Y., Sugiyama, H., and Inokuchi, K. (1998) Novel members of the Vesl/Homer Family of PDZ proteins that bind metabotropic glutamate receptors. *J. Biol. Chem.* **273,** 23969–23975.
135. Beneken, J., Tu, J. C., Xiao, B., Nuriya, M., Yuan, J. P., Worley, P. F., and Leahy, D. J. (2000) Structure of the Homer EVH1 domain-peptide complex reveals a new twist in polyproline recognition. *Neuron* **26,** 143–154.
136. Naisbitt, S., Kim, E., Tu, J. C., et al. (1999) Shank, a novel family of postsynaptic density proteins that binds to the NMDA receptor/PSD-95/GKAP complex and cortactin. *Neuron* **23,** 569–582.
137. Tu, J. C., Xiao, B., Naisbitt, S., et al. (1999) Coupling of mGluR/Homer and PSD-95 complexes by the Shank family of postsynaptic density proteins. *Neuron* **23,** 583–592.
138. Tu, J. C., Xiao, B., Yuan, J. P., et al. (1998) Homer binds a novel proline-rich motif and links group 1 metabotropic glutamate receptors with IP_3 receptors. *Neuron* **21,** 717–726.
139. Kammermeier, P. J., Xiao, B., Tu, J. C., Worley, P. F., and Ikeda, S. R. (2000) Homer proteins regulate coupling of group I metabotropic glutamate receptors to N-type calcium and M-type potassium channels. *J. Neurosci.* **20,** 7238–7245.
140. Roche, K. W., Tu, J. C., Petralia, R. S., Xiao, B., Wenthold, R. J., and Worley, P. F. (1999) Homer 1b regulates the trafficking of Group I metabotropic glutamate receptors. *J. Biol. Chem.* **274,** 25953–25957.
141. Ango, F., Robbe, D., Tu, J. C., Xiao, B., Worley, P. F., Pin, J. P., Bockaert, J., and Fagni, L. (2002) Homer-dependent cell surface expression of metabotropic glutamate receptor type 5 in neurons. *Mol. Cell. Neurosci.* **20,** 323–329.
142. Ciruela, F., Soloviev, M. M., Chan, W. Y., and McIlhenney, R. A. (2000) Homer-1c/Vesl-1L modulates the cell surface targeting of group I metabotropic glutamate receptor type 1α: evidence for an anchoring function. *Mol. Cell. Neurosci.* **15,** 36–50.
143. Tadokoro, S., Tachibana, T., Imanaka, T., Nishida, W., and Sobue, K. (1999) Involvement of unique leucine-zipper motif of PSD-Zip45 (Homer 1c/vesl-1L) in group 1 metabotropic glutamate receptor clustering. *Proc. Natl. Acad. Sci. USA* **96,** 13801–13806.
144. Ango, F., Pin, J.-P., Tu, J. C., Xiao, B., Worley, P. F., Bockaert, J., and Fagni, L. (2000) Dendritic and axonal targeting of type 5 metabotropic glutamate receptor is regulated by homer1 proteins and neuronal excitation. *J. Neurosci.* **20,** 8710–8716.
145. Garcia, E. P., Mehta, S., Blair, L. A., et al. (1998) SAP90 binds and clusters kainate receptors causing incomplete desensitization. *Neuron* **21,** 727–739.
146. Ciruela, F., Escriche, M., Burgueno, J., et al. (2001) Metabotropic glutamate 1α and adenosine A1 receptors assemble into functional interacting complexes. *J. Biol. Chem.* **276,** 18345–18351.
147. Gama, L., Wilt, S. G., and Breitwieser, G. E. (2001) Heterodimerization of calcium sensing receptors with metabotropic glutamate receptors in neurons. *J. Biol. Chem.* **276,** 39053–39059.
148. El Far, O. and Betz, H. (2002) G protein-coupled receptors for neurotransmitter amino acids: C-terminal tails, crowded signalosomes. *Biochem. J.* **365,** 329–336.

II

Molecular Pharmacology of GABA Receptors

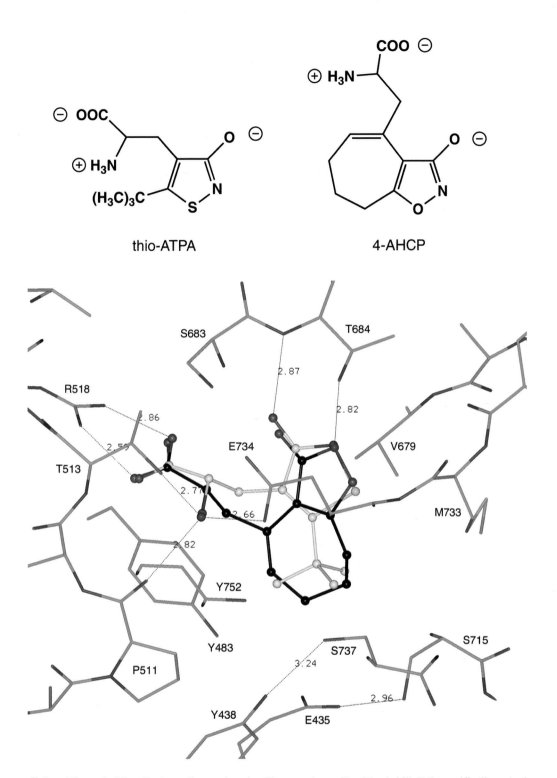

Color Plate 1, Fig. 5. (*see* discussion in Chapter 1, p. 9). (Top) iGluR5-specific ligands 4 AHCP and thio-ATPA. (Bottom) 4 AHCP (dark) and thio-ATPA (cyan) docked to a homology model of iGluR5 by Glide *(13)*.

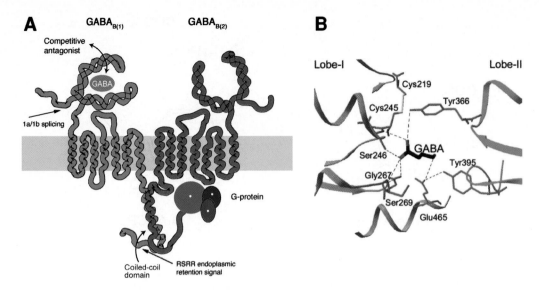

Color Plate 2, Fig. 2. (*see* discussion in Chapter 7, p. 217). (**A**) A schematic illustration of the GABA$_{B(1)}$ and GABA$_{B(2)}$ subunits forming the heterodimeric GABA$_B$ receptor. GABA binds in the cleft of the amino-terminal domain of GABA$_{B(1)}$, which leads to G protein activation through the GABA$_{B(2)}$ subunit. The localization of the coiled-coil domain, the RSRR endoplasmic retention signal, and the GABA$_{B(1a)/(1b)}$ splice site has been noted. Adapted with permission from Marshall et al. *(56)*. (**B**) A model of GABA docked in to the ligand binding site of GABA$_{B(1)}$. The putative interactions between GABA and amino acids of GABA$_{B(1)}$ are shown with dotted lines. Adapted with permission from Kniazeff et al. *(30)*.

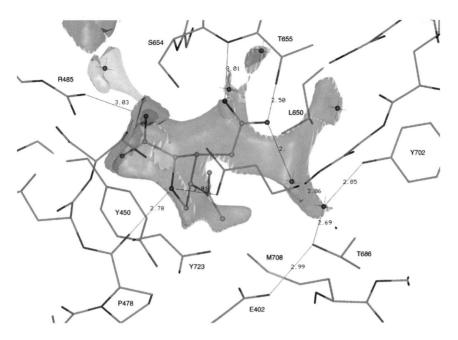

Color Plate 3, Fig. 6. (*see* discussion in Chapter 1, p. 10). Kainate iGluR2-S1S2 complex, including isoenergy contours according to GRID (16)-methyl probe (beige, 23.0 kcal/mol), water probe (cyan, 210.0 kcal/mol), and anionic oxygen probe (magenta, 211.0 kcal/mol).

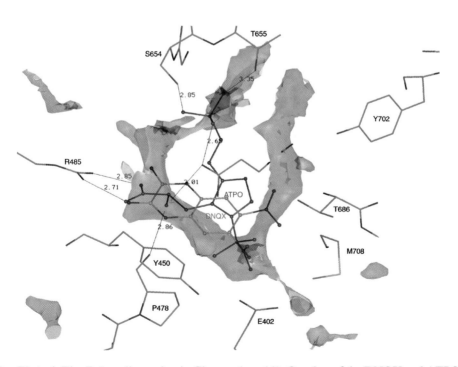

Color Plate 4, Fig. 7. (*see* discussion in Chapter 1, p. 11). Overlay of the DNQX and ATPO–iGluR2-S1S2 complexes including isoenergy contours acording to GRID-methyl probe (beige, 22.9 kcal/mol), water probe (cyan, 28.5 kcal/mol), and hydrogen phosphate anion probe (magenta, 214.4 kcal/mol).

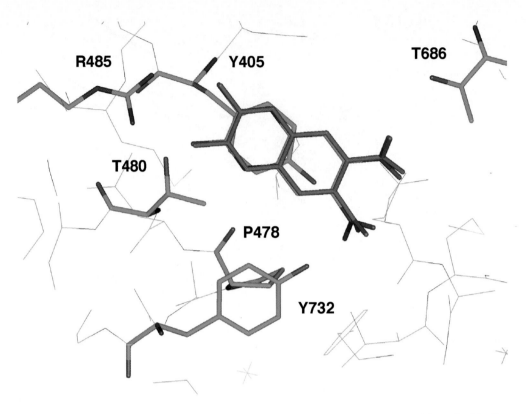

Color Plate 5, Fig. 8. (*see* discussion in Chapter 1, p. 13). DNQX docked into the binding pocket of iGluR2 using Glide *(21)*. The experimentally observed structure is shown in magenta and the atoms of the docked structure are type-coded.

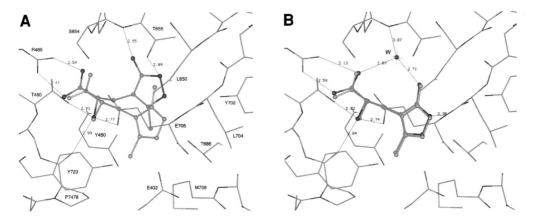

Color Plate 6, Fig. 10. (see discussion in Chapter 1, p. 14). AMPA docked to iGluR2 using Glide *(21)*. **(A)** Excluding water molecules from the calculations; **(B)** Including one water molecule, marked W. The experimentally observed AMPA structure is shown in magenta and the atoms of the docked structure are type-coded.

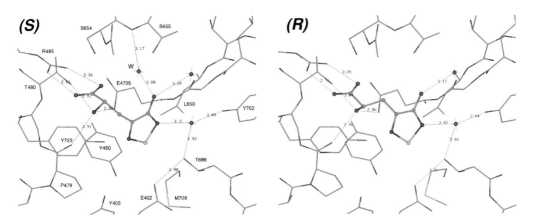

Color Plate 7, Fig. 12. (*see* discussion in Chapter 1, p. 18). *(S)*- and *(R)*-isomers of 2-amino-3-(3-hydroxy-1,2,5-thiadiazol-4-yl)propionic acid (TDPA) docked to iGluR2 *(28)*.

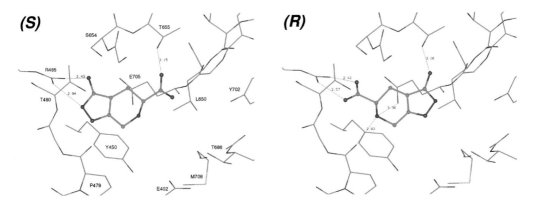

Color Plate 8, Fig. 13. (*see* discussion in Chapter 1, p. 19). *(S)*- and *(R)*-5-HPCA docked to iGluR2, according to Glide *(21)*.

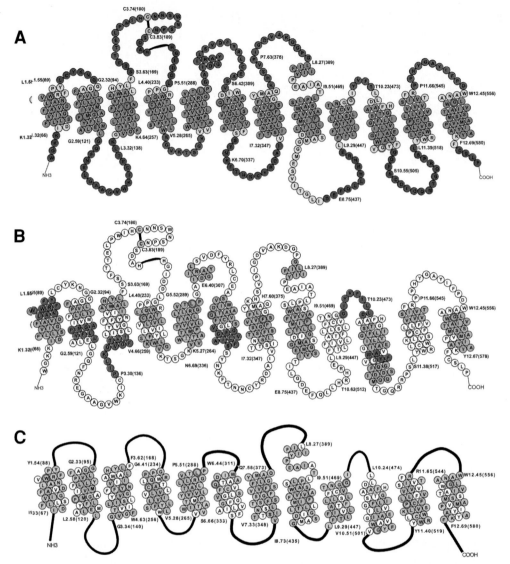

Color Plate 9, Fig. 1. (*see* full caption and discussion in Ch. 12, p. 216). Predicted structural properties of neurotransmitter transporters. The helical nets represent the human dopamine transporter, with the single letter code of each amino acid within a circle. The most conserved residues in the 12 transmembrane segments (index number TM#.50—*see* text) are identified by thick blue lines surrounding the circles. The index nubmers ae shown for selected residues, with human DAT residue numbers in parentheses. In (A) and (B), cysteines in the second extracellular loop thought to form a disulfide bond are shown in yellow. When not shown, residues in the N-and C-termini and in loops are indicated by thick black lines. **(A)** Predicted transmembrane topology. Based on hydrophobicity analysis, cyan residues are predicted to be located within the lipid bilayer, and red residues are predicted to be solvent exposed outside the lipid boundaries. Regions that are not well-defined by the prediction methods are shown in pink. The assignment of residues to the transmembrane segment is based on this hydrophobicity analysis as well as on predictions of secondary structure, lipid accessibility and the summarized experimental data, all of which are discussed below and in the text. **(B)** Secondary structure propensity based on the spatial periodicity of residue properties. Predicted α-helical segments are shown in orange; sequential residues with periodicity consistent with β-strand are shown in green. Segments shown in white did not exhibit an identifiable periodicity. **(C)** Probability of lipid exposure. Positions shown in purple are predicted to lie on the helix-lipid interface, positions in yellow are predicted to face the protein interior, and white regions are undefined.

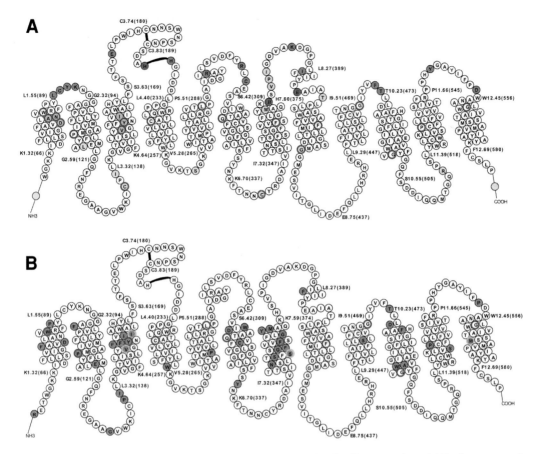

Color Plate 10, Fig. 2. (*see* full caption and discussion in Chapter 12, p. 217). Summary of experimental findings from the neurotransmitter transporter family. As in Fig. 1, the helical nets represent the human dopamine transporter, with the single letter code of each amino acid within a circle and with the index positions shown in blue. Residues not shown in the N- and C-termini are indicated by thick black lines. **(A)** Cysteine accessibility and zinc sites. Endogenous cysteines or substituted cysteines in DAT (or at the aligned positions in other NTs) are shown in orange if they were found to be accessible to impermeant sulfhydryl reagents when applied extracellularly. All accessible residues in the TMs are show in yellow, and residues accessible to impermeant reagents only in membranes or after permeabilization are shown in light yellow. Residues that are protected from reaction by the presence of substrate and/or inhibitor are circled in magenta. Endogenous zinc sites in DAT are shown in bright green, whereas engineered zinc sites are shown in light green. **(B)** Mutations affecting substrate and/or inhibitor recognition. As described in the text, mutations in DAT, or in aligned positions in homologous NTs, that decreased the apparent affinity for substrate or the binding affinity for inhibitors are shown in orange. Also shown in orange are the random mutations in TM7 that decreased or abolished transport and a number of other mutations that affected the function of the transporters (*see* text). Residues shown in orange that were predicted to face lipid in **(C)** are shown surrounded by gold circles.

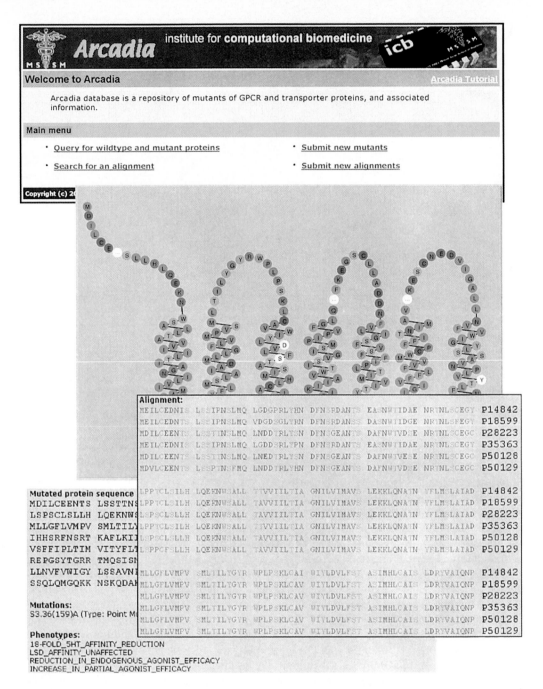

Color Plate 11, Fig. 1. (*see* full caption and discussion in Chapter 13, p. 237). Illustration of the major types of visualization available from the Arcadia Database. The features (snake-like diagram, alignment and mutant details) are superimposed on the home page of the database.

4

Functional Characterization of GABA$_A$ Receptor Ligands In Vitro

Correlation With In Vivo Activity

Bjarke Ebert, Sally Anne Thompson, Signe Í Stórustovu, and Keith A. Wafford

1. INTRODUCTION

In the mammalian brain, the inhibitory neurotransmitter γ-aminobutyric acid (GABA) plays a key role in the control of synaptic activity. Several studies have demonstrated the involvement of GABA and GABA$_A$ receptors in diseases such as anxiety, depression, seizures, schizophrenia, and sleep disorders (recent review: Thomsen and Ebert [1]). Based on this hypothesis, much research has been directed towards an understanding of which receptors are involved in the different disorders and the development of compounds able to alleviate the hypoactivity. Benzodiazepines are in general very effective in several of the mentioned disorders. However, there is a general consensus that unselective benzodiazepines cause too many side effects, and the development of novel drugs with fewer side effects is needed (e.g., [2]). Therefore, over the previous three decades, much research has been focused at the development of ligands for the GABA$_A$ receptor complex selective for certain parts of the brain or certain receptor subtypes.

GABA mediates its effects via opening of a chloride channel formed in the center of an assembly of five different subunit proteins. This chloride channel has the normal characteristics of ligand gated ion channels, in that the activity is voltage-dependent, desensitizes rapidly, and is modulated by several allosteric modulators. One class of modulators, the benzodiazepine site ligands, have been the focus of much research during the last three decades. (Pubmed: 7500 references 1992–2002). The action of benzodiazepines is mediated via the ionotropic GABA$_A$ receptors, which are primarily located at postsynaptic neurons. Upon binding of benzodiazepine agonists, the opening probability of the GABA receptor controlled Cl-channel is increased (3), thereby producing an apparent leftward shift in the dose-response curve to, e.g., GABA and, in case of partial GABA$_A$ receptor agonists like thio-4-PIOL, P4S, and imidazole acetic acid, also an increase the apparent maximal response (4,5). Because benzodiazepines

From: *Molecular Neuropharmacology: Strategies and Methods*
Edited by: A. Schousboe and H. Bräuner-Osborne © Humana Press Inc., Totowa, NJ

in the absence of receptor activation are inactive, the use-dependent activity of benzo-diazepines is potentially very important. Therefore, benzodiazepines, like barbiturates, will act as use-dependent amplifiers of the GABA$_A$-receptor system, and therefore, in principle, only amplify ongoing activity. This was previously thought to be limited to synaptic activity, but recent studies now point towards a degree of tonic activation in some parts of the GABA system *(6–9)*. High levels of receptor activation invariably lead to the desensitization and consequent fading of the GABA$_A$ receptor mediated response. This may lead to a reduction in clinical efficacy of GABAergic drugs, and the introduction of dose escalation and dependence problems limiting the value of long-term treatment, especially with currently available benzodiazepines. In order to deal with these problems, we need to understand how the many physiological consequences of GABAergic activation or potentiation are mediated and whether different effects can be ascribed to the different GABA$_A$ receptor subtypes. The present review will focus on the possibilities for the development of functional selective GABA$_A$ receptor ligands.

2. GABA$_A$ RECEPTOR STRUCTURE AND COMPOSITION

GABA$_A$ receptors are formed as pentameric combinations of different families of subunits. These related families of subunits, termed α (1–6), αb (1–3), γ (1–3), δ, ϵ, π, θ, assemble into a heterogeneous group of functional GABA$_A$ receptors, depending on where in the CNS these subunits are expressed. Most, but not all, GABA$_A$ receptors are formed by 2 α subunits, 2 β subunits, and a γ, ϵ or δ subunit (reviewed by Sieghart et al. *[10]*). The binding site for benzodiazepines is located at the interface between the α and γ subunits *(11–15)*, while the binding site for GABA and other GABA$_A$ receptor agonists is formed at the interface of the α and β subunit *(15–18)*. Modeling based on the recently identified acetylcholine binding protein has provided an excellent basis for a proposed structure of the extracellular portion of the GABA receptor and gives us a first glimpse at how the GABA and benzodiazepine binding sites are formed *(19,20)*.

The free permutation of the subunits leads to more than 40,000 possible combinations. It is likely though that in vivo more than 90% of the receptors are composed of only 20 different combinations *(21)*. This plethora of different subtypes suggest that different functional roles are mediated by distinct and separable receptors, and that these could be individually targeted in order to obtain a drug devoid of side effects. It seems likely that the most abundant receptors represent the important combinations and as such, are most likely to be responsible for major functional consequences when targeted. The specific locations of the less abundant receptor populations may result in very selective effects when modulated.

The clinical consequences of the currently used benzodiazepines range from sedation, muscle relaxation, seizure reduction, anxiolysis, and hypnosis. Clearly, it would be highly desirable to be able to separate some of these effects. In addition, it would be useful to reduce other undesirable consequences such as development of tolerance and dependence, abuse, synergistic interaction with ethanol, and memory impairment (for a comprehensive review: *see [22]*). Animal models for some of the aforementioned conditions, in combination with transgenic mouse technology, have recently led to a deeper understanding of the contribution some of the individual α subunits make to these behaviors.

3. PHYSIOLOGICAL ROLE AND DISTRIBUTION OF SUBUNITS

The binding site for the benzodiazepines is located at the interface of the α and γ subunit. Whereas $\alpha_{1,2,3}$ and α_5 receptors respond to classical benzodiazepine drugs, for example, diazepam, α_4- and α_6-containing receptors do not respond to any clinically used benzodiazepines. Mutation of a histidine in $\alpha_{1,2,3}$, and α_5 to arginine in α_4 and α_6 is a major determinant of this difference in sensitivity towards benzodiazepines *(23)*. Two groups have used recombinant DNA technology to mutate this residue in the α_1 or other α subunits, and investigated the pharmacological consequences of diazepam-insensitive α_1-, α_2-, α_3-, or α_5-containing receptors. Conclusions from these studies suggest that α_1-containing receptors are primarily involved in the sedative effects of diazepam, whereas α_2-containing receptors may be important for the anxiolytic actions *(24–27)*. The cerebellar location of GABA$_A$ receptors containing α_6 subunits suggests a role in motor co-ordination, however, α_6 knockout mice did not seem to show any major ataxia or motor deficits or altered sensitivity on administration of general anaesthetics, barbiturates, or ethanol *(28)*. It is also quite likely that when subunits are deleted in this manner, compensatory changes in other systems may ameliorate some of the phenotypic consequences. There is, therefore, still a number of very important questions regarding the functional implication of different GABA$_A$ receptor combinations that needs to be addressed. Another issue hampering our understanding of the significance of different GABA$_A$ receptor subtypes lies in their cellular distribution. Thus, even in a single cell, receptors at the soma may be different from those located at the dendrites *(29)*, adding to the complexity. Immunocytochemical experiments have also demonstrated that certain receptor subtypes can be located on different parts of the cell; for example, in hippocampal pyramidal neurons the α_2 subunit is highly expressed on the axon initial segment, and α_5 subunits appear to be predominantly extrasynaptic *(30)*. Hence, in one brain region, neurons carrying several different subunit combinations may exist side by side. It is therefore not surprising then, that despite several subtype-selective compounds acting at the benzodiazepine site, a coherent understanding of the specific actions of positive GABA$_A$ receptor modulators is still lacking.

4. IMPACT OF SUBUNIT COMPOSITION ON THE PHARMACOLOGY OBSERVED IN VITRO AND IN VIVO

Studies on compounds interacting directly with the GABA recognition site at the GABA$_A$ receptor and benzodiazepines have shown that it can be very difficult to discover compounds that differ significantly in binding affinity between subtypes. Another mechanism for achieving subtype selectivity is by functional efficacy, and for both GABA and benzodiazepine sites, compounds exist that will discriminate based on receptor subtype *(17,26)*.

The GABA binding site is located at the interface between the α and β subunit *(15,31)*, which, because several members of each family including splice variants exist, raises the possibility that a multitude of different subunit combination-dependent micro-binding domains/receptors may be formed. However, amino acids forming the putative binding domain are conserved within the α and β subunit families, leading to the conclusion that the binding site most likely is conserved irrespective of subunit composition of the GABA$_A$ receptor. In agreement with this hypothesis are studies in which the affinity of a series of agonists, partial agonists, and competitive antagonists has been determined using ^{3}H-muscimol as ligand for the binding site. In these studies, a minor variation (of a factor 7) in the K_D-value of ^{3}H-muscimol was observed. Furthermore, a small variation in the K_i-values for all tested compounds was seen, leading to the conclusion that the affinity of the compounds at this site is independent of subunit composition *(32)*. Based on this, it may be concluded that determining subtype selectivity for GABA$_A$ receptor agonists/antagonist ligands, binding studies are only of very limited value. In contrast, functional studies using electrophysiology at oocytes expressing human GABA$_A$ receptors with predetermined subunit combinations clearly demonstrate major differences in both potency and maximum response with variations of α and β subunits. In these studies, where α subunits were varied with fixed β and γ subunits, compounds like THIP and P4S were low-efficacy partial agonists at one combination and full agonists at another *(17,32,33)*. Illustrating the unpredictability of efficacy, the maximum response of P4S varied from being a full agonist to being a competitive antagonist. There are several important implications of these findings:

1. The simplified terminology in which a certain pharmacological quality (e.g., agonist or antagonist) is ascribed to one compound is meaningless when the activity is highly dependent on the receptor subunit composition.
2. The activity in a tissue is highly dependent on the receptor subtype(s) present, whereas a compound like P4S at $\alpha_4\beta_3\gamma_{2s}$ containing receptor present in the thalamic areas will be an antagonist and at the same time at $\alpha_5\beta_3\gamma_{2s}$ containing receptors present in the hippocampus will be an agonist. The interpretation of in vivo data is therefore very complex, and in some cases impossible, if the knowledge about the subtypes present and pharmacological actions at these receptors is absent. These conclusions are not only relevant for GABA receptors. In the G protein-coupled receptor (GPCR) field of research, it has long been known that receptor and G protein density in the target areas are major determinants of the pharmacological profile, but whether this is the case for ionotropic receptors can only be speculated.

4.1. THIP (Gaboxadol): An Intriguing Example of a GABA$_A$ Agonist Exhibiting a Novel Pharmacological Profile

An illustration of this problem concerns the classical GABA$_A$ agonist THIP. THIP is currently being developed as a novel type of hypnotic under the name Gaboxadol. But why would a GABA$_A$ agonist possess a pharmacological profile superior to the benzodiazepines, which are use-dependent and therefore, in principle, less likely to induce desensitization of the receptors than the constant activation by a GABA$_A$ agonist? Because clinical studies with THIP have indicated that sleep quality improving effects have been obtained at plasma concentrations around 1 μ*M (34,35)* and the potency at $\alpha_1\beta_3\gamma_{2s}$ containing receptors expressed in *Xenopus* oocytes is 238 μ*M (17)*, these data clearly suggest that a significant part of the activity of THIP in a complex system may be mediated via receptors others than the synaptic $\alpha_1\beta_3\gamma_{2s}$ containing receptors. In the frontal cortex, the most abundant synaptically located GABA$_A$ receptors are composed of $\alpha_1\beta_{2/3}\gamma_{2s}$ and $\alpha_3\beta_{2/3}\gamma_{2s}$ subunits *(21)*. In contrast to the synaptic receptors, extrasynaptic receptors in the neocortex contain α_4 and δ subunits. In thalamus and other brain regions where α_4 is more abundant than in neocortex and thus easier to investigate, α_4 and δ subunits are reported to co-localize *(36–39)*. Because $\alpha_4\beta_y\delta$ receptors have proven difficult to express in recombinant systems, this subunit composition has not been characterized in oocytes so far. However, in a very recent study, a stable cell line capable of expressing receptors composed of $\alpha_4\beta_3\delta$ subunits has been created. The pharmacology of GABA, THIP, and muscimol at this receptor subtype was compared to that of $\alpha_4\beta_3\gamma_2$ containing receptors. All agonists exhibited very high potencies at $\alpha_4\beta_3\delta$, GABA being the most potent. In addition, THIP displayed superagonist behavior at this subunit assembly with an E_{max} of approx 160% *(40,41)*. Provided that thalamic receptors containing $\alpha_4\beta_y\delta$ receptors are important for the physiology of sleep, the sleep-improving activity of THIP might be attributed to this functional selectivity observed in recombinant systems. In order to shed further light on this question, THIP and a series of directly acting GABA_A ligands were characterized in the rat cortical-wedge preparation. Within the neocortex, the major neuronal cell types are pyramidal cells and nonpyramidal cells. Most pyramidal cells are projection neurons that are thought to use excitatory amino acids as neurotransmitters. Projection neurons have long axons and mediate the output from neocortex to other cortical areas and to subcortical structures *(42)*. The pyramidal cell bodies are located in layer II–VI *(43)*. From the soma of the cell body, a single apical dendrite rises that ascends vertically towards layer I. In addition, from the cell soma, an array of short dendrites spread laterally. Of particular importance to the cortical-wedge preparation are the corticostriatal projection neurons, which provide a massive glutamatergic input from the neocortex to the striatum *(44)*. In contrast, most nonpyramidal cells are local-circuit neurons, which use GABA as neurotransmitter. These cells have short axons and do not project to other cortical areas or subcortical structures. The physiological function of the GABAergic neurons is to focus and refine the firing pattern of the excitatory projection neurons. Thus, neuronal excitability is under GABAergic control.

Immunocytochemical studies of GABA_A receptor subunit localization have revealed that α_1, β_2, and β_3 subunits are equally distributed throughout the six cortical layers, whereas α_2, α_4, γ_2, and δ are enriched in the outermost layers. In contrast, α_3, α_5, and β_1 subunits are concentrated in the inner layers (α_3 predominantly in layer V–VI, α_5 in IV–VI, and β_1 in IV) *(36)*.

As previously mentioned, the corticostriatal projection neurons provide a dense glutamatergic innervation within the neocortical region used for the cortical-wedge preparation. Therefore, originally, this method was used for pharmacological characterization of compounds interacting with the glutamatergic system as described by Harrison and Simmonds *(45)*. Because application of excitatory amino acids and glutamate analogues to the cortical brain slices gives rise to a depolarization of the brain tissue, Harrison and Simmonds quantified the functional outcome by measuring the height of the peak recorded on a chart recorder. In addition, in the original work, it was described that application Mg^{2+}-free Krebs superfusion medium to the brain slices resulted in development of spontaneous activity (seen as fast spikes) within 30 min in some of the slices. The spike frequency increased upon application of glutamate receptor agonists and vanished upon application of the NMDA receptor selective antagonist, (–)-aminophosphonovaleric acid *(45)*. Thus, the spontaneous activity has NMDA receptor origin. Under physiological conditions where the extracellular concentration of Mg^{2+} is approx 1 mM, extracellular Mg^{2+} ions voltage-dependently block the open NMDA receptor channel, thereby providing a very important inhibition on NMDA receptor activation *(46)*. This inhibitory mechanism is lost when Mg^{2+}-free medium is applied to the cortical slices as the Mg^{2+} present in the brain tissue gradually washes out, giving rise to development of spontaneous activity. The rat cortical-wedge preparation is therefore a much more complex system than the oocytes and the contributions from different types of synapses can be characterized.

THIP and other GABA$_A$ agonists in the rat cortical-wedge preparation study were more than 50 times more potent in the wedge at inhibiting synaptic-field potentials than at the cloned $\alpha_1\beta_3\gamma_{2s}$ GABA receptors expressed in oocytes *(47)*. This observation is in agreement with a study by Kemp et al. *(48)*, using the hippocampal slice showed that, when measured with extracellular electrodes, lower agonist potencies were seen. However, using isolated neurons, Kristiansen et al. *(49)* and Hansen et al. *(50)* showed that the potencies of agonists are much higher than in the cortical preparation, suggesting that at the single neuron, when measuring Cl$^-$ fluxes, the sensitivity towards GABA$_A$ agonists are similar to that seen in to *Xenopus* oocytes. It must be emphasized that whereas oocytes express receptors with a confined subunit composition (in this case $\alpha_1\beta_3\gamma_{2s}$), a wide variety of different subunit assemblies are present in neocortex. *In situ* hybridization and immunoprecipitation techniques have revealed that α_1, β_2, β_3, and γ_{2s} are the most abundant subunits in neocortex, whereas α_2, α_3, α_4, α_5, γ_3, and δ are present in lesser amounts *(36,38,39,51)*. One possible reason why THIP acts much more potently in the rat cortical-wedge preparation may be that the effect of THIP in this system is not mediated primarily via $\alpha_1\beta_3\gamma_{2s}$ containing receptors, although these receptor subunits (along with β_2) are the most predominant in neocortex. A previous study *(17)* in *Xenopus* oocytes expressing GABA$_A$ receptors composed of various combinations of α_1, α_3, α_5, β_1, β_2, β_3, γ_1, γ_2, and γ_3 has revealed that THIP exerts the greatest potency at $\alpha_5\beta_3\gamma_2$ receptors (with EC$_{50}$ an values of 40 μM). At this receptor subtype, THIP acts as a full agonist (showing an E_{max} value of 99%) *(17)*. The higher potency of THIP and the behavior as a full agonist in the rat cortical-wedge preparation may therefore be ascribed to a preferential activation of $\alpha_5\beta_3\gamma_{2/3}$ receptors.

Another possible explanation for the high potency of THIP observed in the rat cortical-wedge preparation is that the effect of THIP to a large extent may arise from activation of extrasynaptically located α_4 containing receptors. At $\alpha_4\beta_3\delta$ containing receptors, muscimol, isoguvacine, and THIP all exhibited very high potencies. In addition, THIP displayed superagonist behavior at this subunit assembly with an E_{max} of 160% *(40,41)*. The physiological role of extrasynaptic receptors is presumably to respond to synaptic spillover of neurotransmitter *(7)* and therefore must be sensitive to low concentrations of transmitter. A preferential localization of $\alpha_4\beta_3\delta$ receptors at extrasynaptic sites in neocortex would imply that this receptor subtype is more readily accessible to exogenous GABA$_A$ receptor ligands, because a diffusion of the ligand into the synaptic cleft prior to interaction with the receptor is not necessary. This, and in particular the fact that GABA$_A$ agonists act very potently at $\alpha_4\beta_3\delta$ receptors, suggests that the effects of GABA and THIP in the rat cortical-wedge preparation are mainly mediated via this receptor subtype, and this could be confirmed by evaluating the extent of modulation by benzodiazepines.

If only a low level of receptor activation is required in order to obtain a full response, one would predict that receptor desensitization would be a smaller problem with THIP, muscimol, and other GABA$_A$ agonists used as hypnotics. In an animal study, Lancel and Langebartels showed that 5-d treatment with THIP produced sustained effects on sleep parameters, whereas a benzodiazepine in a similar paradigm induced tolerance-like effects after only 5 d of treatment *(52)*. Accepting that only functionally relevant receptors should be characterized in oocytes or cell lines raises another question: which receptors are functionally relevant?

As illustrated in the case of THIP, the dogma until recently was that the compound would act via synaptic GABA$_A$ receptors, in which case action at $\alpha_{1/2}\beta_{2/3}\gamma_{2s}$ receptors would be the primary target, and hence would not have predicted that extrasynaptic $\alpha_4\beta_{2/3}\delta$ containing receptors now could well be a critical site of action.

Characterization of THIP at $\alpha_1\beta_2\gamma_{2s}$ receptors is therefore not predictive of the activity in vivo and may be directly misleading when it comes to addressing the potential for interaction with the benzodiazepines.

Further illustrating the complex and not yet understood basis for variations in maximum response as function of subunit composition are data for piperidine-4-sulphonic acid (P4S) at cerebellar granule cells. Cerebellar granule cells contain receptors composed of $\alpha_1\beta_{2/3}\gamma_2$, $\alpha_6\beta_{2/3}\gamma_2$, and $\alpha_1\alpha_6\beta_{2/3}\gamma_2$ subunits *(36,39)*. A comparison of the activity of P4S at these three different combinations reveals that at $\alpha_1\beta_2\gamma_2$ and $\alpha_6\beta_2\gamma_2$ containing receptors, P4S is a partial agonist with a maximal response of 15–40% of that of GABA, whereas at $\alpha_1\alpha_6\beta_2\gamma_2$ containing receptors, the maximum response to P4S is 75% of that of GABA and the potency is lower than that at the other two receptor combinations *(50)*. Not only are these data surprising, they are also in contrast to data for P4S at $\alpha_1/\alpha_5/\alpha_1\alpha_5$ or $\alpha_1/\alpha_3/\alpha_1\alpha_3$ containing receptors, where co-expression of two different α subunits resulted in an α_1-like maximum response and a significantly lower potency than at α_1/α_5 or α_1/α_3 containing receptors *(17)*. Thus in two examples, co-expression of different α subunits results in novel pharmacology (see also *[53]*, where the change appears for benzodiazepines), underlining the complexity of the system. Therefore, using artificial expression systems with predetermined subunit may not reflect the action of ligands at native receptors.

5. CONCLUDING REMARKS

In conclusion, although binding assays have been used for the past few decades to determine selectivity, the basic assumption that functional potency correlates with receptor affinity has been proven worng several times. In a receptor system like the GABA$_A$ receptors, where the binding site is conserved between all receptor combinations characterized so far; affinity measurements are not likely to identify receptor selectivity. Furthermore, because much attention in medicinal research is focused on the development of partial agonists or modulators of the functional consequences of receptor activation, only functional assays will be able to detect these differences. When attempting to identify possible therapeutic compounds that selectively activate one or two of a number of receptor subtypes, it is thus crucial to establish their profile in several different in vitro assays, correlating this with additional data from native tissue where possible. This can then be used in conjunction with in vivo data, as well as using transgenic mice if these are available. Validating the models of subtype selectivity using multiple approaches vastly strengthens our understanding of the role of these individual receptor subtypes and will enable the discovery of safer and more selective therapeutics.

REFERENCES

1. Thomsen, C. and Ebert, B. (2002) Modulators of the GABAA receptor complex: novel therapeutic aspects, in *Glutamate and GABA Receptors and Transporters* (Egebjerg, J., Schousboe, A., and Krogsgaard-Larsen, P., eds.), Taylor & Francis, London, pp. 407–427.

2. Rudolph, U., Crestani, F., and Möhler, H. (2001) GABA(A) receptor subtypes: dissecting their pharmacological functions. *Trends Pharmacol. Sci.* **22,** 188–194.

3. Macdonald, R. L. and Olsen, R. W. (1994) GABAA receptor channels. *Annu. Rev. Neurosci.* **17,** 569–602.

4. Maksay, G., Thompson, S. A., and Wafford, K. A. (2000) Allosteric modulators affect the efficacy of partial agonists for recombinant GABA(A) receptors. *Br. J. Pharmacol.* **129,** 1794–1800.

5. Mortensen, M., Frølund, B., Jørgensen, A., Liljefors, T., Krogsgaard-Larsen, P., and Ebert, B. (2002) Activity of novel 4-PIOL analogues at human alpha(1)beta(2)gamma(2S) GABA(A) receptors-correlation with hydrophobicity. *Eur. J. Pharmacol.* **451,** 125–132.

6. Bai, D., Zhu, G., Pennefather, P., Jackson, M. F., MacDonald, J. F., and Orser, B. A. (2001) Distinct functional and pharmacological properties of tonic and quantal inhibitory postsynaptic currents mediated by gamma-aminobutyric acid(A) receptors in hippocampal neurons. *Mol. Pharmacol.* **59,** 814–824.

7. Brickley, S. G., Cull-Candy, S. G., and Farrant, M. (1996) Development of a tonic form of synaptic inhibition in rat cerebellar granule cells resulting from persistent activation of GABAA receptors. *J. Physiol.* **497(Pt 3),** 753–759.

8. Nusser, Z., Sieghart, W., and Somogyi, P. (1998) Segregation of different GABAA receptors to synaptic and extrasynaptic membranes of cerebellar granule cells. *J. Neurosci.* **18,** 1693–1703.

9. Nusser, Z. and Mody, I. (2002) Selective modulation of tonic and phasic inhibitions in dentate gyrus granule cells. *J. Neurophysiol.* **87,** 2624–2628.

10. Sieghart, W., Fuchs, K., Tretter, V., Ebert, V., Jechlinger, M., Hoger, H., and Adamiker, D. (1999) Structure and subunit composition of GABA(A) receptors. *Neurochem. Int.* **34,** 379–385.

11. Amin, J., Brooks-Kayal, A., and Weiss, D. S. (1997) Two tyrosine residues on the alpha subunit are crucial for benzodiazepine binding and allosteric modulation of gamma-aminobutyric acidA receptors. *Mol. Pharmacol.* **51,** 833–841.

12. Boileau, A. J., Kucken, A. M., Evers, A. R., and Czajkowski, C. (1998) Molecular dissection of benzodiazepine binding and allosteric coupling using chimeric gamma-aminobutyric acidA receptor subunits. *Mol. Pharmacol.* **53,** 295–303.

13. Duncalfe, L. L., Carpenter, M. R., Smillie, L. B., Martin, I. L., and Dunn, S. M. (1996) The major site of photoaffinity labeling of the gamma-aminobutyric acid type A receptor by [3H]flunitrazepam is histidine 102 of the alpha subunit. *J. Biol. Chem.* **271,** 9209–9214.

14. Sigel, E. and Buhr, A. (1997) The benzodiazepine binding site of GABAA receptors. *Trends Pharmacol. Sci.* **18,** 425–429.

15. Smith, G. B. and Olsen, R. W. (1995) Functional domains of GABAA receptors. *Trends Pharmacol. Sci.* **16,** 162–168.

16. Amin, J. and Weiss, D. S. (1993) GABAA receptor needs two homologous domains of the beta-subunit for activation by GABA but not by pentobarbital. *Nature* **366,** 565–569.

17. Ebert, B., Wafford, K. A., Whiting, P. J., Krogsgaard-Larsen, P., and Kemp, J. A. (1994) Molecular pharmacology of gamma-aminobutyric acid type A receptor agonists and partial agonists in oocytes injected with different alpha, beta, and gamma receptor subunit combinations. *Mol. Pharmacol.* **46,** 957–963.

18. Galzi, J. L. and Changeux, J. P. (1994) Neurotransmitter-gated ion channel as unconventional allosteric proteins. *Curr. Opin. Struct. Biol.* **4,** 554–565.

19. Cromer, B. A., Morton, C. J., and Parker, M. W. (2002) Anxiety over GABA(A) receptor structure relieved by AChBP. *Trends Biochem. Sci.* **27,** 280–287.

20. Trudell, J. (2002) Unique assignment of inter-subunit association in GABA(A) alpha 1 beta 3 gamma 2 receptors determined by molecular modeling. *Biochim. Biophys. Acta* **1565,** 91–96.

21. McKernan, R. M. and Whiting, P. J. (1996) Which GABAA-receptor subtypes really occur in the brain? *Trends Neurosci.* **19,** 139–143.

22. Woods, J. H., Katz, J. L., and Winger, G. (1992) Benzodiazepines: use, abuse, and consequences. *Pharmacol. Rev.* **44,** 151–347.

23. Wieland, H. A. and Lüddens, H. (1994) Four amino acid exchanges convert a diazepam-insensitive, inverse agonist-preferring GABAA receptor into a diazepam-preferring GABAA receptor. *J. Med. Chem.* **37,** 4576–4580.

24. Crestani, F., Martin, J. R., Möhler, H., and Rudolph, U. (2000) Resolving differences in GABAA receptor mutant mouse studies. *Nat. Neurosci.* **3,** 1059.

25. Löw, K., Crestani, F., Keist, R., et al. (2000) Molecular and neuronal substrate for the selective attenuation of anxiety. *Science* **290,** 131–134.

26. McKernan, R. M., Rosahl, T. W., Reynolds, D. S., et al. (2000) Sedative but not anxiolytic properties of benzodiazepines are mediated by the GABA(A) receptor alpha1 subtype. *Nat. Neurosci.* **3,** 587–592.

27. Rudolph, U., Crestani, F., Benke, D., Brunig, I., Benson, J. A., Fritschy, J. M., et al. (1999) Benzodiazepine actions mediated by specific gamma-aminobutyric acid(A) receptor subtypes. *Nature* **401,** 796–800.

28. Homanics, G. E., Ferguson, C., Quinlan, J. J., et al. (1997) Gene knockout of the alpha6 subunit of the gamma-aminobutyric acid type A receptor: lack of effect on responses to ethanol, pentobarbital, and general anesthetics. *Mol. Pharmacol.* **51,** 588–596.

29. Brickley, S. G., Cull-Candy, S. G., and Farrant, M. (1999) Single-channel properties of synaptic and extrasynaptic GABAA receptors suggest differential targeting of receptor subtypes. *J. Neurosci.* **19,** 2960–2973.

30. Lüscher, B. and Fritschy, J. M. (2001) Subcellular localization and regulation of GABAA receptors and associated proteins. *Int. Rev. Neurobiol.* **48,** 31–64.

31. Smith, G. B. and Olsen, R. W. (1994) Identification of a [^{3}H]muscimol photoaffinity substrate in the bovine gamma-aminobutyric acidA receptor alpha subunit. *J. Biol. Chem.* **269,** 20380–20387.

32. Ebert, B., Thompson, S. A., Saounatsou, K., McKernan, R., Krogsgaard-Larsen, P., and Wafford, K. A. (1997) Differences in agonist/antagonist binding affinity and receptor transduction using recombinant human gamma-aminobutyric acid type A receptors. *Mol. Pharmacol.* **52,** 1150–1156.

33. Ebert, B., Mortensen, M., Thompson, S. A., Kehler, J., Wafford, K. A., and Krogsgaard-Larsen, P. (2001) Bioisosteric determinants for subtype selectivity of ligands for heteromeric GABA(A) receptors. *Bioorg. Med. Chem. Lett.* **11,** 1573–1577.

34. Faulhaber, J., Steiger, A., and Lancel, M. (1997) The GABAA agonist THIP produces slow wave sleep and reduces spindling activity in NREM sleep in humans. *Psychopharmacology (Berl.)* **130,** 285–291.

35. Madsen, S. M., Lindeburg, T., FØlsgård, S., Jacobsen, E., and Sillesen, H. (1983) Pharmacokinetics of the gamma-aminobutyric acid agonist THIP (Gaboxadol) following intramuscular administration to man, with observations in dog. *Acta Pharmacol. Toxicol. (Copenh.)* **53,** 353–357.

36. Pirker, S., Schwarzer, C., Wieselthaler, A., Sieghart, W., and Sperk, G. (2000) GABA(A) receptors: immunocytochemical distribution of 13 subunits in the adult rat brain. *Neuroscience* **101,** 815–850.

37. Sur, C., Farrar, S. J., Kerby, J., Whiting, P. J., Atack, J. R., and McKernan, R. M. (1999) Preferential coassembly of alpha4 and delta subunits of the gamma-aminobutyric acidA receptor in rat thalamus. *Mol. Pharmacol.* **56,** 110–115.

38. Wisden, W., Herb, A., Wieland, H., Keinanen, K., Lüddens, H., and Seeburg, P. H. (1991) Cloning, pharmacological characteristics and expression pattern of the rat GABAA receptor alpha 4 subunit. *FEBS Lett.* **289,** 227–230.

39. Wisden, W., Laurie, D. J., Monyer, H., and Seeburg, P. H. (1992) The distribution of 13 GABAA receptor subunit mRNAs in the rat brain. I. Telencephalon, diencephalon, mesencephalon. *J. Neurosci.* **12,** 1040–1062.

40. Adkins, C. E., Pillai, G. V., Kerby, J., et al. (2001) alpha4beta3delta GABA(A) receptors characterized by fluorescence resonance energy transfer-derived measurements of membrane potential. *J. Biol. Chem.* **276,** 38934–38939.

41. Brown, N., Kerby, J., Bonnert, T. P., Whiting, P. J., and Wafford, K. A. (2002) Pharmacological characterization of a novel cell line expressing human alpha(4)beta(3)delta GABA(A) receptors. *Br. J. Pharmacol.* **136,** 965–974.

42. Parnavelas, J. G. (1984) Physiological properties of identified neurons, in *Functional Properties of Cortical Cells* (Jones, E. G. and Peters. A., eds.), Plenum Press, New York, NY, pp. 205–239.

43. Zilles, K. (1990) Anatomy of the neocortex: cytoarchitecture and myeloarchitecture, in *The Cerebral Cortex of the Rat* (Kolb, B. and Tees, R. C., eds.), MIT Press, Cambridge, MA, pp. 76–95.

44. Ottersen, O. P., Hjelle, O. P., Osen, K. K., and Laake, J. H. (1995) Amino acid transmitters, in *The Rat Nervous System* (Paxinos, G., ed.), Academic Press, Inc., San Diego, CA, pp. 1017–1037.

45. Harrison, N. L. and Simmonds, M. A. (1985) Quantitative studies on some antagonists of N-methyl-D-aspartate in slices of rat cerebral cortex. *Br. J. Pharmacol.* **84,** 381–391.

46. Mayer, M. L., Westbrook, G. L., and Guthrie, P. B. (1984) Voltage-dependent block by Mg^{2+} of NMDA responses in spinal cord neurones. *Nature* **309,** 261–263.

47. Ebert, B., Storustovu, S., Mortensen, M., and Frølund, B. (2002) Characterization of GABA(A) receptor ligands in the rat cortical wedge preparation: evidence for action at extrasynaptic receptors? *Br. J. Pharmacol.* **137,** 1–8.

48. Kemp, J. A., Marshall, G. R., and Woodruff, G. N. (1986) Quantitative evaluation of the potencies of GABA-receptor agonists and antagonists using the rat hippocampal slice preparation. *Br. J. Pharmacol.* **87,** 677–684.

49. Kristiansen, U., Lambert, J. D., Falch, E., and Krogsgaard-Larsen, P. (1991) Electrophysiological studies of the GABAA receptor ligand, 4-PIOL, on cultured hippocampal neurones. *Br. J. Pharmacol.* **104,** 85–90.

50. Hansen, S. L., Ebert, B., Fjalland, B., and Kristiansen, U. (2001) Effects of GABA(A) receptor partial agonists in primary cultures of cerebellar granule neurons and cerebral cortical neurons reflect different receptor subunit compositions. *Br. J. Pharmacol.* **133,** 539–549.

51. Fritschy, J. M. and Möhler, H. (1995) GABAA-receptor heterogeneity in the adult rat brain: differential regional and cellular distribution of seven major subunits. *J. Comp. Neurol.* **359,** 154–194.

52. Lancel, M. and Langebartels, A. (2000) gamma-aminobutyric Acid(A) (GABA(A)) agonist 4,5,6,7-tetrahydroisoxazolo[4,5-c]pyridin-3-ol persistently increases sleep maintenance and intensity during chronic administration to rats. *J. Pharmacol. Exp. Ther.* **293,** 1084–1090.

53. Verdoorn, T. A. (1994) Formation of heteromeric gamma-aminobutyric acid type A receptors containing two different alpha subunits. *Mol. Pharmacol.* **45,** 475–480.

5

In Vivo Function of GABA$_A$ Receptor Subtypes Unraveled With Mutant Mice

Jean-Marc Fritschy

1. INTRODUCTION

GABA$_A$ receptors mediate most of the fast inhibitory transmission in the brain, and therefore modulate almost every aspect of brain function. In addition, they represent a major site of action for clinically important drugs, including benzodiazepines, barbiturates, neurosteroids, some general anesthetics, as well as drugs of abuse, such as ethanol (1–4). All these drugs act by increasing GABA function by allosteric interaction or by direct action on the channel. Clinically, the significance of GABA$_A$ receptor function is underscored by the multiple neurological and psychiatric diseases for which an alteration in the GABAergic system has been postulated (5), including epilepsy (6–8), anxiety disorders (9), ethanol dependence (10), Huntington's disease (11), Angelman syndrome (12), and schizophrenia (13–15).

The generation of mutant mice, which are either deficient in a particular GABA$_A$ receptor subunit or carry a subunit with a targeted mutation affecting binding of benzodiazepine site ligands, aims at unraveling the in vivo function of GABA$_A$ receptors and at producing animal models of these disorders. In this review, I will summarize major findings derived from the analysis of these mice, focusing on the role of individual subunits for assembly, function, and pharmacology of GABA$_A$ receptors, and on the disease models provided by specific mutations. These studies have shown that considerable compensatory mechanisms are activated in the absence of particular GABA$_A$ receptor subunits, allowing an essentially normal brain function in the absence of a large fraction of GABA$_A$ receptors. These compensatory mechanisms are interesting on their own, but they limit what can be learned about the functional role of the subunits under consideration.

Subheading 2 gives an overview of current concepts on the structural and functional heterogeneity of GABA$_A$ receptors. In Subheading 3, I will focus on results derived from the analysis of GABA$_A$ receptor subunit knockout mice (or from mice with a chromosomal deletion affecting GABA$_A$ receptor subunits), and in Subheading 4, findings derived from mice carrying a point mutation affecting diazepam binding in specific subpopulations of GABA$_A$ receptors will be discussed.

From: *Molecular Neuropharmacology: Strategies and Methods*
Edited by: A. Schousboe and H. Bräuner-Osborne © Humana Press Inc., Totowa, NJ

2. STRUCTURAL AND FUNCTIONAL HETEROGENEITY OF GABA$_A$ RECEPTORS

2.1. GABA$_A$ Receptor Subtypes

GABA$_A$ receptors belong to the superfamily of ligand-gated ion channels, along with nicotinic acetylcholine receptors, glycine receptors, and serotonin type 3 receptors *(16,17)*. GABA$_A$ receptors form multimeric complexes assembled from a family of at least 21 constituent subunits (α_{1-6}, β_{1-4}, γ_{1-4}, δ, ρ_{1-3}, θ, π) *(18,19)*. Their molecular heterogeneity, which is much larger than that of any other ligand-gated ion channel, allows the formation of multiple functionally and pharmacologically distinct GABA$_A$ receptors in the central nervous system (CNS) (reviewed in *[20]*).

The subunit composition and stoichiometry of native GABA$_A$ receptors have not been elucidated. The available evidence favors the existence of pentameric complexes containing 2α, 2β, 1γ subunit variants *(21–24)*. Immunochemical, pharmacological, and functional analyses of GABA$_A$ receptors give convergent results that the majority of GABA$_A$ receptors contain a single type of α and β subunit variant, along with the γ_2 subunit, which is essential for the formation of a benzodiazepine binding site. The $\alpha_1\beta_2\gamma_2$ combination represents the largest population of GABA$_A$ receptors, followed by $\alpha_2\beta_3\gamma_2$ and $\alpha_3\beta_3\gamma_2$. Receptors containing the α_4, α_5, or α_6 subunit, as well as the β_1, γ_1, γ_3, δ, π, and θ subunit, form minor receptor populations. The ρ subunits are primarily expressed in the retina and correspond to the so-called GABA$_C$ receptors *(25)*.

The pharmacological analysis of GABA$_A$ receptors immunoprecipitated with antibodies against specific α subunit variants indicated that each α subunit variant largely corresponds to a distinct GABA$_A$ receptor subtype. The α_1-, α_2-, α_3-, and α_5-GABA$_A$ receptors represent diazepam-sensitive receptors, whereas α_4- and α_6-GABA$_A$ receptors are insensitive to diazepam *(26,27)*. The former are distinguished further by their affinity to zolpidem ($\alpha_1 > \alpha_2 = \alpha_3 >> \alpha_5$) and various β-carbolines ($\alpha_1 > \alpha_2 = \alpha_3$) *(1,20)*. Functionally, the importance of α subunit variants for defining GABA$_A$ receptor subtypes was confirmed by studies showing that the type of α subunit determines ligand affinity and the kinetics of receptor deactivation *(28–30)*. In addition to these studies, it has been shown that the presence of the δ subunit, which is associated mainly with the α_4, α_6, and α_1 subunit in vivo, results in markedly increased agonist affinity and apparent lack of desensitization of the corresponding receptors *(31–33)*.

Immunohistochemical analyses of the distribution of GABA$_A$ receptor subtypes based on the visualization of α subunit variants revealed a region- and neuron-specific distribution pattern that is largely conserved across species *(34–38)*. On the cellular level, even a synapse-specific distribution of GABA$_A$ receptor subtypes has been evidenced, in particular in neurons expressing multiple GABA$_A$ receptor subtypes, such as hippocampal pyramidal neurons. In these cells, a high level of α_1, α_2, and α_5 subunit expression has been reported, along with β_{1-3} and γ_2 subunit *(36,37)*, suggesting that they express at least three main GABA$_A$ receptor subtypes. α_1-GABA$_A$ receptors are located postsynaptically in a majority of somatodendritic synapses and to a lesser extent in the axon initial segment; in contrast, α_2-GABA$_A$ receptors are particularly abundant in the axon initial segment and are only few in somatodendritic synapses *(39,40)*. Finally, α_5-GABA$_A$ receptors have an extrasynaptic localization, being distributed throughout the somatodendritic compartment of hippocampal pyramidal cells without being aggregated at postsynaptic sites *(41,42)*. These observations strongly suggest that individual GABA$_A$ receptor subtypes are present in distinct neuronal circuits. The generation of mutant mice with point mutation affecting benzodiazepine binding in a given GABA$_A$ receptor subtype therefore provides a powerful tool to investigate neuronal circuits mediating a specific behavior (*see* Subheading 4.).

The chromosomal localization of GABA$_A$ receptor subunit genes revealed the existence of several clusters on distinct chromosomes ($\alpha_1/\alpha_6/\beta_2/\gamma_2$ on human 5q, $\alpha_5/\beta_3/\gamma_3$ on 15q, $\alpha_2/\alpha_4/\beta_1/\gamma_1$ on 4p, $\alpha_3/\epsilon/\theta$ on Xq28, ρ_1/ρ_2 on 6q) (reviewed in *[43–45]*). Subunits that form the main GABA$_A$ receptor subtypes are generally part of the same cluster, suggesting that these chromosomal arrangements allow coordinated regulation of subunit expression by shared regulatory sites. However, they are of potential concern for engineering targeted mutations, because alteration of the structure of one gene, for instance by insertion of a neomycin-resistance gene, can affect the expression of the other genes in the cluster, as shown by Uusi-Oukai et al. *(43)* for the $\alpha_1/\alpha_6/\beta_2/\gamma_2$ cluster.

3. ANALYSIS OF KNOCKOUT MICE

3.1. Assembly of GABA$_A$ Receptors

In recombinant expression systems, homomeric GABA$_A$ receptors are expressed at very low levels, with the exception of β_3- and ρ_1-GABA$_A$ receptors, which are readily detected at the cell surface *(46–49)*. In contrast, receptors containing various $\alpha\beta$ subunit combinations form functional Cl$^-$ channels activated by GABA, indicating that receptor assembly and/or membrane targeting readily occur for heteromeric complexes. Inclusion of the γ_2 subunit is an essential requirement for the formation of a benzodiazepine binding site *(50)*, and for proper channel conductance *(32)*. These properties of the γ_2 subunit are valid also for native GABA$_A$ receptors *(51,52)*.

The analysis of mice deficient for a particular α or β subunit variant confirmed that these subunits are required for proper assembly of the corresponding receptor complexes in neurons *(53–59)*. In all these mutants, the gene deletion resulted in a reduction of the number of GABA$_A$ receptors roughly proportional to the abundance of this subunit in wild-type animals. Surprisingly, in all these mice, the resulting phenotype is rather mild; even $\alpha_1^{0/0}$ or $\beta_2^{0/0}$ mice, in which the loss of GABA$_A$ receptors exceeds 50%, are viable and fertile *(54,60)*. In neurons expressing multiple GABA$_A$ receptor subtypes, such as hippocampal pyramidal cells and olfactory bulb granule cells, no replacement of the missing subunit by a homologous variant was observed. Thus, in absence of the α_5 subunit gene, only the corresponding GABA$_A$ receptor subtype was affected, whereas α_2-GABA$_A$ receptors present in the same cells were unaffected *(61)*.

Knockout mice have also revealed further subunit requirements for the assembly of particular GABA$_A$ receptor subtypes. Thus, the absence of the α_6 subunit in cerebellar granule cells results in a complete loss of δ subunit protein, although mRNA expression levels were unaffected *(59,62)*. These results indicate that the δ subunit cannot be incorporated in a receptor lacking α_6 in cerebellar granule cells. Likewise, in $\delta^{0/0}$ mice, the levels of α_4 subunit protein are reduced in all forebrain regions in which the δ subunit is normally abundant, pointing to a preferential co-assembly of these subunits in several neuronal populations *(63)*.

In at least two examples, a compensatory upregulation of certain subunits has been observed in GABA$_A$ receptor subunit knockout mice. First, in $\delta^{0/0}$ mice, expression of the γ_2 subunit was enhanced selectively in brain regions expressing the δ subunit in wild-type *(63–65)*. The newly formed receptors are not aberrant, but represent an increased fraction of a receptor subtype normally present in these neurons. These results suggest that the δ and γ_2 subunits normally compete for assembly in GABA$_A$ receptor subtypes. They raise interesting questions about the physiological significance of this switch in expression, because receptors containing the γ_2 or δ subunit have distinct functional and pharmacological properties *(66,67)*. In particular, it will be important to determine whether the $\alpha_4\beta_x\gamma_2$ receptors formed in the dentate gyrus and thalamus of $\delta^{0/0}$ mice are extrasynaptic or synaptic. The second example concerns the upregulation of the α_2 and α_3 subunits in one line of α_1 subunit-deficient mice, as revealed by Western blotting and pharmacological analysis *(60)*. It is not known, however, whether these subunits are increased in neurons that normally express the α_1 subunit, or in other cell types.

The general rule, namely the lack of compensation of a missing subunit by overexpression of another homologous subunit variant, suggests that GABA$_A$ receptor subtypes are assembled independently from each other and that they are not interchangeable. This conclusion is in line with the demonstration that cell-type specific expression and subcellular targeting of GABA$_A$ receptor subtypes is governed autonomously *(41)*. It also reinforces the notion that GABA$_A$ receptor subtypes are present in distinct neuronal circuits and mediate distinct functions in the CNS.

The reduced GABAergic function resulting from the absence of a given GABA$_A$ receptor subtype is therefore likely compensated for by other, unknown mechanisms to maintain the appropriate level of neuronal excitability and membrane depolarization. Such an adaptation has been demonstrated, for example, in cerebellar granule cells *(68)*, which normally express high levels of α_6-GABA$_A$ receptors. These receptors provide a form of tonic inhibition that disappears in $\alpha_6^{0/0}$ mice. However, the level of granule cell excitability remains unchanged, owing to upregulation of a compensatory voltage-independent K$^+$ conductance *(68)*. The functional and behavioral analysis of GABA$_A$ receptor subunit knockout mice has to take into account the possibility of such adaptations in interpreting any change observed in the mutants.

3.2. Postsynaptic Clustering of GABA$_A$ Receptors

Unlike the α and β subunit variant tested so far, targeted deletion of the γ_2 subunit, which is present in >90% of benzodiazepine-sensitive GABA$_A$ receptors in brain *(69)*, does not result in a deficit of GABA$_A$ receptor binding sites in the brain. However, in view of the critical role of the γ_2 subunit for the formation of a benzodiazepine binding site, mice lacking the γ_2 subunit exhibit over 90% loss of benzodiazepine binding sites at birth *(52)*. No change in the expression of major GABA$_A$ receptor subunits was observed compared to wild-type. Most of these mutants die perinatally, although they show no deficit in brain morphology, or in any of the major organs of the body. Mutant mice surviving up to 12–15 d (less than 10%) exhibit a strong growth deficit, but the development of their brain proceeds as in wild-type, based on the differentiation of cortical layers, maturation of whisker-related structures in the somatosensory system, and differentiation of cerebellar granule cells *(52)*. Most strikingly, immunohistochemical analysis revealed a profound deficit in the number of postsynaptic gephyrin and GABA$_A$ receptor clusters *(70)*, suggesting that impairment of synaptic GABAergic transmission, rather than a lack of GABA$_A$ receptors *per se*, is incompatible with postnatal survival.

The absence of postsynaptic GABA$_A$ receptor and gephyrin clusters was confirmed in primary cultures of cortical and hippocampal neurons *(70)*, indicating that the γ_2 subunit is required for clustering of gephyrin at postsynaptic sites. Downregulation of gephyrin expression by treatment of cultures with antisense oligonucleotides also abolished clustering of GABA$_A$ receptors, providing strong evidence for a bi-directional interaction between gephyrin and the γ_2 subunit for the formation of postsynaptic GABA$_A$ receptor aggregates. These findings were subsequently confirmed by the analysis of gephyrin-deficient mice, which also exhibit a pronounced deficit of GABA$_A$ receptor clustering *(71)*. Interestingly, postsynaptic clustering of dystrophin and β- dystroglycan, which are co-localized with GABA$_A$ receptors and gephyrin in GABAergic synapses of hippocampal neurons, is not affected by the absence of the γ_2 subunit *(72)*.

Replacement of the missing γ_2 subunit by transgenic overexpression of the γ_3 subunit in γ_2 subunit-deficient mice successfully restored clustering of GABA$_A$ receptors and gephyrin, demonstrating that the domains required for interaction with gephyrin are also present on the γ_3 subunit *(73)*. However, this strategy failed to rescue the mutant mice from perinatal lethality, owing either to unidentified dominant-negative effects of the transgenic γ_3 subunit, or to insufficient expression levels.

3.3. Pharmacological Properties of GABA$_A$ Receptor Subtypes

Among the multiple allosteric modulators of GABA$_A$ receptors, neurosteroids, ethanol, and general anesthetics have received considerable attention in GABA$_A$ receptor subunit knockout mice, in view of the possibility that specific GABA$_A$ receptor subtypes represent a target of action for these drugs. Thus, investigations of δ subunit-deficient mice revealed a greatly reduced sensitivity to neurosteroids, accompanied by altered behavioral responses to ethanol, suggesting a causal relationship *(74,75)*. Specifically, $\delta^{0/0}$ mice consume less ethanol in a free-choice test and show reduced withdrawal symptoms following cessation of chronic ethanol exposure, as well as reduced anticonvulsant effects of ethanol. These features suggested that neurosteroids, possibly acting on δ subunit-containing GABA$_A$ receptors, might mediate some effects of ethanol in vivo. The precise localization and function of these receptors in wild-type mice remains to be determined, however, because $\alpha_6^{0/0}$ mice do not show any behavioral alteration to ethanol *(76)*, in spite of the accompanying loss of δ subunit in cerebellar granule cells.

Another potential target of ethanol action on GABA$_A$ receptors has been suggested to be the long splice variant of the γ_2 subunit, γ_{2L}, which carries an additional consensus phosphorylation site by protein kinase C (PKC) *(77)*. However, subsequent analysis of recombinant expression systems led to conflicting results, because ethanol potentiation of GABA$_A$ receptor currents could also be observed for the γ_{2S} subunit *(78,79)*. The generation of $\gamma_{2L}^{0/0}$ mice revealed no functional deficit of GABA$_A$ receptors, and failed to detect any difference in the behavioral effects of ethanol between $\gamma_{2L}^{0/0}$ and wild-type mice, indicating that this splice variant is not required for the modulatory action of ethanol in vitro or in vivo *(80)*. However, $\gamma_{2L}^{0/0}$ mice have been shown to be more sensitive to the sleep-inducing effects of benzodiazepine agonists, possibly owing to an apparent increase of affinity of their $GABA_A$ receptors for benzodiazepine agonists, and a decrease for inverse agonists *(81)*.

Mice deficient in β_3 subunit have been analyzed extensively to determine whether the corresponding $GABA_A$ receptor subtypes mediate some of the actions of general anesthetics *(82,83)*. The results indicated that $\beta_3^{0/0}$ mice are more resistant to the immobilizing (tail clamp) but not obtunding (loss of righting reflex) effects of enflurane and halothane. Mutant mice were also more resistant to etomidate and midazolam, as measured by the sleep time induced by these drugs *(82)*. Although these findings suggest that β_3-$GABA_A$ receptors are a likely target for the immobilizing action of volatile anesthetics, electrophysiological recordings performed in isolated spinal cord failed to reveal any difference in sensitivity between wild-type and mutant mice to enflurane *(83)*. A potentially confounding problem for investigating analgesic action of drugs in $\beta_3^{0/0}$ mice is the fact that these mice display altered sensory thresholds, as shown by their enhanced responsiveness to thermal stimuli and tactile allodynia *(84)*. Furthermore, the antinociceptive effects of baclofen were also reduced in mutants compared to wild-type, suggesting that lack of β_3-$GABA_A$ receptors can affect the action of unrelated analgesic drugs *(84)*. Avoidance of these confounding factors will require the generation of conditional knockout mice, in which the gene deletion is induced in adult animals.

3.4. Role of GABA_A Receptors During Ontogeny

As discussed earlier, a major lesson from the analysis of $GABA_A$ receptor subunit-deficient mice was that the severity of their phenotype does not correlate with the loss of $GABA_A$ receptors induced by the mutation. Therefore, deletion of the γ_2 subunit gene results in a neonatal lethal phenotype, but does not alter the expression of $GABA_A$ receptor subunits or the formation and membrane targeting of functional $GABA_A$ receptors *(52)*. On the opposite site, deletion of the α_1, β_2, or β_3 subunit genes resulted in massive reductions in $GABA_A$ receptors, but produced relatively mild phenotypes, in particular in $\alpha_1^{0/0}$ and $\beta_2^{0/0}$ mutants *(54)*. β_3 subunit knockout mice have cleft palate *(12)*, in line with the results of a chromosomal deletion encompassing the β_3 subunit gene *(85)*, but show no overt anomaly in brain structure at birth. Perinatal mortality of these mice is high, but this does not directly correlate with the presence or absence of cleft palate *(57)*.

Altogether, these results were unexpected because of the speculated importance of $GABA_A$ receptors during embryogenesis (reviewed in *[86]*). However, they are in line with the findings that GAD65/GAD67 double-knockout mice have no obvious structural brain abnormalities at birth *(87)* and that vesicular neurotransmitter release is dispensable for brain development and synaptogenesis, as demonstrated in Munc13 and Munc18-knockout mice *(88,89)*. Likewise, $GABA_B$ receptor knockout mice exhibit no evident brain abnormalities *(90,91)*. These convergent results therefore indicate that functional GABAergic transmission is dispensable for proper embryonic development of the brain. It has to be emphasized, however, that superficial analysis of brain morphology might be insufficient to rule out any abnormalities. For example, the size of the ventromedial hypothalamic nucleus and its content in estrogen-receptor-positive neurons is increased in $\beta_3^{0/0}$ mice, a phenotype that could be replicated in vitro by chronic bicuculline treatment of slice cultures *(92)*. Likewise, the development of several populations of hypothalamic neurons, including cells releasing gonadotropin-releasing hormone, growth hormone-releasing hormone, and somatostatin, is dependent on appropriate GABA$_A$ receptor function, as shown in $\gamma_2^{0/0}$ mice *(93,94)*.

γ_2-Subunit-deficient mice also provided the opportunity to test the functional relevance of the γ_{2S} and γ_{2L} splice variants, by transgenic overexpression of either isoform in $\gamma_2^{0/0}$ mice *(95,96)*. In both instances, mutant mice were undistinguishable from wild-type, and clustering of GABA$_A$ receptors and gephyrin was reinstated. Behaviorally, no differences were noted between mice expressing γ_{2S} and γ_{2L} with regard to anxiety, motor activity, and acute benzodiazepine or ethanol tolerance *(95)*. The additional amino acid segment present in the intracellular loop of the γ_{2L} subunit, including the consensus phosphorylation site for PKC, is therefore not of vital importance.

3.5. Functional Role of GABA$_A$ Receptors

One of the main obstacles to studying the functional role of individual GABA$_A$ receptor subtypes in vivo is the fact that the majority of neurons express multiple populations of receptors. Even at the level of individual synapses, it is almost impossible to rule out that evoked and spontaneous synaptic events are produced by the activation of several GABA$_A$ receptor subtypes. The functional properties of a given receptor subtype can, however, be inferred indirectly in knockout mice, by analyzing the parameters altered in the absence of the subunit of interest. This has been shown for α_1-GABA$_A$ receptors, which are characterized by rapid decay kinetics. In the cerebral cortex and cerebellum, these receptors appear postnatally to become the predominant subtype expressed in adult animals *(97)*. Although a correlation between increased α_1 subunit expression and acceleration of decay kinetics during development has been shown by several groups *(98–100)*, various mechanisms have been proposed to explain this phenomenon, including maturation of GABA uptake systems. Direct evidence that it was owing to the α_1 subunit was provided by the analysis of $\alpha_1^{0/0}$ mice *(101,102)*. Interestingly, the prolongation of decay kinetics observed in cerebral cortex in mutant mice could be reproduced in vitro by α_1 subunit mRNA antisense treatment of organotypic cultures during 1 wk *(102)*. This allowed ruling out potential compensatory factors emerging in knockout mice during embryogenesis that would alter channel kinetics. However, a major difference between the two experimental settings was that the amplitude of spontaneous inhibitory postsynaptic currents in antisense-treated cultures was much reduced, whereas these currents were of similar size in $\alpha_1^{0/0}$ and wild-type mice. This observation indicated that acute blockade of the α_1 subunit mRNA translation results in a reduced number of GABA$_A$ receptors, whereas deletion of the α_1 subunit gene activates mechanisms ensuring the maintenance postsynaptic inhibitory function in spite of the profound loss of receptors *(102)*.

In selected brain regions, a single GABA$_A$ receptor subtype predominates to such an extent that its deletion by inactivation of one of its constituent subunit genes allows analyzing the role of GABA$_A$ receptor-mediated function in the affected neurons. For instance, $\beta_3^{0/0}$ mice provided the opportunity to nearly completely abolish the expression of GABA$_A$ receptors in specific neurons. Thus, in olfactory bulb granule cells, which express a mixture of synaptic and extrasynaptic GABA$_A$ receptors containing the β_3 subunit *(42,61,103)*, the absence of the β_3 subunit results in a nearly complete loss of GABA$_A$ receptors *(104)*. As a consequence, a strong increase in miniature postsynaptic inhibitory currents (mIPSCs) has been observed in olfactory bulb principal cells (mitral cells and tufted cells), along with an increased in the amplitude of θ and γ oscillations recorded in vivo in mutant mice. These changes were reflected by complex alterations in olfactory discrimination performance in $\beta_3^{0/0}$ mice compared to wild-type *(105)*, pointing to a crucial role of GABAergic inhibition in modulating the function of principal cells in the olfactory bulb. A nearly complete loss of $GABA_A$ receptors has likewise been observed in neurons of the thalamic reticular nucleus of $\beta_3^{0/0}$ mice. As a result, a strong increase in oscillatory synchrony was observed in acute thalamic slices *(106)*, indicating that recurrent inhibitory connections within reticular nucleus play an important role in limiting synchronized activity, and likely prevent epileptiform activity in the thalamocortical system. This conclusion was corroborated by a computational analysis *(107)*.

In contrast to these studies focusing on the role of synaptic inhibition, the generation of $\delta^{0/0}$ mice provided insight into the functional significance of extrasynaptic receptors, notably in dentate gyrus granule cells, where $\alpha_4\beta\delta$ receptors mediate tonic inhibition *(105)*. In these cells, the mean tonic current measured in acute slices is about four times larger than the mean phasic currents generated by postsynaptic receptors, underscoring the potential importance of extrasynaptic receptors for the control of neuronal excitability *(105)*. This tonic current was not responsive to modulation by benzodiazepines, in line with the lack of diazepam-sensitivity of α_4-$GABA_A$ receptors. δ-subunit-deficient mice exhibit spontaneous seizures *(75)*. They are also more sensitive to pentylenetretrazole-induced seizures, and this phenotype was correlated with faster decay of evoked and miniature IPSPs in dentate gyrus granule cells *(108)*. These results underscore the role of extrasynaptic $GABA_A$ receptors for controlling the level of neuronal excitability in the dentate gyrus.

Finally, targeted deletion of the α_5 subunit gene provided the opportunity to investigate the role of these $GABA_A$ receptors in adult mice, which show no overt behavioral deficit. As discussed in Subheading 2, these receptors are most abundant in the hippocampus, where they are localized extrasynaptically in the soma and dendrites of pyramidal cells *(41)*. $\alpha_5^{0/0}$ mice exhibit enhanced learning of a hippocampal-dependent task (spatial learning) but no change in anxiety responses *(53)*. The amplitude of spontaneous IPSPs recorded in hippocampal pyramidal cells in a slice preparation was slightly reduced compared to wild-type mice, but it was not determined whether this change corresponds to loss of α_5-$GABA_A$ receptors or to compensatory alterations of remaining subunits *(53)*. Tonic inhibition (likely mediated by extrasynaptic receptors) was not tested in these mice. A selective role of α_5-$GABA_A$ receptors in hippocampal-dependent tasks was also reported in a mouse line carrying a H105R point mutation in the α_5 subunit gene *(42)*, which was designed to render the corresponding $GABA_A$ receptor subtypes diazepam-insensitive (*see* Subheading 4.). Unexpectedly, this mutation caused a profound reduction of α_5-$GABA_A$ receptors selectively in the hippocampal formation and was correlated with altered performance in a delay trace-conditioning task. In contrast, learning of nonhippocampus-dependent behavioral tasks was not affected, pointing to a selective role of extrasynaptic $GABA_A$ receptors for proper hippocampal function *(42)*. Although these behavioral changes were observed in a drug-independent context, alterations of diazepam effects in these mutants are discussed in Subheading 4.

3.6. Disease Models

Two strains of mice with targeted deletions of $GABA_A$ receptor subunits have proven particularly rewarding as models of human neurological and psychiatric diseases. First, $\beta_3{}^{0/0}$ mice, which present with electroencephalographic abnormalities, seizures, and behavioral features that have been described in patients with Angelman syndrome *(12,57)* (for review, *see [109]*). Second, mice heterozygous for the γ_2 subunit gene, which provided evidence for a possible contribution of $GABA_A$ receptors in anxiety disorders *(110)*.

The analysis of $\gamma_2{}^{+/0}$ mice by autoradiography and immunohistochemistry revealed a moderate loss of benzodiazepine binding sites and γ_2 subunit immunoreactivity throughout the brain, indicating that proper expression of this subunit cannot be achieved with a single allele. Most strikingly, the deficit was variable between brain areas, being strongest in the CA1 region of the hippocampus and motor cortex, and weakest in dentate gyrus, somatosensory cortex, and amygdala *(110)*. The reduced expression of the γ_2 subunit in CA1 was correlated with a decrease in the number of postsynaptic $GABA_A$ receptor and gephyrin clusters, suggesting decreased synaptic GABAergic function in this area. Electrophysiologically, the single-channel conductance of $GABA_A$ receptors in cultured neurons from $\gamma_2{}^{+/0}$ mice was altered, with the apparition of a lower conductance (12.6 pS) in addition to the main conductance of 29 pS observed in wild-type *(110)*. This lower conductance most likely corresponds to the formation of receptors containing only α/β subunit combinations.

Behavioral analysis of these mice revealed increased reactivity to naturally aversive stimuli, but no change compared to wild-type in learning or retention of delay or contextual fear conditioning. $\gamma_2{}^{+/0}$ mice were strongly impaired, however, in tests requiring discrimination of an ambiguous stimulus from the conditioned stimulus, as well as in trace fear conditioning, two tasks dependent on hippocampal processing *(110)*. These results therefore demonstrated that a deficit in synaptic $GABA_A$ receptor function, most pronounced in the hippocampus, results in a genetically defined model of trait anxiety that closely reproduces behavioral features of human anxiety disorders.

4. ANALYSIS OF MICE WITH POINT MUTATIONS

As outlined in Subheading 2, the expression of $GABA_A$ receptor subtypes in adult brain exhibits a remarkable region- and neuron-specificity, suggesting that individual subtypes are present in distinct neuronal circuits. Although the functional relevance of having, for instance, an α_1- instead of an α_2-$GABA_A$ receptor in a given type of neuron, remains largely speculative, the specificity of subtype expression is underscored by the remarkable selectivity of action of diazepam in knockin mutant mice carrying "custom-made" diazepam-insensitive $GABA_A$ receptor subtypes *(111)*. This strategy, based on the replacement of a conserved histidine residue by an arginine in α subunit variants forming diazepam-sensitive receptors, allows investigating which $GABA_A$ receptor subtypes mediate the various effects of diazepam, including sedation, anxiolysis, muscle relaxation, and seizure protection, as well as its unwanted side-effects (anterograde amnesia, motor incoordination, ethanol potentiation, and tolerance). These point-mutated mice also provide the opportunity to test in vivo other benzodiazepine site ligands, and to verify the subtype-specificity of novel drugs acting at selective populations of $GABA_A$ receptors (for review, see refs. *[3,111,112]*).

Thus, abolition of diazepam binding on α_2-GABA$_A$ receptors in vivo by exchanging the conserved histidine 101 residue with an arginine results in selective suppression of the anxiolytic action of this drug, whereas anxiolytic drugs acting by other mechanisms are unaffected *(113)*. The same mutation, when introduced in the α_1, α_3, or α_5 subunit, does not influence the anxiolytic action of diazepam. Although the α_2 subunit has a widespread distribution in numerous brain areas, this finding indicates that α_2-GABA$_A$ receptors are strategically located in circuits mediating the anxiolytic action of diazepam.

Likewise, the sedative effect of diazepam is selectively abolished, even at high dose, in mice carrying α_1^{H101R}-GABA$_A$ receptors *(114,115)*. Again, one concludes that brain circuits containing α_1-GABA$_A$ receptors mediate sedation, and that activation of these receptors by diazepam is ineffective for relieving anxiety. In α_1^{H101R} mice, the anticonvulsant action of diazepam is also strongly reduced, whereas its myorelaxant action is retained. α_1-GABA$_A$ receptors appear therefore to be a selective molecular target mediating two of the major actions of diazepam, sedation and seizure protection. This selectivity of drug action was extended to both positive and negative allosteric modulators of the benzodiazepine binding site. Indeed, a recent study reveals a switch in the action of Ro 15-45123 from inverse agonism to agonism in α_1^{H101R}-mice, resulting in decreased baseline locomotion and anticonvulsant action *(116)*. These effects in mutant were exactly opposite to those produced in wild-type mice, indicating that α_1-GABA$_A$ receptors mediate specific behavioral output produced by benzodiazepine site ligands with regard to motor activity and seizure protection, irrespective of the direction of the intrinsic activity of these ligands.

Unlike sedation and anxiolysis, which appear to be mediated by a single GABA$_A$ receptor subtype, the myorelaxant effect of diazepam was reduced, but not abolished, in three lines of mice (α_2^{H101R}, α_3^{H126R}, α_5^{H105R}) *(42,117)*. Although these results are in line with the high level of expression of these three subunits in the spinal cord *(118)*, they suggest that myorelaxation can be produced by increasing GABA$_A$ receptor function in multiple spinal, and possibly supraspinal, circuits. In α_2^{H101R} mice, the myorelaxant action of baclofen was retained, indicating that the loss of diazepam action was not owing to unspecific effects *(117)*.

The selectivity of diazepam action on distinct GABA$_A$ receptor subtypes was confirmed by the analysis of novel drugs, which have a distinct profile of action on GABA$_A$ receptor subtypes. For example, the novel ligand L838417, which binds with high affinity to all diazepam-sensitive GABA$_A$ receptors, but acts as partial agonist only on α_2, α_3, and α_5 subunit-containing receptors, displays anxiolytic properties in rats and enhances exploration of a novel environment in a locomotor activity test *(115)*.

A major issue in the analysis of these knockin mice is the significance of drug-independent changes in GABA$_A$ receptor function or behavior. The analysis of the $\alpha_5^{(H105R)}$ mutants revealed that unexpected effects on GABA$_A$ receptor expression cannot be excluded *(42)*. Although the reasons for the hippocampal deficit in α_5-GABA$_A$ receptors present in these mice is not understood, this example shows that any interpretation of behavioral alteration needs to take potential compensatory changes into account, even when the mutation is very subtle. The possibility to discriminate between drug-induced and drug-independent behavioral changes nevertheless provides a powerful tool for investigating GABA$_A$ receptor function in vivo. In future studies, the generation of time- or site-specific knock-in mutations should allow an even finer dissection of the neuronal circuits mediating the pharmacological effects of diazepam.

REFERENCES

1. Sieghart, W. (1995) Structure and pharmacology of γ-aminobutyric acid$_A$ receptor subtypes. *Pharmacol. Rev.* **47,** 181–234.
2. Dilger, J. P. (2002) The effects of general anaesthetics on ligand-gated ion channels. *Br. J. Anaest.* **89,** 41–51.
3. Mohler, H., Fritschy, J. M., and Rudolph, U. (2002) A new benzodiazepine pharmacology. *J. Pharm. Exp. Ther.* **300,** 2–8.
4. Grobin, A. C., Matthews, D. B., Devaud, L. L., and Morrow, A. L. (1998) The role of GABA$_A$ receptors in the acute and chronic effects of ethanol. *Psychopharmacology* **139,** 2–19.
5. Mohler, H. (2000) Functions of GABA$_A$-receptors: pharmacology and pathophysiology, in *Pharmacology of GABA and Glycine Neurotransmission* (Mohler, H., ed.), Springer-Verlag, Berlin, pp. 101–116.
6. Duncan, J. S. (1999) Positron emission tomography receptor studies in epilepsy. *Rev. Neurol.* **155,** 482–488.
7. Olsen, R. W., DeLorey, T. M., Gordey, M., and Kang, M. H. (1999) GABA receptor function and epilepsy. *Adv. Neurol.* **79,** 499–510.
8. Coulter, D. A. (2001) Epilepsy-associated plasticity in γ-aminobutyric acid-$_A$ receptor expression, function, and inhibitory synaptic properties. *Int. Rev. Neurobiol.* **45,** 237–252.
9. Malizia, A. L. (1999) What do brain imaging studies tell us about anxiety disorders? *J. Psychopharmacol.* **13,** 372–378.
10. Morrow, A. L., VanDoren, M. J., Penland, S. N., and Matthews, D. B. (2001) The role of GABAergic neuroactive steroids in ethanol action, tolerance and dependence. *Brain Res. Rev.* **37,** 98–109.
11. Kunig, G., Leenders, K. L., Sanchez-Pernaute, R., Antonini, A., Vontobel, P., Verhagen, A., and Gunther, I. (2000) Benzodiazepine receptor binding in Huntington's disease: [^{11}C]flumazenil uptake measured using positron. *Ann. Neurol.* **47,** 644–648.
12. DeLorey, T. M., Handforth, A., Anagnostaras, S. G., Homanics, G. E., Minassian, B. A., Asatourian, A., et al. (1998) Mice lacking the β$_3$ subunit of the GABA$_A$ receptor have the epilepsy phenotype and many of the behavioral characteristics of Angelman syndrome. *J. Neurosci.* **18,** 8505–8514.
13. Nutt, D. J. and Malizia, A. L. (2001) New insights into the role of the GABA$_A$-benzodiazepine receptor in psychiatric disorder. *Br. J. Psychiat.* **179,** 390–396.
14. Blum, B. P. and Mann, J. J. (2002) The GABAergic system in schizophrenia. *Int. J. Neuropsychopharmacol.* **5,** 159–179.
15. Lewis, D. A. (2000) GABAergic local circuit neurons and prefrontal cortical dysfunction in schizophrenia. *Brain Res. Rev.* **31,** 270–276.
16. Unwin, N. (1993) Neurotransmitter action: opening of a ligand-gated channel. *Neuron* **10** Suppl., 31–41.
17. Barnard, E. A. (2001) The molecular architecture of GABA$_A$ receptors, in *Pharmacology of GABA and Glycine Neurotransmission,* vol. 150 (Mohler, H., ed.), Springer-Verlag, Berlin, pp. 79–100.
18. Barnard, E. A., Skolnick, P., Olsen, R. W., et al. (1998) International Union of Pharmacology. XV. Subtypes of γ-aminobutyric acid$_A$ receptors: classification on the basis of subunit structure and function. *Pharmacol. Rev.* **50,** 291–313.
19. Whiting, P. J. (1999) The GABA$_A$ receptor gene family: new targets for therapeutic intervention. *Neurochem. Int.* **34,** 387–390.

20. Mohler, H., Fritschy, J. M., Luscher, B., Rudolph, U., Benson, J., and Benke, D. (1996) The GABA$_A$-receptors: from subunits to diverse functions, in *Ion Channels* vol. 4 (Narahashi, T., ed.), Plenum Press, New York, pp. 89–113.

21. Knight, A. R., Stephenson, F. A., Tallman, J. F., and Ramabahdran, T. V. (2000) Monospecific antibodies as probes for the stoichiometry of recombinant GABA$_A$ receptors. *Recept. Channels* **7,** 213–226.

22. Baumann, S. W., Baur, R., and Sigel, E. (2001) Subunit arrangement of γ-aminobutyric acid type A receptors. *J. Biol. Chem.* **276,** 36275–36280.

23. Klausberger, T., Sarto, I., Ehya, N., et al. (2001) Alternate use of distinct intersubunit contacts controls GABA$_A$ receptor assembly and stoichiometry. *J. Neurosci.* **21,** 9124–9133.

24. Farrar, S. J., Whiting, P. J., Bonnert, T. P., and McKernan, R. M. (1999) Stoichiometry of a ligand-gated ion channel determined by fluorescence energy transfer. *J. Biol. Chem.* **274,** 10100–10104.

25. Bormann, J. (2000) The "ABC" of GABA receptors. *Trends Pharmacol. Sci.* **21,** 16–19.

26. Benson, J. A., Löw, K., Keist, R., Mohler, H., and Rudolph, U. (1998) Pharmacology of recombinant γ-aminobutyric acid$_A$ receptors rendered diazepam-insensitive by point-mutated α-subunits. *FEBS Lett.* **431,** 400–404.

27. Wingrove, P. B., Safo, P., Wheat, L., Thompson, S. A., Wafford, K. A., and Whiting, P. J. (2002) Mechanism of α-subunit selectivity of benzodiazepine pharmacology at γ-aminobutyric acid type A receptors. *Eur. J. Pharmacol.* **437,** 31–39.

28. Verdoorn, T. A., Draguhn, A., Ymer, S., Seeburg, P. H., and Sakmann, B. (1990) Functional properties of recombinant rat GABA$_A$ receptors depend upon subunit composition. *Neuron* **4,** 919–928.

29. Hutcheon, B., Morley, P., and Poulter, M. O. (2000) Developmental change in GABA$_A$ receptor desensitization kinetics and its role in synapse function in rat cortical neurons. *J. Physiol.* **522,** 3–17.

30. Devor, A., Fritschy, J. M., and Yarom, Y. (2001) Spatial distribution and subunit composition of GABA$_A$ receptors in the inferior olivary nucleus. *J. Neurophysiol.* **85,** 1686–1696.

31. Burgard, E. C., Tietz, E. I., Neelands, T. R., and Macdonald, R. L. (1996) Properties of recombinant γ-aminobutyric acid$_A$ receptor isoforms containing the α$_5$ subunit subtype. *Mol. Pharmacol.* **50,** 119–127.

32. Fisher, J. L. and Macdonald, R. L. (1997) Single channel properties of recombinant GABA$_A$ receptors containing γ$_2$ or δ subtypes expressed with α$_1$ and β$_3$ subtypes in mouse L929 cells. *J. Physiol.* **505,** 283–297.

33. Adkins, C. E., Pillai, G. V., Kerby, J., et al. (2001) α$_4$β$_3$δ GABA$_A$ receptors characterized by fluorescence resonance energy transfer-derived measurements of membrane potential. *J. Biol. Chem.* **276,** 38934–38939.

34. Schwarzer, C., Berresheim, U., Pirker, S., Wieselthaler, A., Fuchs, K., Sieghart, W., and Sperk, G. (2001) Distribution of the major γ-aminobutyric acid$_A$ receptor subunits in the basal ganglia and associated limbic brain areas of the adult rat. *J. Comp. Neurol.* **433,** 526–549.

35. Waldvogel, H. J., Fritschy, J. M., Mohler, H., and Faull, R. L. M. (1998) GABA$_A$ receptors in the primate basal ganglia: an autoradiographic and a light and electron microscopic immunohistochemical study of the α$_1$ and β$_{2,3}$ subunits in the baboon brain. *J. Comp. Neurol.* **397,** 297–325.

36. Pirker, S., Schwarzer, C., Wieselthaler, A., Sieghart, W., and Sperk, G. (2000) GABA$_A$ receptors: immunocytochemical distribution of 13 subunits in the adult rat brain. *Neuroscience* **101,** 815–850.

37. Fritschy, J. M. and Mohler, H. (1995) GABA$_A$-receptor heterogeneity in the adult rat brain: differential regional and cellular distribution of seven major subunits. *J. Comp. Neurol.* **359,** 154–194.

38. Hornung, J. P. and Fritschy, J. M. (1996) Developmental profile of GABA$_A$-receptors in the marmoset monkey: expression of distinct subtypes in pre- and postnatal brain. *J. Comp. Neurol.* **367,** 413–430.

39. Nusser, Z., Sieghart, W., Benke, D., Fritschy, J. M., and Somogyi, P. (1996) Differential synaptic localization of two major γ-aminobutyric acid type A receptor α subunits on hippocampal pyramidal neurons. *Proc. Natl. Acad. Sci. USA* **93,** 11939–11944.

40. Fritschy, J. M., Weinmann, O., Wenzel, A., and Benke, D. (1998) Synapse-specific localization of NMDA- and GABA_A-receptor subunits revealed by antigen-retrieval immunohistochemistry. *J. Comp. Neurol.* **390,** 194–210.

41. Brünig, I., Scotti, E., Sidler, C., and Fritschy, J. M. (2002) Intact sorting, targeting, and clustering of γ-aminobutyric acid A receptor subtypes in hippocampal neurons in vitro. *J. Comp. Neurol.* **443,** 43–45.

42. Crestani, F., Keist, R., Fritschy, J. M., et al. (2002) Trace fear conditioning involves hippocampal α_5 GABA_A receptors. *Proc. Natl. Acad. Sci. USA* **99,** 8980–8985.

43. Uusi-Oukari, M., Heikkilä, J., Sinkkonen, S. T., et al. (2000) Long-range interactions in neuronal gene expression: evidence from gene targeting in the GABA_A receptor β_2-α_6-α_1-γ_2 subunit gene cluster. *Mol. Cell. Neurosci.* **16,** 34–41.

44. Russek, S. J. (1999) Evolution of GABA_A receptor diversity in the human genome. *Gene* **227,** 213–222.

45. Bailey, M. E. S., Matthews, D. A., Riley, B. P., Albrecht, B. E., Kostrzewa, M., Hicks, A. A., et al. (1999) Genomic mapping and evolution of human GABA_A receptor subunit gene clusters. *Mammal. Gen.* **10,** 839–843.

46. Connolly, C. N., Uren, J. M., Thomas, P., Gorrie, G. H., Gibson, A., Smart, T. G., and Moss, S. J. (1999) Subcellular localization and endocytosis of homomeric γ_2 subunit splice variants of γ-aminobutyric acid type A receptors. *Mol. Cell. Neurosci.* **13,** 259–271.

47. Connolly, C. N., Krishek, B. J., McDonald, B. J., Smart, T. G., and Moss, S. J. (1996) Assembly and cell surface expression of heteromeric and homomeric γ-aminobutyric acid type A receptors. *J. Biol. Chem.* **271,** 89–96.

48. Mihic, S. J. and Harris, R. A. (1996) Inhibition of rho1 receptor GABAergic currents by alcohols and volatile anesthetics. *J. Pharmacol. Exp. Ther.* **277,** 411–416.

49. Amin, J. and Weiss, D. S. (1994) Homomeric ρ1 GABA channels: activation properties and domains. *Recept. Channels.* **2,** 227–236.

50. Pritchett, D. B., Sontheimer, H., Shivers, B. D., Ymer, S., Kettenmann, H., Schofield, P. R., and Seeburg, P. H. (1989) Importance of a novel GABA_A receptor subunit for benzodiazepine pharmacology. *Nature* **338,** 582–585.

51. Lorez, M., Benke, D., Luscher, B., Mohler, H., and Benson, J. A. (2000) Single-channel properties of neuronal GABA_A receptors from mice lacking the γ_2 subunit. *J. Physiol.* **527,** 11–31.

52. Gunther, U., Benson, J., Benke, D., et al. (1995) Benzodiazepine-insensitive mice generated by targeted disruption of the γ_2-subunit gene of γ-aminobutyric acid type A receptors. *Proc. Natl. Acad. Sci. USA* **92,** 7749–7753.

53. Collinson, N., Kuenzi, F. M., Jarolimek, W., et al. (2002) Enhanced learning and memory and altered GABAergic synaptic transmission in mice lacking the α_5 subunit of the GABA_A receptor. *J. Neurosci.* **22,** 5572–5580.

54. Sur, C., Wafford, K. A., Reynolds, D. S., et al. (2001) Loss of the major GABA_A receptor subtype in the brain is not lethal in mice. *J. Neurosci.* **21,** 3409–3418.

55. Kralic, J. E., Korpi, E. R., O'Buckley, T. K., Homanics, G. E., and Morrow, A. (2002) Molecular and pharmacological characterization of GABA_A receptor α_1 subunit knockout mice. *J. Pharmacol. Exp. Ther.* **302,** 1037–1045.

56. Krasowski, M. D., Rick, C. E., Harrison, N. L., Firestone, L., and Homanics, G. E. (1998) A deficit of functional GABA_A receptors in neurons of β_3 subunit knockout mice. *Neurosci. Lett.* **240,** 81–84.

57. Homanics, G. E., Delorey, T. M., Firestone, L. L., et al. (1997) Mice devoid of γ-aminobutyrate type A receptor β_3 subunit have epilepsy, cleft palate, and hypersensitive behavior. *Proc. Natl. Acad. Sci. USA* **94,** 4143–4148.

58. Fritschy, J. M., Benke, D., Johnson, D. K., Mohler, H., and Rudolph, U. (1997) GABA$_A$-receptor α-subunit is an essential prerequisite for receptor formation in vivo. *Neuroscience* **81,** 1043–1053.

59. Nusser, Z., Ahmad, Z., Tretter, V., Fuchs, K., Wisden, W., Seighart, W., and Somogyi, P. (1999) Alterations in the expression of GABA$_A$ receptor subunits in cerebellar granule cells after the disruption of the α6 subunit gene. *Eur. J. Neurosci.* **11,** 1685–1697.

60. Kralic, J. E., O'Buckley, T. K., Khisti, R. T., Hodge, C. J., Homanics, G. E., and Morrow, A. L. (2002) GABA$_A$ receptor α1 subunit deletion alters receptor subtype assembly, pharmacological and behavioral responses to benzodiazepines and zolpidem. *Neuropharmacology* **43,** 685–694.

61. Fritschy, J. M., Johnson, D. K., Mohler, H., and Rudolph, U. (1998) Independent assembly and subcellular targeting of GABA$_A$ receptor subtypes demonstrated in mouse hippocampal and olfactory neurons in vivo. *Neurosci. Lett.* **249,** 99–102.

62. Jones, A., Korpi, E. R., McKernan, R. M., et al. (1997) Ligand-gated ion channel subunit partnerships: GABA$_A$ receptor α6 subunit gene inactivation inhibits delta subunit expression. *J. Neurosci.* **17,** 1350–1362.

63. Peng, Z., Hauer, B., Mihalek, R. M., Homanics, G. E., Sieghart, W., Olsen, R. W., and Houser, C. R. (2002) GABA$_A$ receptor changes in δ subunit-deficient mice: altered expression of α4 and γ2 subunits in the forebrain. *J. Comp. Neurol.* **446,** 179–197.

64. Korpi, E. R., Mihalek, R. M., Sinkkonen, S. T., et al. (2002) Altered receptor subtypes in the forebrain of GABA$_A$ receptor δ subunit-deficient mice: recruitment of γ2 subunits. *Neuroscience* **109,** 733–743.

65. Tretter, V., Hauer, B., Nusser, Z., et al. (2001) Targeted disruption of the GABA$_A$ receptor δ subunit gene leads to an up-regulation of γ2 subunit-containing receptors in cerebellar granule cells. *J. Biol. Chem.* **276,** 10532–10538.

66. Wohlfarth, K. M., Bianchi, M. T., and Macdonald, R. L. (2002) Enhanced neurosteroid potentiation of ternary GABA$_A$ receptors containing the δ subunit. *J. Neurosci.* **22,** 1541–1549.

67. Lin, Y. F., Angelotti, T. P., Dudek, E. M., Browning, M. D., and Macdonald, R. L. (1996) Enhancement of recombinant α1β1γ2$_L$ γ-aminobutyric acid$_A$ receptor whole-cell currents by protein kinase C is mediated through phosphorylation of both β1 and γ2$_L$ subunits. *Mol. Pharmacol.* **50,** 185–195.

68. Brickley, S. G., Revilla, V., Cull-Candy, S. G., Wisden, W., and Farrant, M. (2001) Adaptive regulation of neuronal excitability by a voltage-independent potassium conductance. *Nature* **409,** 88–92.

69. Benke, D., Honer, M., Michel, C., and Mohler, H. (1996) GABA$_A$ receptor subtypes differentiated by their γ-subunit variants: prevalence, pharmacology and subunit architercture. *Neuropharmacology* **35,** 1413–1422.

70. Essrich, C., Lorez, M., Benson, J. A., Fritschy, J. M., and Luscher, B. (1998) Postsynaptic clustering of major GABA$_A$ receptor subtypes requires the γ2 subunit and gephyrin. *Nature Neurosci.* **1,** 563–571.

71. Kneussel, M., Brandstatter, J. H., Laube, B., Stahl, S., Muller, U., and Betz, H. (1999) Loss of postsynaptic GABA$_A$ receptor clustering in gephyrin-deficient mice. *J. Neurosci.* **19,** 9289–9297.

72. Brünig, I., Suter, A., Knuesel, I., Luscher, B., and Fritschy, J. M. (2002) GABAergic presynaptic terminals are required for postsynaptic clustering of dystrophin, but not of GABA$_A$ receptors and gephyrin. *J. Neurosci.* **22,** 4805–4813.

73. Baer, K., Essrich, C., Benson, J. A., Benke, D., Bluethmann, H., Fritschy, J. M., and Luscher, B. (1999) Postsynaptic clustering of γ-aminobutyric acid type A receptors by the γ3 subunit in vivo. *Proc. Natl. Acad. Sci. USA* **96,** 12860–12865.

74. Mihalek, R. M., Banerjee, P. K., Korpi, E. R., and Quinlan, J. J. (1999) Attenuated sensitivity to neuroactive steroids in γ-aminobutyrate type A receptor δ subunit knockout mice. *Proc. Natl. Acad. Sci. USA* **96,** 12905–12910.

75. Mihalek, R. M., Bowers, B. J., Wehner, J. M., Kralic, J. E., VanDoren, M. J., Morrow, A. L., and Homanics, G. E. (2001) GABA$_A$ receptor δ subunit knockout mice have multiple defects in behavioral responses to ethanol. *Alcohol. Clin. Exp. Res.* **25,** 1708–1718.

76. Homanics, G. E., Ferguson, C., Quinlan, J. J., et al. (1997) Gene knockout of the α6 subunit of the γ-aminobutyric acid type A receptor: lack of effect on responses to ethanol, pentobarbital, and general anesthetics. *Mol. Pharmacol.* **51,** 588–596.

77. Wafford, K. A. and Whiting, P. J. (1992) Ethanol potentiation of GABA$_A$ receptors requires phosphorylation of the alternatively spliced variant of the γ$_2$ subunit. *FEBS Lett.* **313,** 113–117.

78. Sigel, E., Baur, R., and Malherbe, P. (1993) Recombinant GABA$_A$ receptor function and ethanol. *FEBS Lett.* **324,** 140–142.

79. Mihic, S. J., Whiting, P. J., and Harris, R. A. (1994) Anaesthetic concentrations of alcohols potentiate GABA$_A$ receptor-mediated currents: lack of subunit specificity. *Eur. J. Pharmacol.* **268,** 209–214.

80. Homanics, G. E., Harrison, N. L., Quinlan, J. J., et al. (1999) Normal electrophysiological and behavioral responses to ethanol in mice lacking the long splice variant of the γ$_2$ subunit of the γ-aminobutyrate type A receptor. *Neuropharmacology* **38,** 253–265.

81. Quinlan, J. J., Firestone, L. L., and Homanics, G. E. (2000) Mice lacking the long splice variant of the γ$_2$ subunit of the GABA$_A$ receptor are more sensitive to benzodiazepines. Pharmacol. *Biochem. Behav.* **66,** 371–374.

82. Quinlan, J., Homanics, G., and Firestone, L. (1998) Anesthesia sensitivity in mice that lack the β$_3$ subunit of the γ-aminobutyric acid type A receptor. *Anesthesiology* **88,** 775–780.

83. Wong, S. M., Cheng, G., Homanics, G. E., and Kendig, J. J. (2001) Enflurane actions on spinal cords from mice that lack the β3 subunit of the GABA$_A$ receptor. *Anesthesiology* **95,** 154–164.

84. Ugarte, S. D., Homanics, G. E., Firestone, L. L., and Hammond, D. L. (2000) Sensory thresholds and the antinociceptive effects of GABA receptor agonists in mice lacking the β$_3$ subunit of the GABA$_A$ receptor. *Neuroscience* **95,** 795–806.

85. Culiat, C. T., Stubbs, L. J., Woychik, R. P., Russell, L. B., Johnson, D. K., and Rinchik, E. M. (1995) Deficiency of the β3 subunit of the type A γ-aminobutyric acid receptor causes cleft palate in mice. *Nature Genetics* **11,** 344–346.

86. Owens, D. F. and Kriegstein, A. R. (2002) Is there more to GABA than synaptic inhibition? *Nature Rev. Neurosci.* **3,** 715–727.

87. Ji, F., Kanbara, N., and Obata, K. (1999) GABA and histogenesis in fetal and neonatal mouse brain lacking both the isoforms of glutamic acid decarboxylase. *Neurosci. Res.* **33,** 187–194.

88. Verhage, M., Maia, A. S., Plomp, J. J., Brussaard, A. B., Heeroma, J. H., Vermeer, H., et al. (2000) Synaptic assembly of the brain in the absence of neurotransmitter secretion. *Science* **287,** 864–869.

89. Varoqueaux, F., Sigler, A., Rhee, J. S., Brose, N., Enk, C., Reim, K., and Rosenmund, C. (2002) Total arrest of spontaneous and evoked synaptic transmission but normal synaptogenesis in the absence of Munc13-mediated vesicle priming. *Proc. Natl. Acad. Sci. USA* **99,** 9037–9042.

90. Prosser, H. M., Gill, C. H., Hirst, W. D., et al. (2001) Epileptogenesis and enhanced prepulse inhibition in GABA$_B$1-deficient Mice. *Mol. Cell. Neurosci.* **17,** 1059–1070.

91. Schuler, V., Lüscher, C., Blanchet, C., et al. (2001) Epilepsy, hyperalgesia, impaired memory, and loss of pre- and postsynaptic GABA$_B$ responses in mice lacking GABA$_B$1. *Neuron* **31,** 47–58.

92. Dellovade, T. L., Davis, A. M., Ferguson, C., Sieghart, W., Homanics, G. E., and Tobet, S. A. (2001) GABA influences the development of the ventromedial nucleus of the hypothalamus. *J. Neurobiol.* **49,** 264–276.

93. Pannell, C., Simonian, S. X., Gillies, G. E., Luscher, B., and Herbison, A. E. (2002) Hypothalamic somatostatin and growth hormone-releasing hormone mRNA expression depend

upon GABA$_A$ receptor expression in the developing mouse. *Neuroendocrinology* **76,** 93–98.

94. Simonian, S. X., Skynner, M. J., Sieghart, W., Essrich, C., Luscher, B., and Herbison, A. E. (2000) Role of the GABA$_A$ receptor γ2 subunit in the development of gonadotropin-releasing hormone neurons in vivo. *Eur. J. Neurosci.* **12,** 3488–3496.

95. Wick, M. J., Radcliffe, R. A., Bowers, B. J., Mascia, M. P., Luscher, B., Harris, R. A., and Wehner, J. M. (2000) Behavioural changes produced by transgenic overexpression of γ$_{2L}$ and γ2$_S$ subunits of the GABA$_A$ receptor. *Eur. J. Neurosci.* **12,** 2634–2638.

96. Baer, K., Essrich, C., Balsiger, S., Wick, M. J., Harris, R. A., Fritschy, J. M., and Luscher, B. (2000) Rescue of γ2 subunit-deficient mice by transgenic overexpression of the GABA$_A$ receptor γ2$_S$ or γ2$_L$ subunit isoforms. *Eur. J. Neurosci.* **12,** 2639–2643.

97. Fritschy, J. M., Paysan, J., Enna, A., and Mohler, H. (1994) Switch in the expression of rat GABA$_A$-receptor subtypes during postnatal development: an immunohistochemical study. *J. Neurosci.* **14,** 5302–5324.

98. Brickley, S. G., Cull-Candy, S. G., and Farrant, M. (1996) Development of a tonic form of synaptic inhibition in rat cerebellar granule cells resulting from persistent activation of GABA$_A$ receptors. *J. Physiol.* **497,** 753–759.

99. Tia, S., Wang, J. F., Kotchabhakdi, N., and Vicini, S. (1996) Developmental changes of inhibitory synaptic currents in cerebellar granule neurons: role of GABA$_A$ receptor α6 subunit. *J. Neurosci.* **16,** 3630–3640.

100. Okada, M., Onodera, K., Van Renterghem, C., Sieghart, W., and Takahashi, T. (2000) Functional correlation of GABA$_A$ receptor α subunits expression with the properties of IPSCs in the developing thalamus. *J. Neurosci.* **15,** 2202–2208.

101. Vicini, S., Ferguson, C., Prybylowski, K., Kralic, J., Morrow, A. L., and Homanics, G. E. (2001) GABA$_A$ receptor α1 subunit deletion prevents developmental changes of inhibitory synaptic currents in cerebellar neurons. *J. Neurosci.* **21,** 3009–3016.

102. Bosman, L. W. J., Rosahl, T. W., and Brussaard, A. B. (2002) Neonatal development of the rat visual cortex: synaptic function of GABA$_A$ receptor α subunits. *J. Physiol.* **545,** 169–181.

103. Miralles, C. P., Li, M., Mehta, A. K., Khan, Z. U., and De Blas., A. L. (1999) Immunocytochemical localization of the β3 subunit of the γ-aminobutyric acid$_A$ receptor in the rat brain. *J. Comp. Neurol.* **413,** 535–548.

104. Nusser, Z., Kay, L. M., Laurent, G., Homanics, G. E., and Mody, I. (2001) Disruption of GABA$_A$ receptors on GABAergic interneurons leads to increased oscillatory power in the olfactory bulb network. *J. Neurophysiol.* **86,** 2823–2833.

105. Nusser, Z., and Mody, I. (2002) Selective modulation of tonic and phasic inhibitions in dentate gyrus granule cells. *J. Neurophysiol.* **87,** 2624–2628.

106. Huntsman, M. M., Porcello, D. M., Homanics, G. E., DeLorey, T. M., and Huguenard, J. R. (1999) Reciprocal inhibitory connections and network synchrony in the mammalian thalamus. *Science* **283,** 541–543.

107. Sohal, V. S., Huntsman, M. M., and Huguenard, J. R. (2000) Reciprocal inhibitory connections regulate the spatiotemporal properties of intrathalamic oscillations. *J. Neurosci.* **20,** 1735–1745.

108. Spigelman, I., Li, Z., Banerjee, P. K., Mihalek, R. M., Homanics, G. E., and Olsen, R. W. (2002) Behavior and physiology of mice lacking the GABA$_A$ receptor δ subunit. *Epilepsia* **43,** 3–8.

109. DeLorey, T. M., and Olsen, R. W. (1999) GABA and epileptogenesis: comparing gabrb3 gene-deficient mice with Angelman syndrome in man. *Epilepsy Res.* **36,** 123–132.

110. Crestani, F., Lorez, M., Baer, K., et al. (1999) Decreased GABA$_A$-receptor clustering results in enhanced anxiety and a bias for threat cues. *Nature Neurosci.* **2,** 833–839.

111. Rudolph, U., Crestani, F., and Mohler, H. (2001) GABA$_A$ receptor subtypes: dissecting their pharmacological functions. *Trends Pharmacol. Sci.* **22,** 188–194.

112. Mohler, H., Crestani, F., and Rudolph, U. (2001) GABA$_A$-receptor subtypes: a new pharmacology. *Curr. Opin. Pharmacol.* **1,** 22–25.

113. Löw, K., Crestani, F., Keist, R., et al. (2000) Molecular and neuronal substrate for the selective attenuation of anxiety. *Science* **290,** 131–134.

114. Rudolph, U., Crestani, F., Benke, D., et al. (1999) Benzodiazepine actions mediated by specific γ-aminobutyric acid$_A$ receptor subtypes. *Nature* **401,** 796–800.

115. McKernan, R. M., Rosahl, T. W., Reynolds, D. S., et al. (2000) Sedative but not anxiolytic properties of benzodiazepines are mediated by the GABA$_A$ receptor α_1 subtype. *Nature Neurosci.* **3,** 587–592.

116. Crestani, F., Assandri, R., Täuber, M., Martin, J. R., and Rudolph, U. (2002) Contribution of the α_1-GABA$_A$ receptor subtype to the pharmacological actions of benzodiazepine site inverse agonists. *Neuropharmacology* **43,** 679–684.

117. Crestani, F., Low, K., Keist, R., Mandelli, M.-J., Mohler, H., and Rudolph, U. (2000) Molecular targets for the myorelaxant action of diazepam. *Mol. Pharmacol.* **59,** 442–445.

118. Bohlhalter, S., Weinmann, O., Mohler, H., and Fritschy, J. M. (1996) Laminar compartmentalization of GABA$_A$-receptor subtypes in the spinal cord. *J. Neurosci.* **16,** 283–297.

6

Characterization of the GABA$_A$ Receptor Recognition Site Through Ligand Design and Pharmacophore Modeling

Bente Frølund, Anne T. Jørgensen, Tommy Liljefors, Martin Mortensen, and Povl Krogsgaard-Larsen

1. INTRODUCTION

4-Aminobutyric acid (GABA), the major inhibitory neurotransmitter in the mammalian central nervous system (CNS), mediates its actions through the ionotropic GABA$_A$ and GABA$_C$ receptors and the metabotropic GABA$_B$ receptors. This classification of the GABA receptors has been developed over a period of 40 years and is primarily based on the pharmacology of a limited number of selective ligands (1). GABA primarily acts through GABA$_A$ receptors, which are built as pentameric assemblies of different families of receptor subunits. Although a large number of subunits have been identified, only a limited number of physiological GABA$_A$ receptors are believed to exist. In addition to the recognition site for GABA, the existence of several modulatory sites for a number of therapeutic agents, including benzodiazepines, barbiturates, neurosteroids, and volatile anaesthetics adds to the complexity of the GABA$_A$ receptor system (2).

The design of selective ligands for the GABA recognition site at the GABA$_A$ receptor is of key importance for the elucidation of structure and function of the different types of GABA$_A$ receptors. Such studies have, so far, been complicated by the lack of knowledge of the topography of the recognition site(s) at the GABA$_A$ receptor complex. The extensive search for selective ligands for the GABA recognition site of the GABA$_A$ receptor complex has provided only a very few different classes of structures, reflecting the strict structural requirements for GABA$_A$ receptor recognition and activation. Within the series of compounds showing agonist activity at the GABA$_A$ receptor site, most of the ligands are structurally derived from the GABA$_A$ receptor agonists muscimol (3), 4,5,6,7-tetrahydroisoxazolo[5,4-*c*]pyridin-3-ol (THIP), or isoguvacine (Fig. 1) (4), which were developed at the initial stage of the project. An even smaller group of structurally different ligands has been identified as selective competitive antagonists for the GABA$_A$ receptor recognition site. The classical GABA$_A$ receptor antagonist, bicuculline, and its quarternized ammonium salt, bicuculline methochloride (BMC), have played a key role in studies of the GABA$_A$ receptors (5). The ary-

From: *Molecular Neuropharmacology: Strategies and Methods*
Edited by: A. Schousboe and H. Bräuner-Osborne © Humana Press Inc., Totowa, NJ

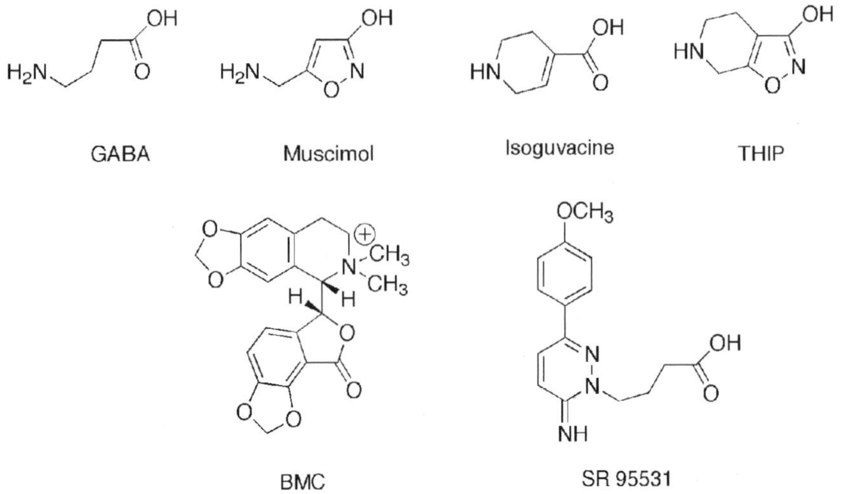

Fig. 1. Structures of GABA, some GABA$_A$ receptor agonists (upper part) and antagonists (lower part).

laminopyridazine derivative of GABA, SR 95531, represents another structural class of GABA$_A$ antagonists (Fig. 1) *(6,7)*.

2. DEVELOPMENT OF LIGANDS FOR THE GABA$_A$ RECEPTOR USING MUSCIMOL AS A LEAD STRUCTURE

Muscimol, a constituent of the mushroom *Amanita muscaria,* has been identified as a potent and selective GABA$_A$ receptor agonist. Owing to the apparent versatility of the 3-isoxazolol moiety of muscimol as a bioisostere to the carboxyl group of GABA, muscimol has been extensively used as a lead in the search for potent and selective ligands acting at the GABA$_A$ receptor *(8,9).* Although a large number of analogs of muscimol have been synthesized, the key step in this drug-design project was the synthesis of the bicyclic analog THIP, a potent and specific GABA$_A$ receptor agonist, and the equally potent and specific amino acid analog isoguvacine. These two compounds are now standard agonists for GABA$_A$ receptor studies.

Reduction of the conformational mobility of flexible compounds is a frequently used approach for studies of the preferred ligand conformation at the receptor target. Therefore, the similarity displayed by muscimol, isoguvacine, and THIP in their pharmacological actions led to the proposal that the structure of isoguvacine and in particular THIP represents the bioactive conformation of muscimol. Based on this assumption, a series of molecular modeling studies have been carried out in order to define pharmacophore models for GABA$_A$ receptor agonists and competitive antagonists *(10,11).* However, the structure-activity relationships of the muscimol analogs and the corresponding analogs of THIP are not straightforward and do not fit into these proposed pharmacophore models, as exemplified by the structure-activity relationships of the 3-isothiazolol analogs of muscimol and THIP (Fig. 2). The replacement of the 3-isoxazolol groups of muscimol and THIP by a 3-isothiazolol unit to give thiomuscimol and thioTHIP has markedly different effects on GABA$_A$ receptor affinity. The affinity

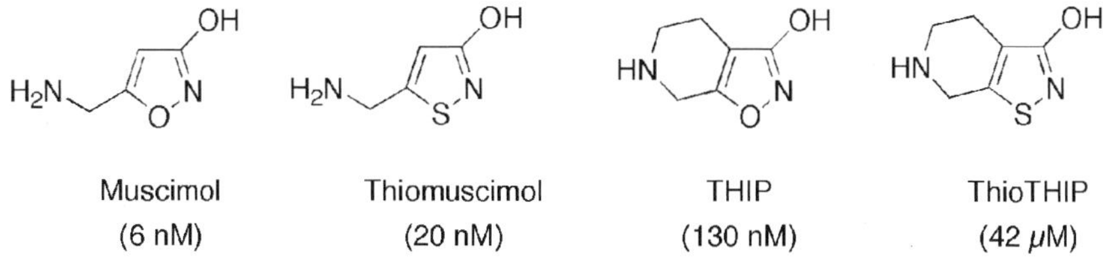

Fig. 2. GABA$_A$ receptor binding data (IC$_{50}$) for muscimol, thiomuscimol, THIP, and thio-THIP.

of thiomuscimol is three times lower than that of muscimol *(9)*, whereas thioTHIP displays more than 300 times lower affinity than THIP *(12)*. These results strongly indicate that the 3-isoxazolol rings in muscimol and THIP do not occupy identical positions in the receptor cavity. Furthermore, introduction of a methyl group in the 4-position of the 3-isoxazolol ring of muscimol severely inhibits interaction with the GABA$_A$ receptor recognition site *(13)*. This effect probably reflects steric repulsion between the methyl group and the receptor, and makes a direct superimposition of the 3-isoxazolol rings of musicmol and THIP questionable.

In order to obtain more information about the receptor-active conformation of muscimol, X-ray structure analysis, and *ab initio* quantum chemical studies have been applied to this compound *(14)*. The flexibility of muscimol exclusively resides in the side chain (O-C-C-N bond) and a calculated potential energy curve for the rotation about this bond of muscimol in its zwitterionic form is shown in Fig. 3. These studies have demonstrated that in order for muscimol to obtain the conformation displayed by THIP, a conformational energy of 8.9 kcal/mol is required, which makes it less likely that the receptor-active conformation of muscimol corresponds to that of THIP. On the contrary, the conformation of THIP seems to mimic a high-energy conformation of muscimol as shown on the energy curve in Fig. 3, where the dihedral angles corresponding to the solid-state conformations of the muscimol zwitterion and the THIP cation are denoted.

To examine this apparent contradiction, a superimposition of muscimol and THIP in their low-energy conformations has been performed (Fig. 4). This superimposition clearly reveals a different position of the 3-isoxazolol moieties of the two compounds and may explain the differences in pharmacology seen for the thio-analogs of muscimol and THIP (Fig. 2).

3. DEVELOPMENT OF A NEW PHARMACOPHORE MODEL FOR GABA$_A$ RECEPTOR LIGANDS

In light of the interest in partial GABA$_A$ receptor agonists as potential therapeutics, structure-activity studies of a number of analogs of the low-efficacy partial GABA$_A$ agonist, 4-PIOL, have been performed *(15)*. In contrast to the thio-analog of muscimol and THIP, the 3-isothiazolol analog of 4-PIOL, thio-4-PIOL, was shown to posses an affinity sixfold higher than that of the parent compound (Fig. 5) *(15)*. As noted previously for muscimol, introduction of alkyl groups into the 4-position of the 3-isoxazolol ring of THIP severely inhibits interaction with the GABA$_A$ receptor. Both 4-methyl-muscimol and 4-methyl-THIP show vanishingly low affinity for the GABA$_A$ receptor-

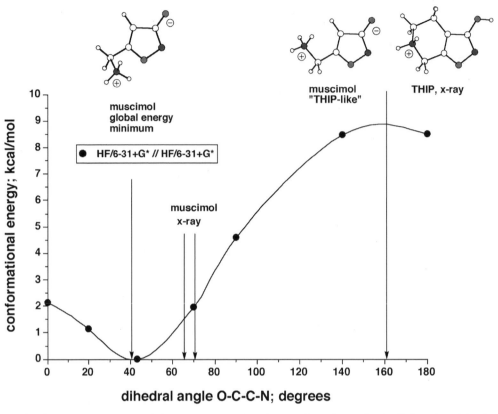

Fig. 3. Calculated potential energy curve for the rotation about the O-C-C-N bond in the muscimol zwitterion using HF/6-31+G* *ab initio* quantum chemical calculations. Dihedral angles corresponding to the global energy minimum of muscimol and the solid-state conformations of the muscimol zwitterion and the THIP cation are denoted by arrows.

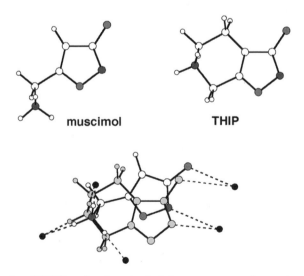

Fig. 4. Muscimol and THIP zwitterions in their low-energy conformations (upper part) and a superimpostion of the two conformations (lower part)

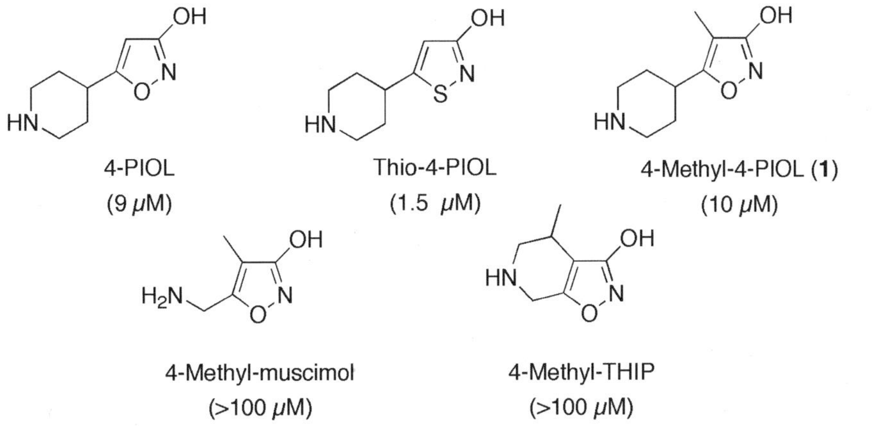

Fig. 5. GABA_A receptor binding data (IC$_{50}$) for 4-PIOL, thio-4-PIOL, 4-methyl-4-PIOL, 4-methyl-muscimol, and 4-methyl-THIP.

binding site *(13,16)*. In contrast, the GABA_A receptor tolerates introduction of a methyl group into the 4-position of the 3-isoxazolol ring in 4-PIOL. The differences in structure-activity relationships observed for these series of compounds indicate that the positions in the GABA_A binding pocket of the 3-isoxazolol rings in muscimol, THIP, and in particular 4-PIOL are different.

In order to rationalize the differences in the structure-activity relationships mentioned previously, a hypothesis for the binding mode of muscimol and 4-PIOL has recently been proposed *(17–19)*. This hypothesis is based on the assumption that an arginine residue in the binding site of the receptor is a suitable binding partner for the 3-isoxazolol anion of muscimol. *Ab initio* quantum chemical calculations have been performed on different bioisosteres of the carboxylate group of GABA, including muscimol, interacting with a guanidinium ion simulating arginine *(20)*. The results from these studies strongly support a bidentate interaction between the anionic part of the GABA_A receptor ligand and an arginine residue in the active site forming two strong hydrogen bonds. Further evidence for the hypothesis has been achieved from subsequent site-directed mutagenesis studies demonstrating a direct implication of an arginine residue in the binding of muscimol and GABA *(21–23)*.

Muscimol and 4-PIOL are assumed to bind to the same binding site, but a direct superimposition of the amino groups and the 3-isoxazolol rings, which make up the pharmacophore elements in these compounds, seems impossible. However, owing to the flexibility of the arginine, this side chain can adopt an extended as well as a bent conformation, which makes it feasible to accommodate muscimol and 4-PIOL in the same binding pocket as illustrated in Fig. 6A.

A molecular least-squares superimposition of the proposed bioactive conformations of muscimol and 4-PIOL in complex with an arginine residue accomplishes a superimposition of the ammonium groups of the two compounds as shown in Fig. 6B *(18)*. According to this superimposition, the 3-isoxazolol rings of muscimol and 4-PIOL do not overlap. This means that the 4-position of the 3-isoxazolol ring of 4-PIOL neither corresponds to the 4-position in the 3-isoxazolol ring of muscimol nor to the corresponding position of THIP during interaction with the receptor recognition site. This

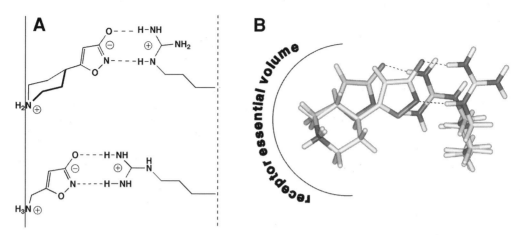

Fig. 6. Hypothesis for the binding of 4-PIOL and muscimol to the GABA$_A$ receptor (**A**). A pharmacophore model for the GABA$_A$ receptor agonists showing the proposed binding modes of muscimol and 4-PIOL and their interactions with two different conformations of an arginine residue (**B**).

hypothesis appears to account for the differences seen for the alkylated analogs and for the thio-analogs of the three compounds under study (Figs. 2 and 5). In order to estimate the receptor essential volume for the GABA$_A$ receptor, a number of structurally related compounds were included in the study *(19)*. Receptor essential volumens were identified in an area around the molecule of muscimol and THIP, an area that also include the piperidine ring of 4-PIOL, suggesting that these structure elements actually do fit into a narrow pocket (Fig. 6B). In contrast, a certain degree of bulk seems to be tolerated around the 4-position of the 3-isoxazolol ring of 4-PIOL.

4. TRANSFORMATION OF PARTIAL AGONISM INTO ANTAGONISM IN 4-PIOL ANALOGS

In order to investigate the steric tolerance of the area around the 4-position of the 3-isoxazolol ring of 4-PIOL, a number of analogs of 4-PIOL with substituents in this position were synthesized *(17,18)*. Substituents of different size and conformational flexibility such as alkyl, phenylalkyl, diphenylalkyl, and naphthylalkyl were explored. The compounds were pharmacologically characterized in receptor binding studies using rat brain membrane preparations and electrophysiologically using whole-cell patch-clamp recordings from cultured cerebral cortical neurones *(17,18)* and two-electrode voltage-clamp recordings from human $\alpha_1\beta_2\gamma_{2s}$ GABA$_A$ receptors expressed in *Xenopus* oocytes *(24)*.

Like 4-PIOL, all of the compounds were shown to be selective for the GABA$_A$ receptor-binding site. The pharmacological data for the compounds are listed in Table 1. Unbranched alkyl groups as substituents in the 4-position of the 3-isoxazolol ring of 4-PIOL such as methyl, ethyl, butyl, hexyl, and octyl groups as well as a benzyl group are tolerated. Compounds **1–7** show affinities for the GABA$_A$ receptor sites comparable to or slightly higher than that of 4-PIOL. Reduction of the phenyl group of compound **7** to a cyclohexyl group, giving compound **6,** was tolerated with only a slight decrease in affinity, indicating that the interaction between the phenyl ring of compound **7** and the receptor is primarily of a hydrophobic character. The optimal length of the alkyl chain connecting the phenyl group and the 3-isoxazolol moiety was found to be three carbon atoms (compound **9.)**

Table 1
Receptor Binding and In Vitro Electrophysiology Data

Compound	R	[³H]muscimol binding[a] K_i(µM)	Electro-physiology[b] IC$_{50}$ (µM)	Compound	R	[³H]muscimol binding[a] K_i(µM)	Electro-physiology[b] IC$_{50}$ (µM)
4-PIOL	H	9.1	110	9	(ring)-(CH$_2$)$_3$-	1.1	0.53
1	CH$_3$-	10	26	10	(diphenyl)CH-CH$_2$-	0.36	0.81
2	CH$_3$CH$_2$-	6.3	10.3	11	(diphenyl)CH-(CH$_2$)$_2$-	0.068	0.02
3	CH$_3$(CH$_2$)$_3$-	7.7	3.0	12	(diphenyl)CH-(CH$_2$)$_3$-	0.70	0.05
4	CH$_3$(CH$_2$)$_5$-	4.5	1.1	13	(2-naphthyl)-CH$_2$-	0.049	0.37
5	CH$_3$(CH$_2$)$_7$-	1.2	0.44	14	(1-naphthyl)-CH$_2$-	0.10	0.48
6	(cyclohexyl)-CH$_2$-	4.9	1.1	15	(9-anthracyl)-CH$_2$-	5.9	1.3
7	(ring)-CH$_2$-	3.8	4.0	16	(biphenyl)-CH$_2$-	0.4	0.71
8	(ring)-(CH$_2$)$_2$-	5.0	4.1	SR 95531		0.074	0.24

[a] Standard receptor binding on rat brain synaptic membranes, $n = 3$
[b] Whole-cell patch-clamp recordings from cerebral cortical neurones, $n = 6$–17 *(17)*.

Addition of an additional phenyl group to the terminal carbon atom of compound **9** gave the 3,3-diphenylpropyl compound, **11,** showing a 16-fold increase in affinity relative to the 3-phenylpropyl analog, **9.** As shown for the phenylalkyl analogs, where the three-carbon chain seems to be optimal for receptor interaction, the 2,2-diphenylethyl as well as the 4,4-diphenylbutyl analog, compounds **10** and **12,** showed reduced affinity for the GABA$_A$ receptor as compared with the 3,3-diphenylpropyl analog, **11.**

A further increase in affinity was obtained by introducing the 2-naphthylmethyl group into the 4-position of the 3-isoxazolol ring of 4-PIOL to give compound **13.** This led to a 78-fold increase in affinity as compared with that of the benzyl analog, **7.** Introduction of the isomeric 1-naphthylmethyl group proved to be less favorable; compound **14** showed a decrease in binding affinity compared to the 2-naphthylmethyl analog **13.** Extension of the aromatic ring system from a 1-naphthylmethyl group in compound **14** to a 9-anthracylmethyl group in compound **15** resulted in a marked reduction in affinity. In contrast, enlarging the aromatic system by attaching a phenyl group to the 4-position of the phenyl group of **7** to give the 4-biphenylmethyl analog **16,** receptor affinity was increased 10-fold relative to that of the benzyl analog **7.**

In the whole-cell patch-clamp studies on cortical neurons, 4-PIOL and selected analogs were characterized as competitive $GABA_A$ receptor antagonists by their ability to induce a parallel shift of the GABA dose-response curves *(17)*. The determined IC_{50} values, shown in Table 1, correspond well with recent data from a voltage-clamp study on recombinant human $\alpha_1\beta_2\gamma_{2s}$ $GABA_A$ receptors expressed in oocytes and show, with a few exceptions, a fairly good correlation to the receptor binding data *(24)*.

4-PIOL has previously been characterized as a partial agonist in whole-cell patch-clamp studies on cerebral cortical neurones and cultured hippocampal neurones *(15,25)*. Only the methyl and ethyl analogs, **1** and **2**, retained detectable ability to induce an agonist effect as shown on cultured cerebral cortical neurones and recombinant receptors *(24)*. In the study of recombinant $GABA_A$ receptors, the weak responses of **1** and **2** were potentiated by simultaneously administered lorazepam and inhibited by the competitive antagonist SR 95531.

As seen from the data in Table 1, the major consequence of replacing the hydrogen in the 4-position of the 3-isoxazolol ring of 4-PIOL with the lipophilic and bulky alkyl or arylalkyl groups is a change in the pharmacology of the compounds from moderately potent low-efficacy partial $GABA_A$ receptor agonist activity to potent antagonist effect. The 2-naphthylmethyl and the 3,3-diphenylpropyl analogs, **13** and **11,** showed an antagonist potency comparable with or markedly higher than that of the standard $GABA_A$ receptor antagonist SR 95531.

5. STRUCTURE-ACTIVITY STUDIES

5.1. Comparative Affinity and Hydrophobicity Studies

Hydrophobicity (log P) is related to the desolvation of the ligand, and it is assumed that the desolvation of the ligands going from water to octanol parallels that of going from water to a cleft or pocket of a receptor. Thus, log P may be an important term in establishing quantitative relationships between structure and activity *(26)*. In order to examine the influence of the lipohilic character of the 4-PIOL analogs on the affinity for the $GABA_A$ receptor, QSAR analysis was performed *(24)*.

To obtain an estimate of the lipophilicity of the compounds, log P values for the substituents in the 4-position of the compounds were calculated according to the method of Ghose and Crippen *(27,28)*. Calculated log P values are listed in the table of Fig 7. A linear correlation between pK_i and log P values was obtained for the data set (R^2 = 0.747, $p < 0.001$) as illustrated in Fig 7. The data set apparently splits up into a large group of compounds showing an excellent correlation (R^2 = 0.947, $p < 0.001$; black line; slope 0.578), two compounds (**11** and **13**) showing higher affinity, and two compounds (**8** and **15**) showing lower affinity than that which can be explained by log P.

Provided that a common binding mechanism of the largest group of compounds is assumed, the highly significant positive correlation between lipophilicity and affinity may reflect the transfer process of the compounds from the aqueous environment to the receptor phase. For the two small groups of compounds, additional interactions in the receptor binding site may be of importance.

5.2. Exploration of the Dimensions of the Receptor Binding Pocket

To study the presumed additional ligand–receptor interactions mentioned earlier in more detail, a conformational analysis was performed on a subset of compounds cho-

Compound	log P
4-PIOL	0
1	1.09
2	1.26
3	2.17
4	3.09
7	2.28
8	2.74
9	3.19
10	3.96
11	4.42
12	4.87
13	3.40
14	3.40
15	4.52
16	3.72

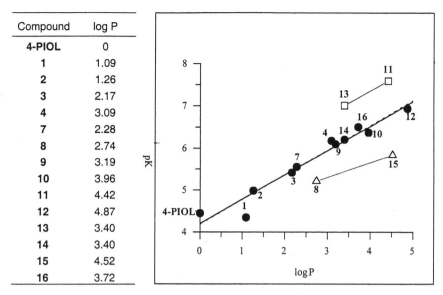

Fig. 7. Estimates of the logarithm of partition coefficient in octanol:water (log P) of 4-PIOL analog substituents obtained by use of Crippen's fragmentation method. Correlation between affinity (pK_i) and lipophilicity (log P) of the substituents of the 4-PIOL analogs. The two gray lines connect the high-affinity compounds (open squares) and the low-affinity compounds (open triangles), respectively, and the black line shows the correlation between the remaining compounds (black circles).

sen on the basis of either high affinity for the GABA$_A$ receptor site and/or informative structural features for the pharmacophore model *(17)*.

Two conformational searches were performed for each molecule using the Monte Carlo multiple minimum method and the MM3* force field in conjunction with the GB/SA hydration model *(29–31)*. The first analysis was performed to search for the global minimum conformation for the free ligand in water. The second conformational search was performed in order to identify possible receptor-bound conformations of the substituents in the 4-position of the 3-isoxazolol ring of 4-PIOL. The latter conformational analysis of the 4-subsituents were performed with the 4-PIOL skeleton conformationally constrained to its proposed bioactive conformation according to the pharmacophore model. Because the bioactive conformation is not necessarily the lowest energy conformation of the molecule, the conformational energy that is required for the molecule to adopt its bioactive conformation was calculated for the compounds. In order to represent a reasonable bioactive conformation, the required energy should be low and there should be no steric conflict with the receptor essential volumes in the pharmacophore model.

The 9-anthracylmethyl analog, **15,** is a member of the group of compounds showing lower affinity for the GABA$_A$ receptor than expected from their log P values (Fig. 7) *(24)*. Conformational analysis of this compound shows a rather high-energy penalty for the most probable bioactive conformation (6.9 kcal/mol). Furthermore, one of the distal aromatic rings in the anthracyl ring system seems to be in electrostatic conflict with a putative electron-rich binding site interacting with the axial piperidyl N$^+$H in the GABA$_A$ receptor binding site (Fig. 8).

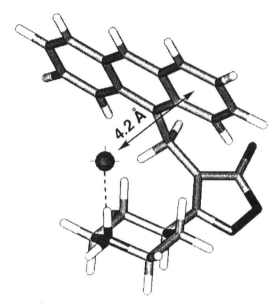

Fig. 8. Proposed bioactive conformation of the 9-anthracylmethyl analog (**15**) displaying electrostatic repulsion between one of the aromatic rings and a putative electron-rich binding site (sphere).

In the proposed bioactive conformation of compound **14**, the 1-naphthylmethyl substituent is approximately orthogonal to the 3-isoxazolol ring and is located on the same side of the ring as the piperidine nitrogen atom (Fig. 9). The unsubstituted aromatic ring is pointing away from the 3-isoxazolol ring, thus avoiding the electrostatic conflict described earlier for the 9-anthracylmethyl analog, **15.** Two possible conformations were found for the isomeric 2-naphthylmethyl analog, **13,** displaying very low and similar conformational energy penalties for binding (0.5 and 0.8 kcal/mol). The two conformations, one shown in Fig. 9, differ only in the orientation of the distal aromatic ring. The conformational analysis of the benzyl analog, **7,** yielded a bioactive conformation in which the phenyl ring closely superimposes the substituted aromatic rings in the two isomeric naphthylmethyl analogs. A superimposition of these compounds, **7,13,** and **14,** respectively, is shown in Fig. 9.

Introduction of a benzyl group in the 4-position of the 3-isoxazolol ring leads to an increase in affinity by a factor less than 3 as compared to that of the methyl analog, **1** (Table 1). In contrast, introduction of a 2- or a 1-naphthylmethyl group increases the affinity by factors of 78 and 38, respectively. A similar effect was observed for the 3,3-diphenylpropyl analogue, **11,** where a marked increase in affinity was seen compared to the corresponding 3-phenylpropyl analog, **9.** Thus, although the affinities for the $GABA_A$ receptor of the diphenyl and naphthyl analogs, **10–14,** are markedly higher than the affinity of the corresponding monophenyl compounds, **7–9,** the energy penalty for converting the low-energy conformation into the supposed bioactive conformation is similar or even higher.

The two high-affinity compounds, the 2-naphthylmethyl and the 3,3-diphenylpropyl analog, **13** and **11,** respectively, constitute the third group in the analysis of the correlation of lipophilicity with affinity, both compounds showing higher affinity for the

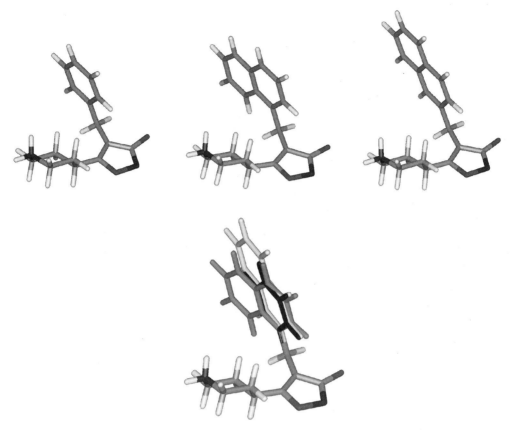

Fig. 9. Proposed bioactive conformations for the benzyl, 1-naphthylmethyl and 2-naphthyl-methyl analogs, **7,14,** and **13,** respectively (upper part). Superimposition of the three compounds in their proposed bioactive conformations (lower part).

GABA$_A$ receptor binding site than expected on the basis of to their log P values (Fig. 7) *(24)*. Because the marked effect on affinity of the additional phenyl ring cannot be rationalized by an increase in lipophilicity or differences in conformational energy, strong specific interaction between the binding cavity of the GABA$_A$ receptor and the unsubstituted aromatic rings may exist.

In order to study these interactions in more detail, a superimposition of the bioactive conformations calculated for the 3,3-diphenylpropyl analog, **11,** and the 2-naphthyl-methyl analog, **13,** was performed (Fig. 10). This superimposition shows that one of the phenyl rings in compound **11** overlaps closely with the distal ring of the 2-naphthyl-methyl analog, **13,** whereas the other phenyl ring of **11** occupies an unexplored region in space in the pharmacophore model. Based on the observed affinities and calculated conformational energies, both of these positions of an aromatic ring seem to be highly favorable for GABA$_A$ receptor binding. Inclusion of the less active 4-biphenylmethyl analog, **16,** in the superimpostion illustrates the considerable dimensions of the proposed receptor cavity (Fig. 10). The cavity may be extending beyond the 2-naphthylmethyl group in compound **13** as demonstrated by the relatively high affinity of the 4-biphenylmethyl analogue **16,** the length of which is 9.7 Å as measured from the methylene carbon to the most distant hydrogen atom of the distal phenyl ring. Furthermore, one of the phenyl

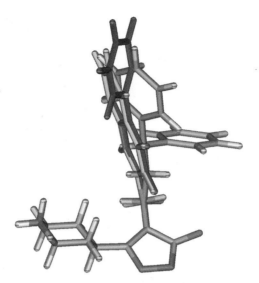

Fig. 10. Superimposition of the 3,3-diphenylpropyl, 2-naphthylmethyl, and 4-biphenyl-methyl analogs, **11,13,** and **16,** respectively, in their proposed bioactive conformations illustrating the large space spanned by the 4-substituent in the 4-position of the 3-isoxazolol rings.

rings in the 3,3-diphenylmethyl analog, **11,** extends 6.4 Å in a direction almost perpendicular to the rings in the three other compounds under study. Because all of the compounds in this study are active, the total volume of the binding pocket has not yet been occupied and it may possible to accomodate even larger substituents.

As previously mentioned, a high correlation was seen between functional antagonist potencies obtained from cloned human receptors and binding affinities determined in rat brain homogenates. Based on the high correlation between binding data obtained in cell lines expressing human GABA$_A$ receptors and rat brain homogenate, it has been concluded, that the inter-species homology in GABA$_A$ receptor sequence and structure is very high. In agreement with earlier observations, this suggests that the potencies of GABA$_A$ antagonists essentially are independent of the subunit composition. In contrast, the potencies of GABA$_A$ receptor agonists have been shown to be highly subunit-dependent *(32).* Along this line, it has been proposed that the GABA$_A$ antagonists bind to and stabilize a distinct inactive receptor conformation. In the case of the 4-substituted 4-PIOL analogs, large substituents increase the affinity of the compounds but concomitantly abolish initiation of a functional response.

Because the GABA binding site in the GABA$_A$ receptor is assumed to be located at the interface between α and β subunits *(33)* it may be speculated that the large cavity accommodating the 4-substituent is located in the space between these subunits. It has been proposed that the mechanism for ligand-induced channel opening in nicotinic acetylcholine receptors involves rotations of the subunits surrounding the ligand binding domain *(34,35).* Belonging to the same superfamily of ligand-gated ion channels, it is likely that the GABA$_A$ and the nicotinic acetylcholine receptors are using the same mechanism for channel opening. Based on this hypothesis, the large substituents in the 4-substituted 4-PIOL analogs may interfere with the conformational transition of the

GABA$_A$ receptor leading to channel opening resulting in antagonistic effects of the compounds *(17).*

6. CONCLUSION

Molecular biology studies have disclosed a high degree of heterogenity of the GABA$_A$ receptor. In order to shed some light on the physiological and pharmacological implications of this receptor heterogenity, the development of selective ligands is of key importance. In the absence of detailed structural informations on the GABA$_A$ receptor complex, the design of ligands for this receptor has, so far, been based on systematic structural, stereochemical and bioisosteric approaches.

Pharmacological data combined with molecular modeling studies have been useful tools to predict the optimal arrangement of the functional groups in the ligands. Using this approach, a new pharmacophore model for the agonist and antagonist binding site(s) in the GABA$_A$ receptor has been developed. Application of this model has clarified some apparent differences in the structure-activity relationships of a series of known GABA$_A$ receptor agonists. In order to further develop the model, a series of new GABA$_A$ receptor ligands has now been synthesized and has provided important information regarding size and dimension of a cavity in the receptorbinding site. Following this line, a better understanding of the structural requirements for binding to and activation or blockade of the GABA$_A$ receptor are obtainable and are likely to be useful for the design of new types of ligands.

REFERENCES

1. Frølund, B., Ebert, B., Kristiansen, U., Liljefors, T., and Krogsgaard-Larsen, P. (2002) GABA$_A$ receptor ligands and their therapeutic potentials. *Curr. Med. Chem.* **2,** 817–832.
2. Chebib, M. and Johnston, G. A. R. (2000) GABA-activated ligand gated ion channels: medicinal chemistry and molecular biology. *J. Med. Chem.* **43,** 1427–1447.
3. Krogsgaard-Larsen, P., Johnston, G. A. R., Curtis, D. R., Game, C. J. A., and McCulloch, R. M. (1975) Structure and biological activity of a series of conformationally restricted analogs of GABA. *J. Neurochem.* **25,** 803–809.
4. Krogsgaard-Larsen, P., Johnston, G. A. R., Lodge, D., and Curtis, D. R. (1977) A new class of GABA agonist. *Nature* **268,** 53–55.
5. Johnston, G. A. R., Beart, P. M., Curtis, D. R., Game, C. J. A., McCulloch, R. M., and MacLachlan, R. M. (1972) Bicuculline methochloride as a GABA antagonist. *Nature (New Biol.)* **240,** 219–220.
6. Wermuth, C. G., Bourguignon, J.-J., Schlewer, G., Gies, J.-P., Schoenfelder, A., Melikian, A., et al. (1987) Synthesis and structure-activity relationships of a series of aminopyridazine derivatives of γ-aminobutyric acid acting as selective GABA-A antagonists. *J. Med. Chem.* **30,** 239–249.
7. Wermuth, C. G. and Biziére, K. (1986) Pyridazinyl-GABA derivatives: a new class of synthetic GABA$_A$ antagonists. *Trends Pharmacol. Sci.* **7,** 421–424.
8. Krogsgaard-Larsen, P., Brehm, L., and Schaumburg, K. (1981) Muscimol, a psychoactive constituent of amanita muscaria, as a medicinal chemical model structure. *Acta Chem. Scand. B* **311,** 0–24.
9. Krogsgaard-Larsen, P., Hjeds, H., Curtis, D. R., Lodge, D., and Johnston, G. A. R. (1979) Dihydromuscimol, thiomuscimol and related heterocyclic compounds as GABA analogues. *J. Neurochem.* **32,** 1717–1724.
10. Galvez-Ruano, E., Aprison, M. H., Robertson, D. H., and Lipkowitz, K. B. (1995) Identifying agonistic and antagonistic mechanisms operative at the GABA receptor. *J. Neurosci. Res.* **42,** 666–673.

11. Buur, J. R., Hjeds, H., Krogsgaard-Larsen, P., and Jørgensen, F. S. (1993) Conformational analysis and molecular modelling of a partial GABA$_A$ agonist and a glycine antagonist related to the GABA$_A$ agonist, THIP. *Drug Des. Discov.* **10,** 213–229.

12. Krogsgaard-Larsen, P., Mikkelsen, H., Jacobsen, P., Falch, E., Curtis, D. R., Peet, M. J., and Leah, J. D. (1983) 4,5,6,7-Tetrahydroisothiazolo[5,4-*c*]pyridin-3-ol and related analogues of THIP. Synthesis and biological activity. *J. Med. Chem.* **26,** 895–900.

13. Krogsgaard-Larsen, P. and Johnston, G. A. R. (1978) Structure-activity studies on the inhibition of GABA binding to rat brain membranes by muscimol and related compounds. *J. Neurochem.* **30,** 1377–1382.

14. Brehm, L., Frydenvang, K., Hansen, L. M., Norrby, P.-O., Krogsgaard-Larsen, P., and Liljefors, T. (1997) Structural features of muscimol, a potent GABA$_A$ receptor agonist. Crystal structure and quantum chemical ab initio calculations. *Struct. Chem.* **8,** 443–451.

15. Frølund, B., Kristiansen, U., Brehm, L., Hansen, A. B., Krogsgaard-Larsen, P., and Falch, E. (1995) Partial GABA$_A$ receptor agonists. Synthesis and in vitro pharmacology of a series of nonannulated analogs of 4,5,6,7-tetrahydroisoxazolo[4,5-c]pyridin-3-ol (THIP). *J. Med. Chem.* **38,** 3287–3296.

16. Haefliger, W., Révész, L., Maurer, R., Römer, D., and Büscher, H.-H. (1984) Analgesic GABA agonists. Synthesis and structure-activity studies on analogues and derivatives of muscimol and THIP. *Eur. J. Med. Chem.* **19,** 149–156.

17. Frølund, B., Jørgensen, A. T., Tagmose, L., Stensbøl, T. B., Vestergaard, H. T., Engblom, C., et al. (2002) A novel class of potent 4-arylalkyl substituted 3-isoxazolol GABA$_A$ antagonists: synthesis, pharmacology and molecular modeling. *J. Med. Chem.* **45,** 2454–2468.

18. Frølund, B., Tagmose, L., Liljefors, T., Stensbøl, T. B., Engblom, C., Kristiansen, U., and Krogsgaard-Larsen, P. (2000) A novel class of potent 3-isoxazolol GABA$_A$ antagonists: design, synthesis, and pharmacology. *J. Med. Chem.* **43,** 4930–4933.

19. Tagmose, L. (2000) A pharmacophore model for GABA$_A$ receptor agonists. Ph.D. Thesis, The Royal Danish School of Pharmacy.

20. Tagmose, L., Hansen, L. M., Norrby, P.-O., and Liljefors, T. (2000) Differences in agonist binding pattern for the GABA$_A$ and the AMPA receptors illustrated by high-level ab initio calculations, in *Molecular Modelling and Prediction of Bioactivity* (Gundertofte, K. and Jørgensen, F. S., eds.), Kluwer Academic/Plenum Publishers, New York, pp. 365–366.

21. Boileau, A. J., Evers, A. R., Davis, A. F., and Czajkowski, C. (1999) Mapping the agonist binding site of the GABA$_A$ receptor: evidence for a β-strand. *J. Neurosci.* **19,** 4847–4854.

22. Hartvig, L., Lükensmejer, B., Liljefors, T., and Dekermendjian, K. (2000) Two conserved arginines in the extracellular N-terminal domain of the GABA$_A$ receptor α_5 subunit are crucial for receptor function. *J. Neurochem.* **75,** 1746–1753.

23. Westh-Hansen, S. E., Witt, M. R., Dekermendjian, K., Liljefors, T., Rasmussen, P. B., and Nielsen, M. (1999) Arginine residue 120 of the human GABA$_A$ receptor α_1, subunit is essential for GABA binding and chloride ion current gating. *NeuroReport* **10,** 2417–2421.

24. Mortensen, M., Frølund, B., Jørgensen, A. T., Liljefors, T., Krogsgaard-Larsen, P., and Ebert, B. (2002) Activity of novel 4-PIOL analogues at human $\alpha_1\beta_2\gamma_{2s}$ GABA$_A$ receptors: correlation with hydrophobicity. *Eur. J. Pharmacol.* **451,** 125–132.

25. Kristiansen, U., Lambert, J. D. C., Falch, E., and Krogsgaard-Larsen, P. (1991) Electrophysiological studies of the GABA$_A$ receptor ligand, 4-PIOL, on cultured hippocampal neurones. *Br. J. Pharmacol.* **104,** 85–90.

26. Leo, A. J. and Hansch, C. (1999) Role of hydrophobic effect in mechanistic QSAR. *Perspect. Drug Discov. Des.* **17,** 1–25.

27. Ghose, A. K. and Crippen, G. M. (1986) Atomic physiochemical parameters for three-dimensional structure-directed quantitative structure-activity relationships: I. Partition coefficients as a measure of hydrophobicity. *J. Comput. Chem.* **4,** 565–577.

28. Ghose, A. K., Pritchett, A., and Crippen, G. M. (1988) Atomic physiochemical parameters for three-dimensional structure-directed quantitative structure-activity relationships: III. Modeling hydrophobic interactions. *J. Comput. Chem.* **9,** 80–90.

29. Chang, G., Guida, W. C., and Still, W. C. (1989) An internal coordinate Monte Carlo method for searching conformational space. *J. Am. Chem. Soc.* **111,** 4379–4386.

30. Still, W. C., Tempczyk, A., Hawley, R. C., and Hendrickson, T. (1990) Semianalytical treatment of solvation for molecular mechanics and dynamics. *J. Am. Chem. Soc.* **112,** 6127–6129.

31. Mohamadi, F., Richards, N. G. J., Guida, W. C., Liskamp, R., Lipton, M., Caufield, C., et al. (1990) MacroModel: an integrated software systems for modeling organic and bioorganic molecules using molecular mechanics. *J. Comput. Chem.* **11,** 440–467.

32. Ebert, B., Thompson, S. A., Suonatsou, K., McKernan, R., Krogsgaard-Larsen, P., and Wafford, K. A. (1997) Differences in agonist/antagonist binding affinity and receptor transduction using recombinant human γ-aminobutyric acid type A receptors. *Mol. Pharmacol.* **52,** 1150–1156.

33. Smith, G. B. and Olsen, R. W. (1995) Functional domains of GABA_A receptors. *Trends Pharmacol. Sci.* **16,** 162–168.

34. Unwin, N. (1993) Nicotinic acetylcholine receptor at 9 A resolution. *J. Mol. Biol.* **229,** 1101–1124.

35. Unwin, N. (1995) Acetylcholine receptor channel imaged in the open state. *Nature* **373,** 37–43.

7

The GABA$_B$ Receptor

From Cloning to Knockout Mice

Bernhard Bettler and Hans Bräuner-Osborne

1. INTRODUCTION

γ-Aminobutyric acid (GABA) is the prevalent inhibitory neurotransmitter in the brain. It exerts it action through ligand-gated Cl⁻ channel (GABA$_A$ and GABA$_C$ receptors) and G protein coupled receptors (GPCR) that inhibit adenylate cyclase (GABA$_B$ receptors) *(1–3)*. GABA$_B$ receptors were first identified in the early 1980s on the basis of pharmacological responses to the agonist baclofen (Fig. 1) and insensitivity to the GABA$_A$ antagonist bicuculline *(4,5)*, but resisted cloning until the late 1990s. Several groups made unsuccessful attempts to isolate the receptor protein by affinity chromatography *(6,7)* or to expression clone the receptor in *Xenopus* oocytes using electrophysiology *(8–10)*. Eventually, medicinal chemistry efforts produced the highly potent radiolabeled GABA$_B$ antagonists [^{125}I]CGP64213 and [^{125}I]CGP71872 (Fig. 1) and provided the necessary tools for expression cloning *(11)*. After screening of two million rat brain cDNA clones using a radioligand binding assay the first GABA$_{B(1)}$ receptor clone was isolated *(12)*. Cloning of the GABA$_{B(1)}$ cDNA was a major milestone in the field as it paved the way for studies on the structure, function, and pharmacology of GABA$_B$ receptors at the molecular level.

2. IN VITRO STUDIES

2.1. Pharmacological Characterization of Recombinant GABA$_B$ Receptors

As expected from early studies on the coupling of GABA$_B$ receptors to inhibition of adenylate cyclase *(1–3)*, the GABA$_{B(1)}$ protein has all the hallmarks of a GPCR. Specifically the receptor belongs to the family C of GPCRs, with similarity to the metabotropic glutamate (mGlu) receptors, which are characterized by a large extracellular amino-terminal domain (ATD) and a 7-transmembrane (7TM) domain (Fig. 2A) *(12)*. Initial cloning efforts revealed two major amino-terminal splice-variants, GABA$_{B(1a)}$ and GABA$_{B(1b)}$, of 960 and 844 amino acids, respectively. The former contain two amino-terminal sushi-repeats, which are protein–protein interaction motifs that are expected to serve as an extracellular targeting signal that dictates subcellular localization (Fig. 2A) *(12)*. Pharmacological analysis of the two GABA$_{B(1)}$ splice vari-

From: *Molecular Neuropharmacology: Strategies and Methods*
Edited by: A. Schousboe and H. Bräuner-Osborne © Humana Press Inc., Totowa, NJ

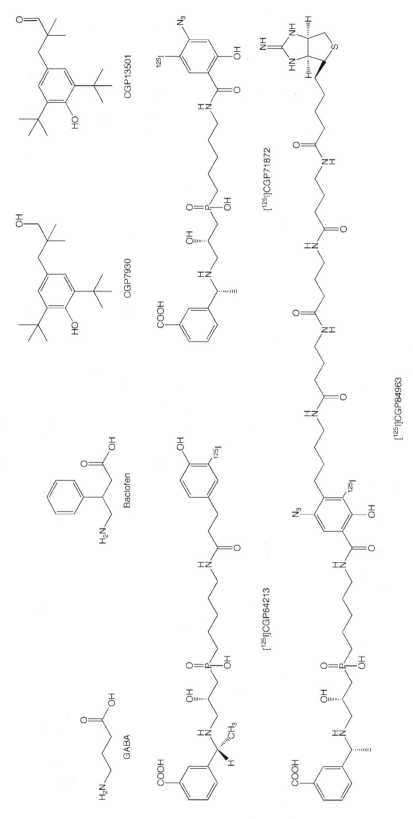

Fig. 1. Chemical structures of ligands used to characterize, clone, and purify the GABA_B receptor. The recently identified positive allosteric modulators CGP7930 and CGP 13501 are expected to broaden the spectrum of therapeutic applications for GABA_B drugs.

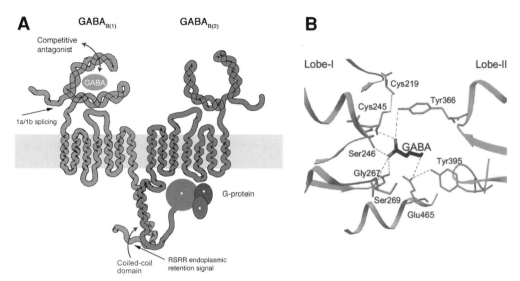

Fig. 2. (A) A schematic illustration of the GABA$_{B(1)}$ and GABA$_{B(2)}$ subunits forming the heterodimeric GABA$_B$ receptor. GABA binds in the cleft of the amino-terminal domain of GABA$_{B(1)}$, which leads to G protein activation through the GABA$_{B(2)}$ subunit. The localization of the coiled-coil domain, the RSRR endoplasmic retention signal, and the GABA$_{B(1a)/(1b)}$ splice site has been noted. Adapted with permission from Marshall et al. *(56)*. **(B)** A model of GABA docked into the ligand binding site of GABA$_{B(1)}$. The putative interactions between GABA and amino acids of GABA$_{B(1)}$ are shown with dotted lines. Adapted with permission from Kniazeff et al. *(30)*.

ants in recombinant binding assays revealed that antagonists bound with the expected affinities, whereas agonists affinities were approx 100-fold lower than those seen with native GABA$_B$ receptors. Furthermore, it was impossible to obtain robust functional responses in cell lines expressing the splice variants *(12)*. Using epitope-tagged receptors and fluorescent microscopy, it was shown that both GABA$_{B(1)}$ splice variants are retained in the endoplasmic reticulum (ER) and thus fail to reach the plasma membrane *(13)*. Given that proper localization to the plasma membrane is a prerequisite for GPCR signaling, this study provided an explanation for the lack of function of GABA$_{B(1a)}$ and GABA$_{B(1b)}$ *(12,13)*.

These puzzling findings were resolved when several groups reported the cloning of a second GABA$_B$ receptor cDNA, GABA$_{B(2)}$ *(14–17)*. The observations that expression of GABA$_{B(1)}$ and GABA$_{B(2)}$ transcripts in the brain is largely overlapping and that the two proteins interact in the yeast two-hybrid system were highly suggestive of a heteromeric receptor. Accordingly, when co-expressing GABA$_{B(1a)}$ or GABA$_{B(1b)}$ with GABA$_{B(2)}$, functional receptors were obtained *(14–18)*. Heteromeric GABA$_{B(1,2)}$ receptors also partially recovered the high-affinity agonist binding-site, although an approx 10-fold lower affinity was still observed in recombinant cells as opposed to brain membranes *(14)*. The reason for this is unclear. A possible explanation is that the amount of G proteins in the heterologous cells is limiting and thus prevents the high-affinity conformation.

As is discussed in greater detail later, it has been impossible to identify pharmacologically distinct subunits. This despite the fact that pharmacological differences were observed in vivo, which suggested the existence of multiple receptor subtypes. Specifically no differences were revealed in binding or functional experiments with native *(19)* or recombinant GABA$_{B(1a)}$ and GABA$_{B(1b)}$ variants *(12,14,18,20)*. It was, however, postulated that the anticonvulsant gabapentin (Fig. 1) is a partial agonist at GABA$_{B(1c)}$ with no activity at GABA$_{B(1b)}$ *(21,22)*. Several groups, including ours, were unable to replicate these findings *(23,24)*. Recently, a transcript named "GABA$_B$-like" (GABA$_{BL}$), which is homologous to GABA$_{B(1)}$ and GABA$_{B(2)}$, was reported *(25)*. Co-expression with GABA$_{B(1)}$ and/or GABA$_{B(2)}$ did not alter pharmacological properties and thus the function of the new gene product remains unknown *(25)*. At present, we are left with only one pharmacological GABA$_B$ receptor subtype, the heteromeric GABA$_{B(1a,2)}$ and GABA$_{B(1b,2)}$. This remains in sharp contrast to the pharmacological diversity described before cloning.

2.2. The Ligand Binding Domain

2.2.1. Orthosteric Ligands

Using chimeric GABA$_{B(1)}$/mGluR1 and truncated GABA$_{B(1)}$ receptors, it was shown that agonists and competitive antagonists bind to the ATD of the GABA$_{B(1)}$ subunit. *(26)*. Further analysis of the GABA$_{B(1)}$ ATD revealed a weak sequence identity with bacterial periplasmic binding proteins, which was exploited to develop computational models of the ligand binding domain based on the 3D X-ray structures of the bacterial proteins *(27–29)*. As described in detail in Chapter 3, the ATD consists of two globular lobes (Lobe I and II) connected with a hinge region that together form a ligand-binding pocket (*see* also Fig. 2A). Extensive mutational analysis of the GABA$_{B(1)}$ ATD, guided by computer models, indicated that S246, S269, D471, and E465 in Lobe I and Y366 in Lobe II are of particular importance for agonist and antagonist binding (Fig. 2B) *(27,28,30)*. Mutation of S247 and Q312 increases the affinity of agonists while decreasing the affinity of antagonists and thus appears to modulate closing of the ligand binding pocket *(27)*. Mutation of several residues selectively affects the binding of certain agonists. For example, the potency of GABA is decreased 30-fold by the S269A mutation whereas the potency of baclofen remains unaltered by the mutation *(31,32)*. As will be discussed in Subheading 2.2.2., this correlates with a potentation of GABA binding by Ca^{2+}, which is not observed with baclofen *(32)*.

The GABA$_{B(2)}$ subunit does not bind radiolabeled GABA$_B$ agonists or antagonists and is not functional when expressed alone *(14–18)*. It does, however, contain the large ATD, which binds ligands in GABA$_{B(1)}$, the mGlu, and calcium sensing receptors (*see* Chapter 3). It is thus surprising that all GABA$_B$ radioligands fail to bind to GABA$_{B(2)}$. A possible explanation for this finding is that an alternative, as yet unknown, ligand binds to the ATD of GABA$_{B(2)}$. However, based on a phylogenetic analysis of GABA$_B$ receptors from various species, this appears to be a remote possibility only *(30)*.

As was described in Chapter 3, an enormous wealth of information has been obtained by crystallization and X-ray structure determination of the ATD of mGlu1. It is expected that an equal leap in knowledge will occur once the X-ray structure of the ATD of GABA$_{B(1)}$ will be available, with or without co-crystallization of the GABA$_B$(2) ATD. So far no crystalline GABA$_B$ ATD fragment suitable for X-ray structure determination has been reported. However, high-affinity biotinylated radiolabeled photoaffinity antagonists such as [125I]CGP84963 were developed (Fig. 1), which now can be used for receptor affinity chromatography *(11)*.

2.2.2. Allosteric Modulators

As noted in the previous section, Ca^{2+} allosterically modulates the potency of GABA but not baclofen. Presumably Ca^{2+} interacts with S269 *(31,32)* and stabilizes the active closed conformation of the ATD *(32,33)*. The EC$_{50}$ for the Ca^{2+} potentiation is 37 µ*M (32)*, which is considerably lower than the potency of Ca^{2+} at the mGlu receptors (low m*M* range). It is likely that the Ca^{2+} site on the GABA$_B$ receptor is saturated under normal physiological conditions. However, it is possible that Ca^{2+} modulation plays a role in pathophysiological situations characterized by a significant drop in extracellular Ca^{2+}.

Recently, a series of positive allosteric modulators was described *(34)*. These compounds, CGP7930 and CGP13501 (Fig. 1), are small organic molecules and thus attractive from a drug-development perspective. Like Ca^{2+}, these compounds do not activate the receptor directly, but they increase the potency and affinity of classical GABA$_B$ agonists including GABA *(34)*. The compounds do not displace binding of radiolabeled orthosteric ligands and therefore do not bind to the GABA binding site. So far the allosteric binding site has not been elucidated, but positive allosteric modulators at mGlu1 and calcium-sensing receptors were shown to bind within the 7TM domain *(35,36)*. This suggests that CGP7930 and CGP13501 similarly bind in the 7TM domain of GABA$_{B(1)}$ and/or GABA$_{B(2)}$, but other binding sites are not ruled out *(37)*. As is described later, positive allosteric modulators may well turn out to be therapeutically superior to classical agonists, such as baclofen.

2.3. G protein Coupling

GABA$_B$ receptors need to heteromerize in order to form functional receptors. The roles of the individual subunits in the receptor complex are just beginning to emerge. GABA$_{B(1)}$ contains the orthosteric ligand-binding site and is retained in the ER in the absence of GABA$_{B(2)}$. However GABA$_{B(2)}$ is not merely a chaperone ensuring proper trafficking of the receptor complex to the plasma membrane. Recent data indicate that this subunit is mainly responsible for coupling to the G protein. This was initially demonstrated by use of chimeric GABA$_{B(1)}$ and GABA$_{B(2)}$ receptors. Receptors that only contain the 7TMs of GABA$_{B(2)}$ subunits are functional, whereas the reverse configuration with all the 7TMs from GABA$_{B(1)}$ yields nonfunctional receptors *(38)*. This observation was confirmed and extended by exchanging the intracellular loops between the GABA$_{B(1)}$ and GABA$_{B(2)}$ subunits *(39,40)*, or by mutation of individual residues within the intracellular loops *(41,42)*. Intriguingly, mutation of a single residue in the i3 loop *(41)* or of three residues in the i2 loop *(42)* of the GABA$_{B(2)}$ subunit leads to a complete loss of function while mutation of the corresponding residues of the GABA$_{B(1)}$ subunit has no effect on receptor function.

2.4. Receptor Trafficking

GABA$_{B(1)}$ and GABA$_{B(2)}$ subunits need to heteromerize in order to form a functional receptor. GABA$_{B(1)}$, when expressed alone, is retained in the ER *(13)*, but co-expression with GABA$_{B(2)}$ leads to proper trafficking of the receptor complex to the plasma membrane *(14–17)*. Both GABA$_{B(1)}$ and GABA$_{B(2)}$ contain alpha-helical domains in their intracellular carboxy-temini that undergo a coiled-coil interaction (Fig. 2A). It has been shown that GABA$_{B(1)}$ features a RSRR motif adjacent to the alpha-helical coil, which is responsible for the ER retention by an as yet unknown mechanism *(43,44)*. Formation of the coiled-coil interaction by heteromerization appears to mask the RSRR motif allowing release of the GABA$_{B(1)}$ subunit from the ER and surface trafficking of the heteromeric complex. GABA$_{B(1)}$ subunits with an inactivated RSRR motif reach the plasma membrane but remain nonfunctional. This has led to the conclusion that the ER retention signal functions as a trafficking checkpoint ensuring correct assembly of heteromeric GABA$_B$ receptors *(43–45)*. Interestingly, co-expressed truncated GABA$_{B(1)}$ and GABA$_{B(2)}$ subunits devoid of the coiled-coil domains remain functional, which indicates that the C-terminal domains do not contain the only interaction motif in the heteromer *(44,45)*. It is tempting to speculate that the ATDs also form a dimerization interface as observed with the mGlu receptors (Chapter 3).

2.5. Interaction With Protein Kinases and Other Intracellular Proteins

Phosphorylation of intracellular loops and the carboxy-termini of GPCRs by protein kinases (e.g., protein kinase A [PKA] and protein kinase C [PKC]) generally leads to receptor inactivation. Both GABA$_{B(1)}$ and GABA$_{B(2)}$ contain PKA and PKC consensus phosphorylation motifs. Surprisingly, it was recently reported that PKA phosphorylation of S892 in GABA$_{B(2)}$ positively modulates receptor function by enhancing the membrane stability of the heteromer *(46)*.

Using the yeast-two hybrid system, the carboxy-termini of GABA$_{B(1)}$ and GABA$_{B(2)}$ subunits were also shown to interact with a plethora of other intracellular proteins, including members of the 14-3-3 family *(47)* and activation transcription factor 4 (ATF4) *(48–50)*. These interactions are potentially interesting as they could modulate heteromerization and intracellular signaling pathways, and perhaps even be responsible for the subtype heterogeneity observed in animal tissues. Investigation of the effects of these interactions has just started and no clear picture has emerged yet.

3. GABA$_B$(1) KNOCKOUT MICE AND IN VIVO STUDIES

Baclofen, the prototypical GABA$_B$ receptor agonist, is clinically used as a muscle relaxant to treat spasticity in spinal injury and amyotrophic lateral sclerosis (ALS) patients. Because GABA$_B$ receptors are ubiquitous in the nervous system, a primary goal in pharmaceutical GABA$_B$ research is to target novel drugs selectively to defined receptor populations, e.g., pre- vs postsynaptic or activated vs inactive receptors. This is expected to reduce baclofen's side effects and expand its currently limited clinical use. Therefore it is important to understand the extent of the molecular and functional diversity in the GABA$_B$ receptor system. On the basis of pharmacological differences, receptor heterogeneity between pre- and postsynaptic, as well as between auto- and heteroreceptors, on inhibitory and excitatory terminals, respectively, has been claimed *(51)*. As mentioned earlier, molecular studies only identified two GABA$_B$ genes, each encoding several splice variants *(52)*. Pharmacological differences between recombinant GABA$_{B(1a,2a)}$ and GABA$_{B(1b,2a)}$ receptors were claimed *(21,53)*. In contrast, mixing and matching of splice variants in a number of other laboratories, including ours, did not uncover any significant pharmacological differences, either in ligand-binding experiments or in functional assay systems *(14,19,23,24)*. Similarly, the search for additional GABA$_B$ receptor-related subunits in DNA databases did not precipitate any pharmacological receptor subtypes *(54)*. Intriguingly, the expression pattern of the cloned GABA$_B$ subunits matches the brain distribution of GABA$_B$ binding sites *(55)*. Moreover, the heteromeric GABA$_{B(1a,2)}$ and GABA$_{B(1b,2)}$ receptors activate all well-characterized GABA$_B$ effector pathways in transfected cells *(56)*. These findings, together with the fact that cloning studies failed to identify additional GABA$_B$ sub-types, prompted speculations that the cloned subunits are responsible for all known GABA$_B$-mediated effects. However, as long as pharmacological evidence against receptor subtypes is derived solely from recombinant work, definite conclusions are not warranted. The possibility remains that GABA$_B$ receptor subtypes result from asso-ciation of either subunit with other, possibly structurally unrelated proteins, which in turn affect pharmacological properties. Given these speculations on receptor subtypes, it is imperative to understand to which pharmacological properties and receptor func-tions the cloned GABA$_B$ subunits can contribute in vivo. To address this question, sev-eral laboratories disabled the gene for GABA$_{B(1)}$ in mice *(57,58)*. Only the mutant mice generated in the Balb/c genetic background *(57)* turned out to be viable. Mice gener-ated on other genetic backgrounds die within a 3–4 wk after birth, thus precluding extensive behavioral analysis *(58,59)*. Strain differences in viability are not uncommon *(60)*. The overt phenotype of all GABA$_{B(1)-/-}$ mouse strains includes spontaneous epileptic seizures and these seizures may be suppressed to some extent in the Balb/c genetic background. A reduced seizure activity, in turn, may rescue mice from lethality. Strikingly, in all biochemical, electrophysiological, and behavioral paradigms studied, GABA$_{B(1)-/-}$ animals demonstrate a complete lack of detectable GABA$_B$ responses. These experiments indicate that no other protein can compensate for GABA$_{B(1)}$ and that this subunit is absolutely essential for the functioning of pre- and postsynaptic GABA$_B$ receptors. In agreement with these findings, the GABA$_{B(2)}$ protein is heavily downregulated in GABA$_{B(1)-/-}$ mice. This requirement of GABA$_{B(1)}$ for stable expres-sion of GABA$_{B(2)}$ indicates that most GABA$_{B(2)}$ protein is associated with GABA$_{B(1)}$, in support of biochemical *(61)* and recombinant studies that demonstrate that GABA$_{B(2)}$ does not form a receptor in its own right *(38)*. In conclusion, GABA$_{B(1)}$ knockout stud-ies do not provide evidence for pharmacologically or functionally distinct GABA$_B$ receptor subtypes and are in line with release experiments that challenged the notion of subtypes early on *(62)*.

GABA$_B$ receptors are known to influence many physiological processes, including, e.g., memory formation, thermoregulation, motor and sensory control, or gastrointesti-nal functions. When studied in classical GABA$_B$ paradigms, viable Balb/c GABA$_{B(1)}$-deficient mice exhibit defects in many of these processes *(57)*. Besides a clear impair-ment of passive avoidance performance, reflecting impaired learning or memory for-mation, GABA$_{B(1)-/-}$ mice are hyperalgesic, and exhibit reduced body temperature and increased locomotor activity. This indicates that GABA$_B$ receptors are tonically active and that, as has been proposed for many years, GABA$_B$ drugs could be used to modu-late excessive neuronal firing characteristic of epilepsy and pain, or perhaps help to bolster memory. Strikingly, the locomotor phenotype of Balb/c GABA$_{B(1)-/-}$ mice resembles the one seen with dopamine transporter knockout mice that are similarly aroused by novelty and respond with hyperlocomotion to a new environment *(63)*. The hyperactive behavior of mice lacking dopamine transporters was shown to be related to a hyperdopaminergic state, reminiscent of human Attention-Deficit Hyperactivity Disorder (ADHD) *(64)*. GABA$_B$ receptors are known to alter dopaminergic transmission in the brain *(65)*. It therefore seems reasonable to assume that a loss of tonic GABA$_B$ control in GABA$_{B(1)-/-}$ mice triggers a similar hyperdopaminergic state as in dopamine transporter knockout mice.

Despite the obvious expression of GABA$_B$ receptors in many peripheral organs, such as heart, spleen, lung, liver, intestine, stomach, and urinary bladder, no overt peripheral phenotype has been described for GABA$_{B(1)}$-deficient mice. However, as in the central nervous system (CNS), knockout studies demonstrate that the GABA$_{B(1)}$ subunit is an essential requirement for GABA$_B$ receptor function in the enteric and peripheral nervous system (PNS) *(66)*.

The demonstration of the presence of a tonic brake on memory impairment, locomotion, and nociception is of particular importance for the management of disease. Allosteric GABA$_B$ drugs *(34,67)* rely on tonic receptor activity to discriminate between activated and nonactivated states. GABA$_B$ drugs that exploit the GABA$_B$ tone and only work on the subset of activated GABA$_B$ receptors may therefore be able to dissociate the unwanted side-effects seen with classical GABA$_B$ agonists like baclofen (*see* below).

Given that GABA$_{B(1)}$ harbors the binding site for GABA, it was suggested that the role of GABA$_{B(2)}$ is restricted to serving as a chaperone, thus enabling GABA$_{B(1)}$ to traffic to the plasma membrane. However, in recombinant experiments where the C-terminal tail of GABA$_{B(2)}$ is used to deliver wild-type GABA$_{B(1)}$ to the cell surface, homomeric GABA$_{B(1)}$ is not sufficient for receptor function *(44)*. It follows that if GABA$_B$ receptor subtypes exist, they must incorporate the GABA$_{B(1)}$ subunit plus a protein that can substitute, both as a chaperone and in coupling to the appropriate effector systems, for GABA$_{B(2)}$. It therefore is most exciting to determine whether or not GABA$_B$ functions persist in GABA$_{B(2)-/-}$ mice.

4. THE RELATIONSHIP BETWEEN GABA$_B$ AND γ-HYDROXYBUTYRATE (GHB) RECEPTORS

Because GABA$_B$ subtypes could be exploited therapeutically, it is not only crucial to determine their molecular diversity, but also to elucidate how they relate to the receptors for GHB, a popular recreational drug of abuse. GHB is a metabolite of GABA that is present at micromolar concentration in the brain. GHB is concentrated in specific brain regions where high-affinity binding sites and/or possibly receptors are located. To a lesser extent, GHB is also found in peripheral tissues such as heart, kidney, liver, and muscle. Although it was initially assumed that GHB has no biological activity, it is now proposed to function as a neurotransmitter or neuromodulator *(68–70)*. The physiological role of endogenous GHB remains obscure. Behavioral effects are usually only observed when endogenous GHB levels are increased, either pathologically or by exogenous administration of GHB. Patients suffering from GHB aciduria, a congenital enzyme defect causing GHB accumulation, exhibit psychomotor retardation, delayed or absent speech, hypotonia, ataxia, hyporeflexia, seizures, and EEG abnormalities. How GHB leads to these pathological changes is unclear. Likewise, when administered exogenously, GHB freely penetrates into the brain and induces a large spectrum of behavioral responses. GHB is described as producing a feeling of euphoria and a "high," which likely is linked to its dopaminergic effects. These mood-elevating properties, together with sedative, relaxing, and muscle growth-promoting properties, lead to illicit use and abuse of GHB, with intense Internet marketing contributing to the rapid spread of use of the drug *(71,72)*. Some clinical features of GHB aciduria are reproduced in regular GHB users, with high doses (100–400 mg/kg) causing EEG hypersynchronization, absence-like seizures, sedation, sleep, and coma. When combined with other substances, especially alcohol, GHB can be fatal and therefore has become known as "coma killer." GHB is currently also one of the most frequently used acquaintance sexual assault ("date rape") drugs *(72)*. Perpetrators choose these drugs because they act rapidly, produce disinhibition, cause relaxation of voluntary muscles, and cause the victim to have lasting anterograde amnesia for events that occurred under the influence of the drug. In March 2000, the U.S. Department of Justice and the U.S. Drug Enforcement Agency classified GHB in Schedule I of the Controlled Substances Act. Restrictions were placed on the availability of its main synthetic precursor, γ-butyrolactone (GBL).

Findings that GHB shows therapeutic benefits sparked interest in GHB in the pharmaceutical industry. The arguments regarding GHB's medical uses are under much debate, and it is impossible to conglomerate the often conflicting data into a reliable summary. In the past, GHB was clinically used as an anesthetic. Recently, it became obvious that GHB normalizes sleep patterns in narcoleptic patients *(70,73)*. Moreover, the regulating properties of GHB on dopaminergic pathways are a cause for the intense interest in synthetic ligands that act specifically at GHB receptors. Preliminary preclinical and clinical data suggest that GHB is useful in the therapy of alcoholism, nicotine, and opiate dependencies *(74)*. In view of these therapeutic prospects and public health concerns, it is important to understand the molecular events underlying the effects of GHB.

Although high-affinity [^{3}H]GHB binding sites in the brain were described, the molecular mechanism of action of GHB remains elusive, mostly owing to obvious GABA$_B$ receptor-mediated effects. GHB was demonstrated to bind to native and recombinant GABA$_B$ receptors, although with significantly lower affinity than to the high-affinity binding sites *(75)*. Besides the differences in binding affinity, several other lines of evidence support that native high-affinity [^{3}H]GHB binding sites and GABA$_B$ receptors are distinct. First, the distribution and ontogenesis of [^{3}H]GHB binding sites and GABA$_B$ receptors are different. Second, NCS-382 (6,7,8,9-tetrahydro-5-[H]benzocycloheptene-5-ol-4-ylidene acetat), the only available GHB receptor antagonist, has no activity on GABA$_B$ receptors *(76,77)*. There is nevertheless mounting evidence that in vivo, GHB can mediate some of its effects via GABA$_B$ receptors *(78)*. Experiments using recombinant receptors support this view. In *Xenopus* oocytes, GHB activates heterologous GABA$_{B(1a,2a)}$ and GABA$_{B(1b,2a)}$ receptors with an EC$_{50}$ of approx 5 mM and a maximal stimulation of 69% when compared to L-baclofen *(75)*. The results indicate that GHB is a weak partial agonist at the GABA binding-site of cloned GABA$_B$ receptors. This EC$_{50}$ is clearly too high to explain a high-affinity GHB binding site on recombinant GABA$_B$ receptors. Therefore, under normal conditions, endogenous brain levels of GHB are too low to significantly activate GABA$_B$ receptors *(75,79)*. Under abuse conditions, however, it is expected that GHB reaches concentrations high enough to activate GABA$_B$ receptors. It this respect, it is interesting to note that some of the clinical signs of GHB intoxication resemble those of a baclofen overdose. If endogenous GHB is involved in active signaling, high-affinity GHB receptors, likely related to brain [^{3}H]GHB binding sites, must mediate these effects. In line with this, several reports suggest that high-affinity [^{3}H]GHB binding sites are coupled to G proteins *(80,81)*. Similarly another report claims that specific GHB receptors modulate Ca^{2+} channels *(82)*. The GHB receptor antagonist NC382 extends the lifespan of mice with genetically induced GHB aciduria, further suggesting a $GABA_B$ receptor-independent GHB signaling *(83)*. It was also suggested that GHB activates a GHB recognition site related to, although separate from, $GABA_B$ receptors, thereby forming a presynaptic $GABA_B$/GHB receptor complex *(84)*. Such a scenario is conceivable with the recent demonstration that distantly related G protein coupled receptors can form heteromeric complexes *(85)*. Although many of the aforementioned studies would suggest the existence of specific GHB receptors, this proposal has also been challenged because GHB stimulates the release of GABA or can be metabolized to GABA *(86,87)*. $GABA_B$ receptor-mediated effects of GHB may be secondary to such mechanisms and could wrongly suggest the existence of a specific G protein coupled GHB receptor. For example, a conversion of only 01.% of GHB into GABA would be sufficient to explain many published "GHB-specific" effects.

It appears impossible to reconcile all the aforementioned findings into a common scheme. However, a scenario that accommodates most of the available data is that low doses of administered GHB (5–50 mg/kg) activate specific GHB receptors, whereas high doses (100–400 mg/kg) additionally recruit $GABA_B$ receptors. The availability of $GABA_{B(1)}$-deficient mice *(57,58)*, which are devoid of any functional $GABA_B$ receptors, now provide the opportunity to study GHB effects in the absence of coincident $GABA_B$ effects. Analysis of these mice will make it possible to unequivocally identify whether biochemical and behavioral effect relate to [^{3}H]GHB binding sites. Preliminary data presented at the Fifth International $GABA_B$ Symposium (Paris, July 2002) suggest that such mice exhibit unchanged [^{3}H]GHB binding, but do not show the typical muscle relaxation in response to GHB application. Therefore at least some of the behavioral responses of GHB appear to be mediated by $GABA_B$ receptors. It clearly remains controversial whether GHB binding-sites mediate G protein signaling, or whether they are, for instance, part of an uptake system.

5. CONCLUSIONS AND OUTLOOK

The $GABA_B$ field is currently in an exciting stage of rapid development, but many important questions remain unresolved. For example it is unknown which structural features dictate pre- vs postsynaptic $GABA_B$ receptor localization. It is still unclear whether pharmacological subtypes of $GABA_B$ receptors exist. Related to this, we do not know what the specific roles of the prominent amino-terminal splice variants $GABA_{B(1a)}$ and $GABA_{B(1b)}$ are and whether compounds specifically targeted to them are likely to have distinct effects. Very little is still known about $GABA_B$ receptor internalization, recycling, and control of surface-expression levels. It is still unclear how $GABA_B$ receptors relate to the GHB binding-sites, and whether these sites mediate some known actions of GHB. As the mouse genome becomes more accessible to experimental manipulation, it is becoming feasible to address some of these questions using gene targeting. It is therefore to be expected that significant progress on these issues will be made in the forthcoming years.

The discovery of $GABA_B$ receptor genes from 1997 to 1998 has not only yielded new insights into the structural and functional properties of G protein coupled receptors, but also renewed interest in GABA$_B$ receptors as therapeutic targets. Because the GABA$_B$ receptor system is distributed throughout the nervous system, the therapeutic promise of manipulation depends on pharmacological subtypes that can be selectively targeted. A large body of literature suggests the existence of such pharmacological GABA$_B$ receptor subtypes *(51)*. Moreover, because it has become virtually axiomatic that neurotransmitter receptor systems are composed of pharmacologically distinct subgroups, it was to be expected that GABA$_B$ receptors form a family of related genes. Therefore the discovery that GABA$_B$ receptors arise from the expression of only two genes comes as a surprise and, for the time being, limits the number of possible receptor subtypes. Most people will agree that the cloned GABA$_{B(1a,2)}$ and GABA$_{B(1b,2)}$ have identical pharmacological properties and that they do not reproduce the heterogeneity described for native receptors *(56)*. Whether the pharmacological diversity described for native receptors relates to differences in the effector systems or to the existence of additional, as yet unidentified, GABA$_B$ receptor-associated proteins remains a key issue in the field. The discovery that G protein coupled receptors can form heteromeric complexes with distinct pharmacology and properties raises fascinating possibilities concerning the plasticity and diversity of these receptors *(85)*. An obvious implication is that association between GABA$_B$ receptor subunits and other G protein coupled receptors could be the source of pharmacological diversity. GABA$_B$ receptors were therefore expressed together with the structurally most closely related mGlu1-8 receptors *(44)*. However, no significant pharmacological or functional differences were observed in the presence of co-expressed mGlu proteins, looking at radioligand binding, surface trafficking, and functional assay systems. It looks as if we have to await GABA$_{B(2)}$ knockout studies to clarify whether GABA$_{B(1)}$ can form functional receptors in the absence of GABA$_{B(2)}$. For now, GABA$_{B(1a,2)}$ and GABA$_{B(1b,2)}$ receptors are the only distinct target sites that can be exploited. The design of drugs capable of distinguishing between the GABA$_{B(1a)}$ and GABA$_{B(1b)}$ sites may prove to be difficult. As mentioned earlier, the GABA$_{B(1a)}$ and GABA$_{B(1b)}$ subunits only differ in the amino-terminal sushi repeats. Given the current lack of subunit-specific drugs, the dissection of the in vivo functions of GABA$_{B(1a,2)}$ and GABA$_{B(1a,2)}$ receptors will probably first be achieved using genetic means. The development of GABA$_{B(1a)}$- and GABA$_{B(1b)}$-deficient mice is crucial, because it allows the elaboration of a link between molecularly distinct GABA$_B$ receptor populations and the phenotypes observed in GABA$_{B(1)-/-}$ mice.

The most widely exploited clinical response to the GABA$_B$ agonist baclofen is its muscle-relaxant effect. This action appears largely owing to a reduction in neurotransmitter release onto motoneurons in the ventral horn of the spinal cord. GABA$_B$ receptors also represent attractive drug targets in the pharmacotherapy of various neurological and psychiatric disorders, including neuropathic pain, anxiety, depression, absence epilepsy, and drug addiction. However, the (side) effects of agonists, principally sedation, tolerance, and muscle relaxation, limit their utility for the treatment of many of these diseases. Novel GABA$_B$ drugs that largely lack these components are therefore much sought after. A clear breakthrough represents the development of allosteric GABA$_B$ receptor modulators *(34,67)*. These novel compounds do not activate GABA$_B$ receptors on their own, but they enhance the physiological effects of GABA in the brain, just as the benzodiazepines (Valium) do at GABA$_A$ receptors. The screening for such allosteric GABA$_B$ receptor drugs has only been made possible by the development of functional assay systems based on recombinant receptors. The novel allosteric drugs are expected to have an improved side-effect profile as compared to GABA$_B$ agonists, which act indiscriminately on all GABA$_B$ receptors. Unlike agonists, these drugs should also prevent the development of tolerance, because they avoid prolonged receptor activation leading to desensitization and internalization. If allosteric modulators retain the therapeutic potential of baclofen, while being devoid of any side effects and muscle-relaxant properties, they will considerably broaden the spectrum of therapeutic applications for GABA$_B$ drugs.

A development in the field of drug discovery is the emerging concept that GABA$_B$ effects and GHB effects may turn out to be same. Although it is clear from molecular studies that GHB can activate GABA$_B$ receptors, truly compelling evidence for a specific G protein signaling through high-affinity GHB binding sites is still lacking. By the same token, a growing preclinical and clinical literature shows therapeutic efficacy for GHB and GABA$_B$ compounds in the same indications. As an example, both GHB and GABA$_B$ receptor agonists promote abstinence and reduce the use of cocaine, heroin, alcohol, and nicotine *(74,88)*.

REFERENCES

1. Wojcik, W. J. and Neff, N. H. (1984) Gamma-aminobutyric acid B receptors are negatively coupled to adenylate cyclase in brain and in the cerebellum these receptors may be associated with granule cells. *Mol. Pharmacol.* **25,** 24–28.
2. Hill, D. R. (1985) GABA$_B$ receptor modulation of adenylate cyclase activity in rat brain slices. *Br. J. Pharmacol.* **84,** 249–257.
3. Karbon, E. W. and Enna, S. J. (1985) Characterization of the relationship between gamma-aminobutyric acid B agonists and transmitter-coupled cyclic nucleotide-generating systems in rat brain. *Mol. Pharmacol.* **27,** 53–59.
4. Bowery, N. G., Hill, D. R., Hudson, A. L., Doble, A., Middlemiss, D. N., Shaw, J. and Turnbull, M. J. (1980) (–)Baclofen decreases neurotransmitter release in the mammalian CNS by an action at a novel GABA receptor. *Nature* **283,** 92–94.
5. Hill, D. R. and Bowery, N. G. (1981) ^{3}H-Baclofen and ^{3}H-GABA bind to bicuculline-insensitive GABA$_B$ sites in rat brain. *Nature* **290,** 149–152.
6. Nakayasu, H., Nishikawa, M., Mizutani, H., Kimura, H., and Kuriyama, K. (1993) Immunoaffinity purification and characterization of gamma-aminobutyric acid (GABA)$_B$ receptor from bovine cerebral cortex. *J. Biol. Chem.* **268,** 8658–8664.
7. Facklam, M. and Bowery, N. G. (1993) Solubilization and characterization of GABA$_B$ receptor binding sites from porcine brain synaptic membranes. *Br. J. Pharmacol.* **110,** 1291–1296.
8. Sekiguchi, M., Sakuta, H., Okamoto, K., and Sakai, Y. (1990) GABA$_B$ receptors expressed in *Xenopus* oocytes by guinea pig cerebral mRNA are functionally coupled with Ca^{2+}-dependent Cl$^-$ channels and with K$^+$ channels, through GTP-binding proteins. *Brain Res. Mol. Brain Res.* **8,** 301–309.
9. Taniyama, K., Takeda, K., Ando, H., Kuno, T., and Tanaka, C. (1991) Expression of the GABA$_B$ receptor in *Xenopus* oocytes and inhibition of the response by activation of protein kinase C. *FEBS Lett.* **278,** 222–224.
10. Woodward, R. M. and Miledi, R. (1992) Sensitivity of *Xenopus* oocytes to changes in extracellular pH: possible relevance to proposed expression of atypical mammalian GABA$_B$ receptors. *Mol. Brain Res.* **16,** 204–210.
11. Froestl, W., Bettler, B., Bittiger, H., Heid, J., Kaupmann, K., Mickel, S. J. and Strub, D. (2001) Ligands for expression cloning and isolation of GABA$_B$ receptors. *Farmaco* **56,** 101–105.

12. Kaupmann, K., Huggel, K., Heid, J., et al. (1997) Expression cloning of GABA_B receptors uncovers similarity to metabotropic glutamate receptors. *Nature* **386,** 239–246.

13. Couve, A., Filippov, A. K., Connolly, C. N., Bettler, B., Brown, D. A., and Moss, S. J. (1998) Intracellular retention of recombinant GABA_B receptors. *J. Biol. Chem.* **273,** 26361–26367.

14. Kaupmann, K., Malitschek, B., Schuler, V., et al. (1998) GABA_B-receptor subtypes assemble into functional heteromeric complexes. *Nature* **396,** 683–687.

15. Jones, K. A., Borowsky, B., Tamm, J. A., et al. (1998) GABA_B receptors function as a heteromeric assembly of the subunits GABA_BR1 and GABA_BR2. *Nature* **396,** 674–679.

16. White, J. H., Wise, A., Main, M. J., et al. (1998) Heterodimerization is required for the formation of a functional GABA_B receptor. *Nature* **396,** 679–682.

17. Kuner, R., Kohr, G., Grunewald, S., Eisenhardt, G., Bach, A., and Kornau, H. C. (1999) Role of heteromer formation in GABA_B receptor function. *Science* **283,** 74–77.

18. Bräuner-Osborne, H. and Krogsgaard-Larsen, P. (1999) Functional pharmacology of cloned heteromeric GABA_B receptors expressed in mammalian cells. *Br. J. Pharmacol.* **128,** 1370–1374.

19. Malitschek, B., Rüegg, D., Heid, J., et al. (1998) Developmental changes in agonist affinity at GABA_BR1 receptor variants in rat brain. *Mol. Cell. Neurosci.* **12,** 56–64.

20. Kaupmann, K., Schuler, V., Mosbacher, J., et al. (1998) Human GABA_B receptors are differentially expressed and regulate inwardly rectifying K^+ channels. *Proc. Natl. Acad. Sci. USA* **95,** 14991–14996.

21. Ng, G. Y., Bertrand, S., Sullivan, R., et al. (2001) Gamma-aminobutyric acid type B receptors with specific heterodimer composition and postsynaptic actions in hippocampal neurons are targets of anticonvulsant gabapentin action. *Mol. Pharmacol.* **59,** 144–152.

22. Bertrand, S., Ng, G. Y., Purisai, M. G., et al. (2001) The anticonvulsant, antihyperalgesic agent gabapentin is an agonist at brain gamma-aminobutyric acid type B receptors negatively coupled to voltage-dependent calcium channels. *J. Pharmacol. Exp. Ther.* **298,** 15–24.

23. Jensen, A. A., Mosbacher, J., Elg, S., et al. (2002) The anticonvulsant gabapentin (neurontin) does not act through gamma-aminobutyric acid-B receptors. *Mol. Pharmacol.* **61,** 1377–1384.

24. Lanneau, C., Green, A., Hirst, W. D., Wise, A., Brown, J. T., Donnier, E., et al. (2001) Gabapentin is not a GABA_B receptor agonist. *Neuropharmacology* **41,** 965–975.

25. Calver, A. R., Davies, C. H., and Pangalos, M. (2002) GABA_B receptors: from monogamy to promiscuity. *Neurosignals* **11,** 299–314.

26. Malitschek, B., Schweizer, C., Keir, M., et al. (1999) The N-terminal domain of gamma-aminobutyric acid_B receptors is sufficient to specify agonist and antagonist binding. *Mol. Pharmacol.* **56,** 448–454.

27. Galvez, T., Parmentier, M. L., Joly, C., et al. (1999) Mutagenesis and modeling of the GABA_B receptor extracellular domain support a venus flytrap mechanism for ligand binding. *J. Biol. Chem.* **274,** 13362–13369.

28. Galvez, T., Prezeau, L., Milioti, G., et al. (2000) Mapping the agonist-binding site of GABA_B type 1 subunit sheds light on the activation process of GABA_B receptors. *J. Biol. Chem.* **275,** 41166–41174.

29. Bernard, P., Guedin, D., and Hibert, M. (2001) Molecular modeling of the GABA/GABA_B receptor complex. *J. Med. Chem.* **44,** 27–35.

30. Kniazeff, J., Galvez, T., Labesse, G., and Pin, J. P. (2002) No ligand binding in the GB2 subunit of the GABA_B receptor is required for activation and allosteric interaction between the subunits. *J. Neurosci.* **22,** 7352–7361.

31. Jensen, A. A., Madsen, B. E., Krogsgaard-Larsen, P., and Brüuner-Osborne, H. (2001) Pharmacological characterization of homobaclofen on wild type and mutant GABA_B1b receptors coexpressed with the GABA_B2 receptor. *Eur. J. Pharmacol.* **417,** 177–180.

32. Galvez, T., Urwyler, S., Prezeau, L., et al. (2000) Ca^{2+} requirement for high-affinity gamma-aminobutyric acid (GABA) binding at $GABA_B$ receptors: involvement of serine 269 of the $GABA_B$R1 subunit. *Mol. Pharmacol.* **57,** 419–426.

33. Pin, J. P., Parmentier, M. L., and Prezeau, L. (2001) Positive allosteric modulators for gamma-aminobutyric acid$_B$ receptors open new routes for the development of drugs targeting family 3 G protein-coupled receptors. *Mol. Pharmacol.* **60,** 881–884.

34. Urwyler, S., Mosbacher, J., Lingenhoehl, K., et al. (2001) Positive allosteric modulation of native and recombinant gamma-aminobutyric acid$_B$ receptors by 2,6-di-tert-butyl-4-(3-hydroxy-2,2-dimethyl-propyl)-phenol (CGP7930) and its aldehyde analog CGP13501. *Mol. Pharmacol.* **60,** 963–971.

35. Knoflach, F., Mutel, V., Jolidon, S., et al. (2001) Positive allosteric modulators of metabotropic glutamate 1 receptor: characterization, mechanism of action, and binding site. *Proc. Natl. Acad. Sci. USA* **98,** 13402–13407.

36. Hauache, O. M., Hu, J., Ray, K., Xie, R., Jacobson, K. A., and Spiegel, A. M. (2000) Effects of a calcimimetic compound and naturally activating mutations on the human Ca^{2+} receptor and on Ca^{2+} receptor/metabotropic glutamate chimeric receptors. *Endocrinology* **141,** 4156–4163.

37. Jensen, A. A., Greenwood, J. R., and Bräuner-Osborne, H. (2002) The dance of the clams: twists and turns in the family C GPCR homodimer. *Trends Pharmacol. Sci.* **23,** 491–493.

38. Galvez, T., Duthey, B., Kniazeff, J., et al. (2001) Allosteric interactions between GB1 and GB2 subunits are required for optimal $GABA_B$ receptor function. *EMBO J.* **20,** 2152–2159.

39. Margeta-Mitrovic, M., Jan, Y. N., and Jan, L. Y. (2001) Function of GB1 and GB2 subunits in G protein coupling of $GABA_B$ receptors. *Proc. Natl. Acad. Sci. USA* **98,** 14649–14654.

40. Havlickova, M., Prezeau, L., Duthey, B., Bettler, B., Pin, J. P., and Blahos, J. (2002) The intracellular loops of the GB2 subunit are crucial for G protein coupling of the heteromeric γ-aminobutyrate B receptor. *Mol. Pharmacol.* **62,** 343–350.

41. Duthey, B., Caudron, S., Perroy, J., Bettler, B., Fagni, L., Pin, J. P., and Prezeau, L. (2002) A single subunit (GB2) is required for G protein activation by the heteromeric $GABA_B$ receptor. *J. Biol. Chem.* **277,** 3236–3241.

42. Robbins, M. J., Calver, A. R., Filippov, A. K., Hirst, W. D., Russell, R. B., Wood, M. D., et al. (2001) $GABA_B$2 is essential for G protein coupling of the $GABA_B$ receptor heterodimer. *J. Neurosci.* **21,** 8043–8052.

43. Margeta-Mitrovic, M., Jan, Y. N., and Jan, L. Y. (2000) A trafficking checkpoint controls $GABA_B$ receptor heterodimerization. *Neuron* **27,** 97–106.

44. Pagano, A., Rovelli, G., Mosbacher, J., et al. (2001) C-terminal interaction is essential for surface trafficking but not for heteromeric assembly of $GABA_B$ receptors. *J. Neurosci.* **21,** 1189–1202.

45. Calver, A. R., Robbins, M. J., Cosio, C., et al. (2001) The C-terminal domains of the $GABA_B$ receptor subunits mediate intracellular trafficking but are not required for receptor signaling. *J. Neurosci.* **21,** 1203–1210.

46. Couve, A., Thomas, P., Calver, A. R., et al. (2002) Cyclic AMP-dependent protein kinase phosphorylation facilitates $GABA_B$ receptor-effector coupling. *Nat. Neurosci.* **25,** 25.

47. Couve, A., Kittler, J. T., Uren, J. M., Calver, A. R., Pangalos, M. N., Walsh, F. S., and Moss, S. J. (2001) Association of $GABA_B$ receptors and members of the 14-3-3 family of signaling proteins. *Mol. Cell. Neurosci.* **17,** 317–328.

48. Nehring, R. B., Horikawa, H. P., El Far, O., Kneussel, M., Brandstatter, J. H., Stamm, S., et al. (2000) The metabotropic $GABA_B$ receptor directly interacts with the activating transcription factor 4. *J. Biol. Chem.* **275,** 35185–35191.

49. White, J. H., McIllhinney, R. A., Wise, A., Ciruela, F., Chan, W. Y., Emson, P. C., et al. (2000) The $GABA_B$ receptor interacts directly with the related transcription factors CREB2 and ATFx. *Proc. Natl. Acad. Sci. USA* **97,** 13967–13972.

50. Vernon, E., Meyer, G., Pickard, L., Dev, K., Molnar, E., Collingridge, G. L., and Henley, J. M. (2001) $GABA_B$ receptors couple directly to the transcription factor ATF4. *Mol. Cell. Neurosci.* **17,** 637–645.

51. Bonanno, G. and Raiteri, M. (1993) Multiple GABA_B receptors. *Trends Pharmacol. Sci.* **14**, 259–261.

52. Billinton, A., Ige, A. O., Bolam, J. P., White, J. H., Marshall, F. H., and Emson, P. C. (2001) Advances in the molecular understanding of GABA_B receptors. *Trends Neurosci.* **24**, 277–282.

53. Leaney, J. L. and Tinker, A. (2000) The role of members of the pertussis toxin-sensitive family of G proteins in coupling receptors to the activation of the G protein-gated inwardly rectifying potassium channel. *Proc. Natl. Acad. Sci. USA* **97**, 5651–5656.

54. Robbins, M. J., Charles, K. J., Harrison, D. C., and Pangalos, M. N. (2002) Localisation of the GPRC5B receptor in the rat brain and spinal cord. *Brain Res. Mol. Brain Res.* **106**, 136.

55. Bischoff, S., Leonhard, N., Reymann, N., Schuler, V., Kaupmann, K., and Bettler, B. (1997) Distribution of the GABA_BR1 mRNA in rat brain. Comparison with the GABA_B binding sites. *Soc. Neurosci.* **23**, 954.

56. Marshall, F. H., Jones, K. A., Kaupmann, K., and Bettler, B. (1999) GABA_B receptors: the first 7TM heterodimers. *Trends Pharmacol. Sci.* **20**, 396–399.

57. Schuler, V., Luscher, C., Blanchet, C., et al. (2001) Epilepsy, hyperalgesia, impaired memory, and loss of pre- and postsynaptic GABA_B responses in mice lacking GABA_B(1). *Neuron* **31**, 47–58.

58. Prosser, H. M., Gill, C. H., Hirst, W. D., et al. (2001) Epileptogenesis and enhanced prepulse inhibition in GABA_B1-deficient mice. *Mol. Cell. Neurosci.* **17**, 1059–1070.

59. Mitchell, K. J., Pinson, K. I., Kelly, O. G., et al. (2001) Functional analysis of secreted and transmembrane proteins critical to mouse development. *Nat. Genet.* **28**, 241–249.

60. Pearson, H. (2002) Surviving a knockout blow. *Nature* **415**, 8–9.

61. Benke, D., Honer, M., Michel, C., Bettler, B., and Mohler, H. (1999) Gamma-aminobutyric acid type B receptor splice variant proteins GBR1a and GBR1b are both associated with GBR2 in situ and display differential regional and subcellular distribution. *J. Biol. Chem.* **274**, 27323–27330.

62. Waldmeier, P. C., Wicki, P., Feldtrauer, J. J., Mickel, S. J., Bittiger, H., and Baumann, P. A. (1994) GABA and glutamate release affected by GABA_B receptor antagonists with similar potency: no evidence for pharmacologically different presynaptic receptors. *Br. J. Pharmacol.* **113**, 1515–1521.

63. Spielewoy, C., Biala, G., Roubert, C., Hamon, M., Betancur, C., and Giros, B. (2001) Hypolocomotor effects of acute and daily d-amphetamine in mice lacking the dopamine transporter. *Psychopharmacology* **159**, 2–9.

64. Viggiano, D., Grammatikopoulos, G., and Sadile, A. G. (2002) A morphometric evidence for a hyperfunctioning mesolimbic system in an animal model of ADHD. *Behav. Brain Res.* **130**, 181–189.

65. Waldmeier, P. C. (1991) The GABA_B antagonist, CGP 35348, antagonizes the effects of baclofen, gamma-butyrolactone and HA 966 on rat striatal dopamine synthesis. *Naunyn Schmiedebergs Arch. Pharmacol.* **343**, 173–178.

66. Sanger, G. J., Munonyara, M. L., Dass, N., Prosser, H., Pangalos, M. N., and Parsons, M. E. (2002) GABA_B receptor function in the ileum and urinary bladder of wildtype and GABA_B1 subunit null mice. *Auton. Autacoid Pharmacol.* **22**, 147–154.

67. Kerr, D. I., Ong, J., Puspawati, N. M., and Prager, R. H. (2002) Arylalkylamines are a novel class of positive allosteric modulators at GABA_B receptors in rat neocortex. *Eur. J. Pharmacol.* **451**, 69–77.

68. Bernasconi, R., Mathivet, P., Bischoff, S., and Marescaux, C. (1999) Gamma-hydroxybutyric acid: an endogenous neuromodulator with abuse potential? *Trends Pharmacol. Sci.* **20**, 135–141.

69. Maitre, M., Andriamampandry, C., Kemmel, V., Schmidt, C., Hode, Y., Hechler, V., and Gobaille, S. (2000) Gamma-hydroxybutyric acid as a signaling molecule in brain. *Alcohol* **20**, 277–283.

70. Nicholson, K. L. and Balster, R. L. (2001) GHB: a new and novel drug of abuse. *Drug Alcohol. Depend.* **63,** 1–22.

71. Galloway, G. P., Frederick, S. L., Staggers, F. E., Jr., Gonzales, M., Stalcup, S. A., and Smith, D. E. (1997) Gamma-hydroxybutyrate: an emerging drug of abuse that causes physical dependence. *Addiction* **92,** 89–96.

72. Schwartz, R. H., Milteer, R., and LeBeau, M. A. (2000) Drug-facilitated sexual assault ('date rape'). *South Med. J.* **93,** 558–561.

73. Tunnicliff, G. and Raess, B. U. (2002) Gamma-hydroxybutyrate (orphan medical). *Curr. Opin. Invest. Drugs* **3,** 278–283.

74. Gallimberti, L., Spella, M. R., Soncini, C. A., and Gessa, G. L. (2000) Gamma-hydroxybutyric acid in the treatment of alcohol and heroin dependence. *Alcohol* **20,** 257–262.

75. Lingenhoehl, K., Brom, R., Heid, J., Beck, P., Froestl, W., Kaupmann, K., Bettler, B., and Mosbacher, J. (1999) Gamma-hydroxybutyrate is a weak agonist at recombinant GABA$_B$ receptors. *Neuropharmacology* **38,** 1667–1673.

76. Castelli, M. P., Mocci, I., Pistis, M., et al. (2002) Stereoselectivity of NCS-382 binding to gamma-hydroxybutyrate receptor in the rat brain. *Eur. J. Pharmacol.* **446,** 1–5.

77. Mehta, A. K., Muschaweck, N. M., Maeda, D. Y., Coop, A., and Ticku, M. K. (2001) Binding characteristics of the gamma-hydroxybutyric acid receptor antagonist [^{3}H](2*E*)-(5-hydroxy-5,7,8,9-tetrahydro-6H-benzo[a][7]annulen-6-ylidene) ethanoic acid in the rat brain. *J. Pharmacol. Exp. Ther.* **299,** 1148–1153.

78. Jensen, K. and Mody, I. (2001) GHB depresses fast excitatory and inhibitory synaptic transmission via GABA$_B$ receptors in mouse neocortical neurons. *Cereb. Cortex* **11,** 424–429.

79. Bernasconi, R., Mathivet, P., Otten, U., Bettler, B., Bischoff, S., and Marescaux, C. (2002) Part of gamma-hydroxybutyrate pharmacological actions are mediated by GABAB receptors, in *Gamma-Hydroxybutyrate: Pharmacological and Functional Aspects* (Tunnicliff, G. and Cash, C. D., eds), Taylor & Francis, New York, pp. 28–63.

80. Snead, O. C., 3rd. (2000) Evidence for a G protein-coupled gamma-hydroxybutyric acid receptor. *J. Neurochem.* **75,** 1986–1996.

81. Ratomponirina, C., Hode, Y., Hechler, V., and Maitre, M. (1995) Gamma-hydroxybutyrate receptor binding in rat brain is inhibited by guanyl nucleotides and pertussis toxin. *Neurosci. Lett.* **189,** 51–53.

82. Kemmel, V., Taleb, O., Perard, A., Andriamampandry, C., Siffert, J. C., Mark, J., and Maitre, M. (1998) Neurochemical and electrophysiological evidence for the existence of a functional gamma-hydroxybutyrate system in NCB-20 neurons. *Neuroscience* **86,** 989–1000.

83. Gupta, M., Greven, R., Jansen, E. E., et al. (2002) Therapeutic intervention in mice deficient for succinate semialdehyde dehydrogenase (gamma-hydroxybutyric aciduria). *J. Pharmacol. Exp. Ther.* **302,** 180–187.

84. Snead, O. C., 3rd. (1996) Relation of the [^{3}H]gamma-hydroxybutyric acid (GHB) binding site to the gamma-aminobutyric acid$_B$ (GABA$_B$) receptor in rat brain. *Biochem. Pharmacol.* **52,** 1235–1243.

85. Angers, S., Salahpour, A., and Bouvier, M. (2002) Dimerization: an emerging concept for G protein-coupled receptor ontogeny and function. *Annu. Rev. Pharmacol. Toxicol.* **42,** 409–435.

86. Hechler, V., Ratomponirina, C., and Maitre, M. (1997) Gamma-hydroxybutyrate conversion into GABA induces displacement of GABA$_B$ binding that is blocked by valproate and ethosuximide. *J. Pharmacol. Exp. Ther.* **281,** 753–760.

87. Gobaille, S., Hechler, V., Andriamampandry, C., Kemmel, V., and Maitre, M. (1999) Gamma-hydroxybutyrate modulates synthesis and extracellular concentration of gamma-aminobutyric acid in discrete rat brain regions in vivo. *J. Pharmacol. Exp. Ther.* **290,** 303–309.

88. Cousins, M. S., Roberts, D. C., and de Wit, H. (2002) GABA$_B$ receptor agonists for the treatment of drug addiction: a review of recent findings. *Drug Alcohol. Depend.* **65,** 209–220.

III MOLECULAR PHARMACOLOGY OF TRANSMITTER TRANSPORTERS

8

Structure and Molecular Characterization of the Substrate-Binding Sites of the Dual-Function Glutamate Transporters

Baruch I. Kanner, Michael P. Kavanaugh, and Lars Borre

1. INTRODUCTION

In the central nervous system (CNS), glutamate is the predominant excitatory neurotransmitter. Glutamate transporters remove the transmitter from the cleft and maintain its extracellular concentrations below neurotoxic levels *(1–5)*. In addition, at some synapses glutamate transporters play an important role in limiting the duration of synaptic excitation *(6–9)*. Glutamate uptake is an electrogenic process *(10,11)* in which the transmitter is cotransported with three sodium ions and a proton *(3,12)*, followed by countertransport of a potassium ion *(13–15)*. In addition to this coupled flux, glutamate transporters mediate a thermodynamically uncoupled chloride flux, activated by two of the molecules they transport: sodium and glutamate *(16–18)*. This indicates the existence of a tight link between gating of the anion conductance and permeation of glutamate. It has been suggested that this capacity for enhancing chloride permeability could alter neuronal excitability *(17)*.

2. THE GLUTAMATE TRANSPORTER FAMILY: CLONING AND PURIFICATION

Several neurotransmitter transporters were successfully cloned based on the assumption that they are related to the GABA *(19)* and norepinephrine *(20)* transporters *(21–23)*. This approach, however, was unsuccessful for the glutamate transporter. Three homologous eukaryotic glutamate transporters were cloned using different approaches: GLAST-1 *(24)*, GLT-1 *(25)*, and EAAC-1 *(26)*. Their human homologs, termed EAAT1, -2, and -3, respectively, were cloned later *(27)*. In the brain, GLAST-1 and GLT-1 appear to be of glial origin *(24,25,28,29)*, whereas EAAC-1 is neuronal *(26,30)*. The subsequently cloned EAAT4 *(17)* and EAAT5 *(18)* appear to be localized to neurons and the retina, respectively.

These eukaryotic glutamate transporters form a family that displays approx 50% overall identity and approx 60% overall similarity and is distinct from the GABA and norepinephrine transporter family. The eukaryotic glutamate transporters are also

From: *Molecular Neuropharmacology: Strategies and Methods*
Edited by: A. Schousboe and H. Bräuner-Osborne © Humana Press Inc., Totowa, NJ

related to the proton-coupled glutamate transporter glt-P from *Escherichia coli* and other bacteria *(31)* and the dicarboxylate transporter dct-A of *Rhizobium meliloti (32)*. The identity between these prokaryotic and eukaryotic glutamate transporters is approx 25–30%. Furthermore, the glutamate transporter family also includes the sodium-dependent exchangers, termed ASCT-1 *(33,34)* and ASCT-2 *(35,36),* that do not transport dicarboxylic acids, but rather the neutral amino acids alanine, serine, and cysteine.

Prior to its cloning, GLT-1 was reconstituted and purified and shown to have a relative molecular mass of 64 kDa, which agrees well both with the value of 65 kDa of the purified and deglycosylated transporter *(28,37)* and its 573 amino acids determined from its nucleotide sequence *(25)*. Reconstituted GLT-1 has been shown to form dimers, trimers, and higher molecular-weight homomers in the absence of reducing agents *(38)*. The possibility of homomeric assemblies is corroborated by a recent freeze fracture, which shows pentameric assemblies *(39)*.

3. SECONDARY STRUCTURE AND PROXIMITY RELATIONS OF THE CARBOXYL TERMINAL HALF

Hydropathy plots are relatively straightforward at the amino terminal side of the protein. The three different groups that originally cloned GLAST-1, GLT-1, and EAAC-1 have predicted six transmembrane α-helices at very similar positions *(24–26)*. On the other hand, there is much more ambiguity at the carboxyl terminal half, where zero *(24)*, two *(25)*, or four *(26)*, α-helices have been predicted. In recent years, attempts have been made to determine experimentally the topology of the glutamate transporters. The studies have focused on the highly conserved carboxyl terminal part of the glutamate transporters and have involved determination of accessibility to cysteines that have been introduced at selected positions by site-directed mutagenesis *(40–42)*. This assay revealed the surprising topology of the carboxyl terminal part shown in Fig. 1 and identifies two reentrant loops (I and II), two transmembrane domains (7 and 8) long enough to span the membrane as α-helices as well as an outward-facing hydrophobic linker *(40–42)*. Another study arrives at a somewhat different model, including the assignment of transmembrane domain 7 as a reentrant loop *(43)*. These models disagree only on the accessibility of one of the engineered cysteines to impermeable sulfhydryl reagents *(42,44)*.

The accessibility studies of GLT-1 have shown that cysteine residues introduced at positions 364 and 440 located in reentrant loops I and II, respectively, react with the impermeable sulfhydryl reagent MTSET added from the extracellular side *(42,45)*. Substrates and nontransportable analogs are partially protected against the modification of cysteines introduced at these two positions. Therefore we have suggested that positions 364 and 440 may be close in the three-dimensional structure of the protein *(42)*. To test this prediction and to obtain the first information regarding the tertiary structure of the carboxyl-terminal half of the glutamate transporters, we set out to determine proximity relations between the different structural elements in this region. Two cysteine pairs, A412C/V427C and A364C/S440C, were identified, which behave as if they are close enough to form intramolecular disulfides and therefore must be close in space in the transporter monomer *(46)*. The disulfide formation could be inferred from functional assays because exposure of the cysteine double mutants to oxidative conditions resulted in inhibition of uptake *(46)* and inhibition of the coupled

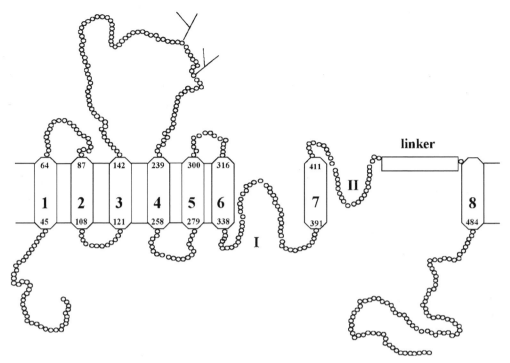

Fig. 1. Topology of glutamate transporters. The topology was determined by introducing cysteines at selected positions in a transporter devoid of endogenous cysteines and then probing their accessibility. This experimental approach revealed the first reentrant loops (I and II) in a transporter protein as well as a highly hydrophobic region (linker) linking reentrant loop II to transmembrane domain 8.

and uncoupled currents (unpublished observations). The proximity of these residues provides evidence that the two oppositely oriented reentrant loops are spatially close to one another *(46)*. The recent crystal structures of an aquaglyceroporin *(47)* and an aquaporin *(48)* revealed that these proteins also contain two reentrant loops of opposite orientation, which are close in space and line the substrate permeation pathway *(47,48)*. Although the precise molecular details of substrate permeation in glutamate transporters is likely to differ it is interesting to note that the glutamate analog DHK can protect against MTSET modification at all four positions; A364C, A412C, V427C, and S440C *(42,45* and unpublished observations).

4. COUPLED AND UNCOUPLED FLUXES

Glutamate uptake is an electrogenic process *(10,11)* in which the transmitter is cotransported with three sodium ions and one proton *(3,12)* followed by the countertransport of one potassium ion *(13–15)*. The sequential scheme shown in Fig. 2A depicts this. The resulting current is termed stoichiometrically coupled transport current, coupled current, or stoichiometric current. Because the exact binding order of sodium, amino acid, and protons remains an active area of research, they have been collapsed into one binding step. In addition to this coupled glutamate uptake, the glutamate transporters also mediate a thermodynamically uncoupled chloride flux that is

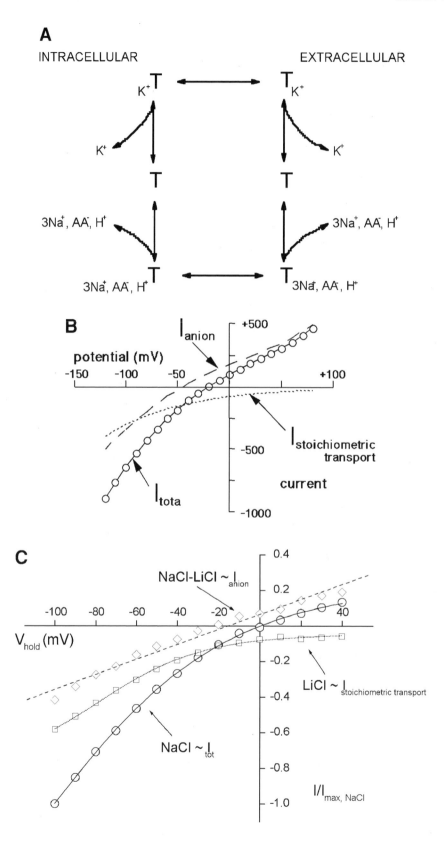

Fig. 2. Transport cycle: coupled vs uncoupled fluxes. **(A)** Inward transport of glutamate (i.e., uptake) is initiated by the binding of 3Na$^+$, 1H$^+$, and 1 acidic amino acid (1AA$^-$) to the extracellular side; translocation of the complex to the intracellular side; and subsequent release of the 3Na$^+$, 1H$^+$, and 1AA$^-$ followed by binding of 1K$^+$ on the intracellular side. The binding of potassium re-orients the transporter and a new cycle can begin. Because these substrates interact with the transporter with a defined stoichiometry, uptake is a thermodynamically coupled process and the resulting current is variously termed stoichiometrically coupled transport current, coupled current, or stoichiometric current. The exact binding order or Na$^+$, H$^+$, and glutamate$^-$ remains an active area of research and has therefore been lumped into one step. **(B)** Na$^+$ and the AA$^-$ also gate a Cl$^-$ conductance. Strikingly, a change in Cl$^-$ concentration does not change [^{3}H]-AA$^-$ uptake *(16–18)* and glutamate does not permeate through the anion pathway *(49)*. The anion current is variously termed stoichiometrically uncoupled current or uncoupled current. Their proposed interrelation is shown; the total current (I$_{total}$), is the sum of the stoichiometric transport current and the uncoupled anion current (I$_{anion}$), observed in sodium chloride under normal physiological conditions. The current-voltage relation shows that the stoichiometric transport current remains inward in the tested potential range while I$_{total}$ reverses; this relation differs from subtype to subtype and with AA$^-$. The model also shows that it is possible to selectively measure the stoichiometrically coupled cycle by clamping the potential at E$_{Cl}$, a fact utilized in the experiments shown in Fig. 3A. **(C)** We recently characterized EAAC1 (EAAT3) with respect to sodium selectivity and found that lithium is able to support AA$^-$ uptake but in its presence the AA$^-$ cannot gate the anion conductance *(53)*. Indeed, the subtracted current (NaCl – LiCl) isolates a nearly ohmic current (I$_{Cl}$; stippled line), reversing close to E$_{Cl}$ and resembling the modeled I$_{anion}$.

activated by two of the molecules they transport, sodium and glutamate *(16–18)*. The simulation in Fig. 2B shows that the total current (I$_{tot}$)—obtained by subtracting the current in presence of substrate from that in its absence—in fact is a sum of an inwardly rectifying stoichiometric transport current and a nearly ohmic uncoupled anion current (I$_{anion}$). Different glutamate transporters exhibit different reversal potentials because the ratio of coupled to uncoupled currents differs between the various transporters (EAAT2 > EAAT3 > EAAT1 > EAAT4 ~ EAAT5) and also with the substrates *(16,49)*. The family relation between ASCT1 and ASCT2 and the glutamate transporters also manifests itself by the facts that EAAT3 is able to transport cysteine *(50)* and that exchange by ASCT1 and ASCT2 can activate an uncoupled chloride flux *(51,52)* with the same permeability sequence to that of the eukaryotic glutamate transporters (SCN$^-$ > NO$_3^-$ > I$^-$ > Cl$^-$) *(49)*.

Although the selectivity for different amino acids as well as the ability of anions to substitute for chloride has been investigated, the cation selectivity is less well-characterized. In a recent study, we investigated the ability of lithium to replace sodium in EAAC-1. To our surprise, we found that lithium can replace sodium in the coupled uptake, but not in its capacity to gate the sodium and glutamate-dependent anion current *(53)*. Indeed, subtracting net currents in sodium from those obtained in lithium isolates the nearly ohmic anion current (Fig. 2C). Three important facts emerge from Fig. 2B and C; the total current is not a direct measure of the turnover rate, the coupled current can be selectively determined at E$_{Cl}$, and an outward current seen at potentials more positive than E$_{Cl}$ does not reflect reversal of the transporter under normal physiological conditions. The differing cation dependence of the net radioactive AA$^-$ flux and the substrate-induced anion conductance also suggests that the conformation gating the anion conductance is different from that during substrate translocation *(53)*.

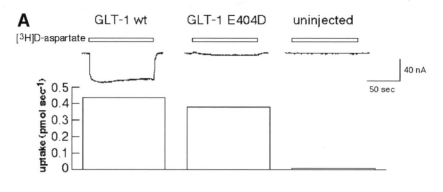

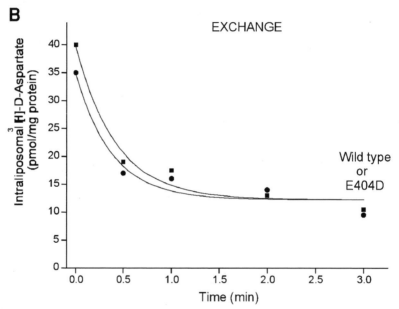

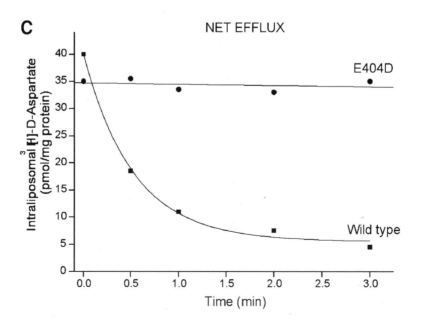

Fig. 3. Net flux vs exchange. (A) By perfusing a wild-type expressing oocyte, clamped at the reversal potential of chloride, with radioactive [³H]-AA⁻ an inward current reflecting the coupled cycle is observed and the subsequent scintillation counting shows that the [³H]-AA⁻ also is transported. In contrast, the mutant E404D shows no coupled current but it is able to accumulate radioactive AA⁻. **(B)** and **(C)** Reconstitution of transporters provides a way to control both intracellular and extracellular media compositions and any change in net-flux vs exchange can be investigated. (B) shows that wild-type and E404D are both able to exchange intracellular [³H]-AA⁻ with extracellular nonradioactive AA⁻, resulting in a decrease of intracellular [³H]-AA⁻. (C) shows that extraliposomal potassium promotes net efflux in the wild-type but not in E404D.

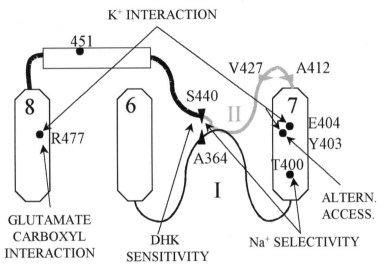

Fig. 4. Topology and proximity in the carboxyl terminal half with the position of identified substrate interacting residues indicated. Topological model of the carboxyl-terminal half, beginning at transmembrane domain 6, with the proximity of the two cysteine-pairs S440C/A364C and V427C/A412C, indicated by pointed triangles. Transmembrane domains 6, 7, and 8 and reentrant loop I are drawn in the plane of the paper (black line). Residues 412-440 of reentrant loop II are drawn behind the plane (gray line) and the continuation of this loop is drawn out of the plane (thick black line) until transmembrane domain 8 is reached. Residues that are involved in interactions with the substrates are indicated, as well as residue 451 in the linker. When a cysteine is introduced at this position, its chemical modification leaves the uncoupled anion conductance intact but abolishes the coupled uptake *(54)*. Besides residue 404, residue 403 is also a determinant of potassium interaction *(15,60)*. Moreover, residue 403 is alternatingly accessible from either side of the membrane *(61)*. Residue 440 is involved in the interaction with the non-transportable substrate analog dihydrokainate (DHK) *(45)*. This latter residue and threonine-400 are determinants of the sodium specificity *(45,53)*.

If this idea is correct, it should be possible to selectively perturb one of the two fluxes mediated by the glutamate transporters. Indeed, analysis of a mutant with a cysteine residue introduced in the hydrophobic linker region, I421C (corresponding to position 451 in GLT-1, *see* Fig. 4) reveals that the coupled uptake in this mutant is very sensitive to sulfhydryl reagents *(54)*. Strikingly, the substrate-induced anion conductance is not affected at all by the sulfhydryl reagents *(54)*. Similar observations have been made on single cysteine mutants in $EAAT_1$ at position corresponding to 447 *(55)* and 450 *(56)* of GLT-1.

The sulfhydryl-modified I421C allowed us to characterize the anion current in isolation of the coupled pathway. Under conditions where the uncoupled current is dominant and entry into the coupled pathway is blocked by sulfhydryl reagents, the anion current could be increased further by more than fourfold *(54)*. This behavior is consistent with the idea that the two processes, which both are activated by sodium and glutamate, are coupled and exist in a dynamic equilibrium *(54)*. Notably, the anion currents in the modified I421C have a hyperbolic dependence on aspartate and sigmoid dependence on sodium *(54)* indicating that the stoichiometries of the coupled and uncoupled processes are similar. Whereas the affinity for sodium was unchanged, we found that the affinity for aspartate was lowered by almost 10-fold as compared with the affinity when measuring both processes *(54)*. Evidence has been presented that entry/exit into/from the coupled pathway occurs slightly faster than the uncoupled pathway *(57,58)*. We find it tempting to speculate that under normal physiological conditions, the relatively higher aspartate affinity, exhibited by the coupled process, controls the equilibrium and brings the transporter(s) into the coupled pathway before entering the uncoupled pathway.

The use of lithium to isolate the anion current is not feasible for all transporters because lithium cannot support uptake in GLT-1 *(59)*. This raises the interesting question of which residues are responsible for the interaction with lithium (see next section).

5. STRUCTURE–FUNCTION RELATIONS OF THE GLUTAMATE TRANSPORTERS

5.1. Defective Potassium Coupling Locks the Transporter in the Exchange Mode

Because of the sequential nature of the transport process (*see* Fig. 2A), a mutation that affects the interaction of the transporter with potassium is expected to abolish net flux but not exchange. Because the uncoupled anion flux is activated from the exchange part of the cycle, the uncoupled current is also expected to remain intact. Because potassium is positively charged, conserved negatively charged or polar amino acids are likely to be important for its binding and translocation. Indeed, two adjacent amino acid residues of GLT-1, tyrosine-403 and glutamate-404, located in transmembrane domain 7 (Fig. 4), appear to be involved in potassium binding *(15,60)*. Moreover, residue 403 is alternatingly accessible from either side of the membrane *(61)*. Oocytes expressing either GLT-1 or E404D were clamped at –30mV (~E_{cl}), where only the coupled current is measured, and perfused with radiolabeled D-aspartate (Fig. 3A). The absence of coupled current suggests that the [^{3}H]-D-aspartate uptake by E404D may represent exchange with the internal pool of unlabeled glutamate and aspartate *(15)*. This was proven when GLT-1 and E404D were reconstituted into liposomes where the intracellular medium composition can be controlled. Both GLT-1 and E404D are able to perform exchange of intraliposomal [^{3}H]-D-aspartate with unlabeled extraliposomal D-aspartate (Fig. 3B). Strikingly, only the wild-type GLT-1 is able to perform net-efflux, a process that requires interaction with potassium (Fig. 3C).

Although E404D is conserved in all the mammalian glutamate transporters, it is of interest to note that it is neither conserved in ASCT1 nor ASCT2, which performs potassium-independent exchange *(51)*.

5.2. Sodium/Lithium Discrimination

Glutamate and sodium/lithium-induced conformational changes in the GLT-1 transporter have been detected by the altered accessibility of trypsin-sensitive sites to the protease *(59)*. These experiments in GLT-1 shows that lithium can occupy at least one of the sodium ion binding sites, but lithium by itself cannot support coupled transport *(59)*. Therefore, at least one of the sodium binding sites in GLT-1 discriminates between sodium and lithium. As described earlier, this contrasts with EAAC-1, where lithium is able to support uptake. It should therefore be possible to identify residues that are responsible for the sodium/lithium selectivity difference between EAAC-1 and GLT-1.

When serine-440 (Fig. 4) is changed to glycine in GLT-1, the mutant is able to catalyse coupled transport in the presence of lithium, as if the specific sodium site now has become more promiscuous; the transporter can now accept lithium at all three sites *(45)*. Importantly, the corresponding position in EAAC-1 is occupied by a glycine (G410) and the reciprocal mutant G410S has lost the ability of lithium to support uptake while sodium remains able to do so *(53)*. In addition, we also showed that, in another EAAC-1 mutant, T370S, sodium, but not lithium, is able to support uptake *(53)*. This threonine corresponds to T400 in GLT-1, see Fig. 4. The results obtained from mutating the two positions, 370 and 410 in EAAC-1, indicates that we indeed have been able to identify two residues that control the interaction with lithium.

5.3. Glutamate Binding Site

We have identified an arginine residue (R447) of EAAC-1 located in transmembrane domain 8 that controls the binding of the γ-carboxyl group of aspartate and glutamate *(62)*. This residue corresponds to R477 in GLT-1 (*see* Fig. 4). Because EAAT3, the human homolog of EAAC-1, is able to support cysteine transport *(50)*, we reasoned that EAAC-1 also is able to do so, which indeed is the case *(62)*. When R447 is mutated to neutral or negative amino acid residues, it completely abolishes transport of L-glutamate and D- and L-aspartate, but remarkably leaves cysteine transport unaffected *(62)*. Thus, it appears that this residue controls the binding of the γ-carboxyl group of glutamate *(62)*. However, R447 also controls the interaction with potassium, and its phenotype is reminiscent of the exchange mutant E404D *(62)*. A hypothetical scheme of how R447 can participate alternatingly in potassium and glutamate interaction has been published elsewhere *(63)*.

6. CONCLUDING REMARKS

Considerable progress has been achieved in our understanding of the structural basis of glutamate transporter function. The unique topology as well as the identification of amino acid residues involved in the interaction of the transporter with glutamate and the two cosubstrates, sodium and potassium, form the basis for exciting new experiments that help us gain further insights into the molecular mechanism of coupled and uncoupled transport.

ACKNOWLEDGMENTS

The work presented here was supported by the Bernard Katz Minerva Center for Cell Biophysics, and by grants from the U.S.–Israel Binational Science Foundation; the

Basic Research Foundation administered by the Israel Academy of Sciences and Humanities; the NINDS; National Institutes of Health; and the Federal Ministry of Education, Science, Research and Technology of Germany and its International Bureau at the Deutsches Zentrum für Luft und Raumfahrt. L.B. is a recipient of a Danish Research Agency fellowship.

REFERENCES

1. Kanner, B. I. and Schuldiner, S. (1987) Mechanism of transport and storage of neurotransmitters. *CRC Crit. Rev. Biochem.* **22,** 1–38.
2. Nicholls, D. and Attwell, D. (1990) The release and uptake of excitatory amino acids. *Trends Pharmacol. Sci.* **11,** 462–468.
3. Zerangue, N. and Kavanaugh, M. P. (1996) Flux coupling in a neuronal glutamate transporter. *Nature* **383,** 634–637.
4. Rothstein, J. D., Dykes-Hoberg, M., Pardo, C. A., Bristol, L. A., Jin, L., Kuncl, R. W., et al. (1996) Knockout of glutamate transporters reveals a major role for astroglial transport in excitotoxicity and clearance of glutamate. *Neuron* **16,** 675–686.
5. Tanaka, K., Watase, K., Manabe, T., et al. (1997) Epilepsy and exacerbation of brain injury in mice lacking the glutamate transporter GLT-1. *Science* **276,** 1699–1702.
6. Mennerick, S. and Zorumski, C. F. (1994) Glial contributions to excitatory neurotransmission in cultured hippocampal cells. *Nature* **368,** 59–62.
7. Tong, G. and Jahr, C. E. (1994) Block of glutamate transporters potentiates postsynaptic excitation. *Neuron* **13,** 1195–1203.
8. Otis, T. S., Wu, T. C., and Trussell, L. O. (1996) Delayed clearance of transmitter and the role of glutamate transporters at synapses with multiple release sites. *J. Neurosci.* **16,** 1634–1644.
9. Diamond, J. S. and Jahr, C. E. (1997) Transporters buffer synaptically released glutamate on a submillisecond time scale. *J. Neurosci.* **17,** 4672–4687.
10. Kanner, B. I. and Sharon, I. (1978) Active transport of L-glutamate by membrane vesicles isolated from rat brain. *Biochemistry* **17,** 3949–3953.
11. Brew, H. and Attwell, D. (1987) Electrogenic glutamate uptake is a major current carrier in the membrane of axolotl retinal glial cells. *Nature* **327,** 707–709.
12. Levy, L. M., Warr, D., and Attwell, D. (1998) Stoichiometry of the glial glutamate transporter GLT-1 expressed inducibly in a Chinese hamster ovary cell line selected for low endogenous Na^+-dependent glutamate uptake. *J. Neurosci.* **18,** 9620–9628.
13. Kanner, B. I. and Bendahan, A. (1982) Binding order of substrates to the sodium and potassium ion coupled L-glutamic acid transporter from rat brain. *Biochemistry* **21,** 6327–6330.
14. Pines, G. and Kanner, B. I. (1990) Counterflow of L-glutamate in plasma membrane vesicles and reconstituted preparations from rat brain. *Biochemistry* **29,** 11209–11214.
15. Kavanaugh, M. P., Bendahan, A., Zerangue, N., Zhang, Y., and Kanner, B. I. (1997) Mutation of an amino acid residue influencing potassium coupling in the glutamate transporter GLT-1 induces obligate exchange. *J. Biol. Chem.* **272,** 1703–1708.
16. Wadiche, J. I., Amara, S. G., and Kavanaugh, M. P. (1995) Ion fluxes associated with excitatory amino acid transport. *Neuron* **15,** 721–728.
17. Fairman, W. A., Vandenberg, R. J., Arriza, J. L., Kavanaugh, M. P., and Amara, S. G. (1995) An excitatory amino-acid transporter with properties of a ligand-gated chloride channel. *Nature* **375,** 599–603.
18. Arriza, J. L., Eliasof, S., Kavanaugh, M. P., and Amara, S. G. (1997) Excitatory amino acid transporter 5, a retinal glutamate transporter coupled to a chloride conductance. *Proc. Natl. Acad. Sci USA* **94,** 4155–4160.
19. Guastella, J., Nelson, N., Nelson, H., et al. (1990) Cloning and expression of a rat brain GABA transporter. *Science* **249,** 1303–1306.

20. Pacholczyk, T., Blakely, R. D., and Amara, S. G. (1991) Expression cloning of a cocaine- and antidepressant-sensitive human noradrenaline transporter. *Nature* **350,** 350–353.

21. Uhl, G. R. (1992) Neurotransmitter transporters (plus): a promising new gene family. *Trends Neurosci.* **15,** 265–268.

22. Schloss, P., Mayser, W., and Betz, H. (1992) Neurotransmitter transporters. A novel family of integral plasma membrane proteins. *FEBS Lett.* **307,** 76–78.

23. Amara, S. G. and Kuhar, M. J. (1993) Neurotransmitter transporters: recent progress. *Ann. Rev. Neurosci.* **16,** 73–93.

24. Storck, T., Schulte, S., Hofmann, K., and Stoffel, W. (1992) Structure, expression, and functional analysis of a Na⁺-dependent glutamate/aspartate transporter from rat brain. *Proc. Natl. Acad. Sci. USA* **89,** 10955–10959.

25. Pines, G., Danbolt, N. C., Bjoras, M., et al. (1992) Cloning and expression of a rat brain L-glutamate transporter. *Nature* **360,** 464–467.

26. Kanai, Y. and Hediger, M. A. (1992) Primary structure and functional characterization of a high-affinity glutamate transporter. *Nature* **360,** 467–471.

27. Arriza, J. L., Fairman, W. A., Wadiche, J. I., Murdoch, G. H., Kavanaugh, M. P., and Amara, S. G. (1994) Functional comparisons of three glutamate transporter subtypes cloned from human motor cortex. *J. Neurosci.* **14,** 5559–5569.

28. Danbolt, N. C., Storm-Mathisen, J., and Kanner, B. I. (1992) A (Na⁺ + K⁺) coupled L-glutamate transporter purified from rat brain is located in glial cell processes. *Neuroscience* **51,** 295–310.

29. Lehre, K. P., Levy, L. M., Ottersen, O. P., Storm–Mathisen, J., and Danbolt, N. C. (1995) Differential expression of two glial glutamate transporters in the rat brain: quantitative and immunocytochemical observations. *J. Neurosci.* **15,** 1835–153.

30. Rothstein, J. D., Martin, L., Levey, A. I., et al. (1994) Localization of neuronal and glial glutamate transporters. *Neuron* **13,** 713–725.

31. Tolner, B., Poolman, B., Wallace, B., and Konings, W. N. (1992) Revised nucleotide sequence of the gltP gene, which encodes the proton-glutamate-aspartate transport protein of Escherichia coli K-12. *J. Bacteriol.* **174,** 2391–2393.

32. Jiang, J., Gu, B., Albright, L. M., and Nixon, B. T. (1989) Conservation between coding and regulatory elements of Rhizobium meliloti and Rhizobium leguminosarum dct genes. *J. Bacteriol.* **171,** 5244–5253.

33. Shafqat, S., Tamarappoo, B. K., Kilberg, M. S., Puranam, R. S., McNamara, J. O., Guadano-Ferraz, A., and Fremeau, R. T. (1993) Cloning and expression of a novel Na⁺-dependent neutral amino acid transporter structurally related to mammalian Na⁺/glutamate cotransporters. *J. Biol. Chem.* **268,** 15351–15355.

34. Arriza, J. L., Kavanaugh, M. P., Fairman, W. A., Wu, Y.-N., Murdoch, G. H., North, R. A., and Amara, S. G. (1993) Cloning and expression of a human neutral amino acid transporter with structural similarity to the glutamate transporter gene family. *J. Biol. Chem.* **268,** 15329–15332.

35. Utsunomiya-Tate, N., Endo, H., and Kanai, Y. (1996) Cloning and functional characterization of a system ASC-like Na⁺-dependent neutral amino acid transporter. *J. Biol. Chem.* **271,** 14883–14890.

36. Kekuda, R., Prasad, R. D., Fei, R.-J., et al. (1996) Cloning of the sodium-dependent, broad-scope, neutral amino acid transporter Bᵒ from a human placental choriocarcinoma cell line. *J. Biol. Chem.* **271,** 18657–18661.

37. Danbolt, N. C., Pines, G., and Kanner, B. I. (1990) Purification and reconstitution of the sodium- and potassium-coupled glutamate transport glycoprotein from rat brain. *Biochemistry* **29,** 6734–6740.

38. Haugeto, Ø., Ullensvang, K., Levy, L. M., Chaudhry, F. A., Honoré, T., Nielsen, M., et al. (1996) Brain glutamate transporter proteins form homomultimers. *J. Biol. Chem.* **271,** 27715–27722.

39. Eskandari, S., Kreman, M., Kavanaugh, M. P., Wright, E. M., and Zampighi, G. A. (2000) Pentameric assembly of a neuronal glutamate transporter *Proc. Natl. Acad. Sci. USA* **97,** 8641–8646.

40. Grunewald, M., Bendahan, A., and Kanner, B. I. (1998) Biotinylation of single cysteine mutants of the glutamate transporter GLT-1 from rat brain reveals its unusual topology. *Neuron* **21,** 623–632.

41. Slotboom, D. J., Sobczak, I., Konings, W. N., and Lolkema, J. S. (1999) A conserved serine-rich stretch in the glutamate transporter family forms a substrate-sensitive reentrant loop. *Proc. Natl. Acad. Sci. USA* **96,** 14282–14287.

42. Grunewald, M. and Kanner, B. I. (2000) The accessibility of a novel reentrant loop of the glutamate transporter GLT-1 is restricted by its substrate. *J. Biol. Chem.* **275,** 9684–9689.

43. Seal, R. P., Leighton, B. H., and Amara, S. G. (2000) A model for the topology of excitatory amino acid transporters determined by the extracellular accessibility of substituted cysteines. *Neuron* **25,** 695–706.

44. Seal, R. P. and Amara, S. G. (1998) A reentrant loop domain in the glutamate carrier EAAT1 participates in substrate binding and translocation. *Neuron* **21,** 1487–1498.

45. Zhang, Y. and Kanner, B. I. (1999) Two serine residues of the glutamate transporter GLT-1 are crucial for coupling the fluxes of sodium and the neurotransmitter. *Proc. Natl. Acad. Sci. USA* **96,** 1710–1715.

46. Brocke, L., Bendahan, A., Grunewald, M., and Kanner, B. I. (2002) Proximity of two oppositely oriented reentrant loops in the glutamate transporter GLT-1 identified by paired cysteine mutagenesis. *J. Biol. Chem.* **277,** 3985–3992.

47. Fu, D., Libson, A., Miercke, L. J., Weitzman, C., Nollert, P., Krucinski, J., and Stroud, R. M. (2000) Structure of a glycerol-conducting channel and the basis for its selectivity. *Science* **290,** 481–486.

48. Murata, K., Mitsuoka, K., Hirai, T., et al. (2000) Structural determinants of water permeation through aquaporin-1. *Nature* **407,** 595–605.

49. Wadiche, J. and Kavanaugh, M. P. (1998) Macroscopic and microscopic properties of a cloned glutamate transporter/chloride channel. *J. Neurosci.* **18,** 7650–7661.

50. Zerangue, N. and Kavanaugh, M. P. (1996) Interaction of L-cysteine with a human excitatory amino acid transporter. *J. Physiol.* **493,** 419–423.

51. Zerangue, N. and Kavanaugh, M. P. (1996) ASCT-1 is a neutral amino acid exchanger with chloride channel activity. *J. Biol. Chem.* **271,** 27991–27994.

52. Broer, A., Wagner, C., Lang, F., and Broer, S. (2000) Neutral amino acid transporter ASCT2 displays substrate-induced Na^+ exchange and a substrate-gated anion conductance. *Biochem. J.* **346,** 705–710.

53. Borre, L. and Kanner, B. I. (2001) Coupled, but not uncoupled, fluxes in a neuronal glutamate transporter can be activated by lithium ions. *J. Biol. Chem.* **276,** 40396–40401.

54. Borre, L., Kavanaugh, M. P., and Kanner, B. I. (2002) Dynamic equilibrium between coupled and uncoupled modes of a neuronal glutamate transporter. *J. Biol. Chem.* **277,** 13501–13507.

55. Seal, R. P., Shigeri, Y., Eliasof, S., Leighton, B. H., and Amara, S. G. (2001) Sulfhydryl modification of V449C in the glutamate transporter EAAT1 abolishes substrate transport but not the substrate-gated anion conductance. *Proc. Natl. Acad. Sci. USA* **98,** 15324–15329.

56. Ryan, R. M. and Vandenberg, R. J. (2002) Distinct conformational states mediate the transport and anion channel properties of the glutamate transporter EAAT-1. *J. Biol. Chem.* **277,** 13494–13500.

57. Otis, T. S. and Kavanaugh, M. P. (2000) Isolation of current components and partial reaction cycles in the glial glutamate transporter EAAT2 *J. Neurosci.* **20,** 2749–2757.

58. Grewer, C., Watzke, N., Wiessner, M., and Rauen, T. (2000) Glutamate translocation of the neuronal glutamate transporter EAAC1 occurs within milliseconds *Proc. Natl. Acad. Sci. USA* **97,** 9706–9711.

59. Grunewald, M. and Kanner, B. I. (1995) Conformational changes monitored on the glutamate transporter GLT-1 indicate the existence of two neurotransmitter-bound states. *J. Biol. Chem.* **270,** 17017–17024.

60. Zhang, Y., Bendahan, A., Zarbiv, R., Kavanaugh, M. P., and Kanner, B. I. (1998) Molecular determinant of ion selectivity of a (Na^+ + K^+)-coupled rat brain glutamate transporter. *Proc. Natl. Acad. Sci. USA* **95,** 751–755.

61. Zarbiv, R., Grunewald, M., Kavanaugh, M. P., and Kanner, B. I. (1998) Cysteine scanning of the surroundings of an alkali-ion binding site of the glutamate transporter GLT-1 reveals a conformationally sensitive residue. *J. Biol. Chem.* **273,** 14231–14237.

62. Bendahan, A., Armon, A., Madani, N., Kavanaugh, M. P., and Kanner, B. I. (2000) Arginine 447 plays a pivotal role in substrate interactions in a neuronal glutamate transporter. *J. Biol. Chem.* **275,** 37436–37442.

63. Kanner, B. I. and Borre, L. (2002) The dual-function glutamate transporters: structure and molecular characterisation of the substrate binding sites. *BBA* **1555,** 92–95.

9

Allosteric Modulation of Glutamate Transporters

Robert J. Vandenberg and Renae M. Ryan

1. INTRODUCTION

Glutamate is the predominant excitatory neurotransmitter in the mammalian central nervous system (CNS) and mediates a wide variety of physiological functions, but the failure to adequately control extracellular glutamate concentrations can lead to excessive excitatory neurotransmission and cell death. Glutamate transporters contribute to maintaining extracellular glutamate concentrations and thereby play an important role in regulating the dynamics of excitatory neurotransmission. In this chapter we will briefly discuss the structure and function of glutamate transporters and then focus on how the activity of glutamate transporters can be allosterically regulated by endogenous factors and also pharmacological agents and how this may impact on glutamatergic neurotransmission. This will be addressed in terms of what compounds have been identified as allosteric regulators, the structural specificity of the interactions, where the binding sites are located on the transporters, and how ligand binding to these sites may allosterically influence transporter function.

2. MOLECULAR BIOLOGY OF GLUTAMATE TRANSPORTERS

Five subtypes of glutamate transporters have been identified and, although different terminology has been used by various investigators, we shall use the Excitatory Amino Acid Transporter (EAAT) terminology (1–3) with EAAT1 equivalent to GLAST1 (4,5), EAAT2 equivalent to GLT1 (6), and EAAT3 equivalent to EAAC1 (7). EAAT2 is the most abundant glutamate transporter and is found predominantly in glial cells throughout the brain (2,6). EAAT1 is also found predominantly in glial cells, with the greatest levels observed in the cerebellum (8) and the retina (9). EAAT3 (7), EAAT4, and EAAT5 are expressed in neurones, with EAAT4 found in highest levels in the Purkinje cells of the cerebellar molecular layer (10,11) and EAAT5 within the retina (1).

The amino acid sequences of the glutamate transporters show a high degree of similarity with between 40–60% of amino acid residues identical between subtypes. At present, the three-dimensional (3D) structure of the transporters is unknown and indirect methods based on amino acid sequence hydropathy plots and amino acid accessibility methods have been employed to predict the transmembrane topology of the transporters. Two similar models developed by the groups of Amara (12,13) and Kanner

From: *Molecular Neuropharmacology: Strategies and Methods*
Edited by: A. Schousboe and H. Bräuner-Osborne © Humana Press Inc., Totowa, NJ

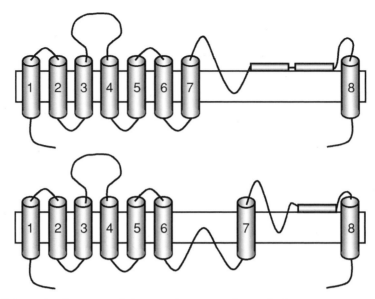

Fig. 1. Schematic diagrams of the membrane topology of glutamate transporters as proposed by the groups of Amara (top) and that of Kanner (bottom). In both cases, the carboxy terminal and amino terminals of the protein are intracellular. In the top figure, the segment after transmembrane domain 7 enters into the membrane but does not pass through the membrane and the extra two segments before transmembrane domain 8 are likely to be associated with the extracellular surface of the membrane. In the bottom figure, there is an additional re-enterant loop between transmembrane domains 6, and 7. The regions where the two models differ is highly conserved and plays an important role in glutamate translocation through the transporter and as such it is likely to undergo considerable conformational changes during the transport process.

(14,15) have been proposed and both incorporate a series of α-helical transmembrane domains with reentrant loops within the highly conserved carboxy-terminal domain (*see* Fig. 1). The re-entrant loops seem to play very important roles in binding of glutamate and the various co- and countertransported ions.

3. FUNCTIONAL PROPERTIES OF GLUTAMATE TRANSPORTERS

At inactive synapses, extracellular concentrations of glutamate are in the low micromolar to nanomolar range, but stimulation of glutamate release from presynaptic terminals will cause extracellular glutamate concentrations to rise transiently to millimolar concentrations. In contrast, intracellular glutamate concentrations are relatively stable in the millimolar range *(16,17)*. Thus, glutamate transporters pump glutamate into the cell under a wide range of concentration gradients, which requires a considerable input of energy into the system. This energy is derived from the co- and countertransport of ions moving down their electro-chemical gradients. The exact stoichiometry of these transporters has been debated for over 30 yr *(18–20)*, but recent studies on EAAT3 *(21)* and EAAT2 (GLT-1) *(22,23)* agree that glutamate is transported into the cell with three Na^+ and one H^+, followed by the countertransport of one K^+ (Fig. 2). Theoretically, this coupling can support a transmembrane glutamate concentration gradient exceeding 10^6

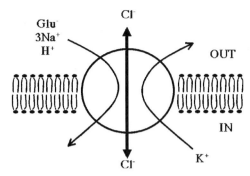

Fig. 2. Schematic diagram of the stoichometry of ion flux coupling and the chloride channel activity of glutamate transporters. Glutamate is coupled to the co-transport of 3 Na^+, $1K^+$, and the countertransport of 1 K^+. In addition, glutamate and Na^+ binding to the transporter activates an uncoupled chloride flux through the transporter.

under equilibrium conditions, and allows the transporters to continue removing glutamate over a wide range of ionic conditions.

Application of glutamate to cells expressing glutamate transporters generates a number of distinct conductances. The glutamate transport conductance (or the stoichiometric conductance) mentioned earlier results in the net transfer of two positive charges into the cell, allowing glutamate transport to be monitored as a membrane current *(24)*. The application of glutamate and Na^+ to cells expressing glutamate transporters also activates a thermodynamically uncoupled anion conductance through the transporter *(3,25,26)* and recently a third glutamate independent leak conductance has been characterized *(27,28)*. Finally, application of arachidonic acid to the EAAT4 transporter activates a conductance carried by protons (*see* Subheading 6.1.).

4. THE UNCOUPLED CHLORIDE CONDUCTANCE OF THE EAATS

Traditionally, cotransporters and ion channels have been thought of as being structurally and functionally distinct. Neurotransmitter transport is a process whereby the substrates are pumped actively against their transmembrane electrochemical gradients. This is possible because the transport process is thermodynamically coupled to the passive transport of ions. In contrast, ion channels are considered to be pores that allow the dissipation of pre-existing electrochemical gradients *(29)*. Glutamate transporters are an example of a protein that can act as both a transporter and an ion channel. In addition, several other neurotransmitter transporters have recently been shown to conduct ions in a process that is thermodynamically uncoupled from the transport of neurotransmitter, and includes transporters for: GABA, dopamine, noradrenaline, proline, and serotonin *(30–36)*. Sodium-dependent glutamate binding to the cloned glutamate transporter proteins activates an anion conductance that has the following selectivity sequence: $SCN^- > ClO_4^- > NO_3^- > I^- > Br^- > Cl^- > F^- >>$ gluconate *(25,26,28)*.

Although all five glutamate transporter subtypes support uncoupled chloride conductances, there are a number of distinct differences between the transporter subtypes. The current generated by EAAT4 and EAAT5 is predominantly owing to chloride ions, and in the case of L-aspartate transport by EAAT4, it has been estimated that up to 95%

of the current observed at –60 mV is owing to chloride *(3)*. The contribution of chloride to the total current of EAAT2 and EAAT3 is negligible, whereas different substrates affect the proportion of the current owing to chloride via EAAT1. Aspartate transport by EAAT1 allows a large chloride flux, whereas the chloride conductance activated by glutamate is much smaller *(26)*. These differences in the relative proportions of transport and uncoupled chloride conductances for the different transporters results in quite distinct current voltage relationships. In the cases of EAAT4 and EAAT5, under standard physiological conditions the glutamate elicited conductance reverses direction close to the reversal potential for chloride ions, whereas for EAAT2 and EAAT3 the current does not reverse at potentials less than +80 mV.

The existence of two distinct functions within the one membrane protein raises a number of interesting questions about transporter functions, such as: what is the relative importance of the two functions? Can they be allosterically regulated under physiological conditions or manipulated by pharmacological methods? What is the structural basis for the dual functions? In the next section, we shall attempt to address some of these issues and also to present some of the current understanding of the structural basis for these functional states.

The physiological relevance of the uncoupled anion conductance is not fully understood, but it has been suggested that the chloride conductance may act as a voltage clamp *(25,37)*. The two positive charges that move into the cell with each transport cycle will depolarize the cell membrane *(38)* and reduce the driving force of glutamate into the cell. If chloride ions enter the cell during glutamate transport, a more hyperpolarized membrane potential will result, and thus an optimal rate of glutamate transport will be maintained. A similar suggestion by Billups and colleagues *(25)* predicts the uncoupled anion conductance activated during glutamate transport could clamp the pre-synaptic terminal at a negative potential. This could reduce further exocytotic release by decreasing the probability that action potentials arriving at the terminal will successfully depolarize the membrane *(25)*. Grant and Dowling *(39)* investigated the role of the transporter-associated chloride conductance in ON bipolar cells of the retina and demonstrated that this conductance is responsible for the light response of the ON bipolar cells *(39)*.

The relationship between the different conductances of the glutamate transporter is not well-understood. The structural basis for these conductances and whether the anions and glutamate permeate the same pore of the transporter protein is not known. Sonders and Amara *(40)* proposed two models: a single pathway for substrate and the uncoupled anion movement (Fig. 3A), or multiple permeation pathways in a single transporter molecule (Fig. 3B) *(40)*. This idea will be discussed further in relation to what is known about the quaternary structure of glutamate transporters.

5. QUATERNARY STRUCTURE

A number of different transporter proteins have been shown to exist as oligomers, including the glucose transporter GLUT1 *(41)*, the serotonin transporter *(42)*, and the GABA transporter GAT1 *(43)*. The quaternary structure of the glutamate transporters is unknown, but the variable electrophoretic mobility of the glutamate transporters with higher molecular-mass species *(44)* suggests that these proteins might also form oligomeric complexes. When fresh brain tissue is directly homogenized and solubi-

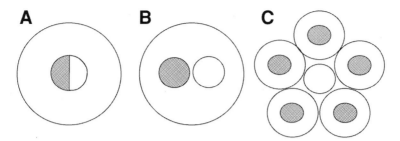

Fig. 3. Single- and multiple-pore models of glutamate transporters. The glutamate transloca-tion pore is represented as the hatched area and the pore responsible for chloride permeation is represented as a clear symbol. In the first two models (**A** and **B**), each transporter subunit is capable of functioning as both a transporter and an ion channel. In the first model (A), glutamate and chloride ions permeate the same pore region. In the second model (B), glutamate and chlo-ride ions permeate separate pores within the one protein "subunit." In the third model (**C**), the glutamate transporter subunits associate as a pentamer and whilst each subunit is capable of transporting glutamate, the formation of the chloride channel only arises as a consequence of the pentamer structure (adapted from Eskandari et al. *[48]*). At this stage, there is no definitive evidence that favors one model over another and other potential arrangements and interactions between subunits are possible.

lized in sodium dodecyl sulfate (SDS) and the proteins immediately separated on SDS-polyacrylamide gel electrophoresis (PAGE) in the absence of reducing agents, gluta-mate transporters exist as monomers *(45)*. This suggests that the subunits are noncovalently attached. When fresh brain membranes are exposed to chemical cross-linkers in a reduced environment, and then solubilized in SDS, oligomer bands appear on the gels *(46,47)*. Large molecular-weight complexes corresponding to dimers and trimers have also been observed for the GLT1 transporters purified and reconstituted in liposomes, which also indicates that transporters can form oligomeric complexes. However, the different glutamate transporter subtypes do not appear to form oligomeric complexes with each other, which suggests that the complexes observed in rat brain membranes are likely to be homo-oligomeric complexes *(46)*.

A recent freeze-fracture electron micrograph study of EAAT3 expressed in *Xenopus laevis* oocytes suggests that the transporters exist as a pentameric assembly with a cross-sectional area of 48 ± 5 nm^2 in the plasma membrane *(48)*. In this study, the authors proposed a functional model for glutamate transporters based on the observed pentameric structure. It was suggested that each transporter "subunit" conducts cou-pled glutamate transport, while the association of multiple subunits allows the forma-tion of a central chloride channel (Fig. 3C). This model can be related to the multiple permeation pathway model proposed by Sonders and Amara *(40)* (*see* Fig. 3b) where the central pore between the subunits forms a distinct pathway for the chloride ions. In the absence of suitable crystals for structure determination, it is not possible to verify this function model, but from a series of recent mutagenesis studies it has been demon-strated that the two components of transport can be independently regulated and may have separate molecular determinants *(49–51)*. This would be consistent with the model predicted by Eskandari et al. *(48)*.

6. ALLOSTERIC MODULATORS OF GLUTAMATE TRANSPORT

There are a large number of therapeutic drugs and drugs of abuse that act on neurotransmitter transporters, such as fluoxetine, cocaine, or amphetamines, and in most cases these drugs prevent or reduce the rate of clearance of the neurotransmitter from the synapse, which leads to elevations in the transmitter concentrations. However, the excitotoxic potential of excessive glutamate concentrations limits the utility of drugs that may block glutamate transporters. As such, there are no clinically useful glutamate transport blockers. An alternative approach to developing therapeutically useful drugs that manipulate glutamate concentrations may be to identify processes that lead to enhanced transporter function. For the most part, it would be expected that drugs that act to enhance transporter function would do so in an allosteric manner by acting at sites on the transporter that are not intimately involved in glutamate translocations. In the next section, we shall discuss the mechanisms of action of the endogenous modulators arachidonic acid, zinc, and reactive oxygen species (ROS); how they interact with the transporters; and how they may be expected to influence synaptic transmission (*see* Fig. 4). A better understanding of these process may lead to the development of novel drugs that allosterically upregulate transporter function. A number of other compounds have also been reported to modulate the activity of glutamate transporters and include protons *(52)*, ammonium *(53)*, phorbol esters *(54,55)*, and glutamate transporter-associated proteins *(56,57)*. However, at this stage the mechanisms of action of these compounds at the molecular level are only superficially characterized and therefore we shall not discuss these modulatory processes in any further detail.

6.1. Arachidonic Acid

Arachidonic acid is a cis-polyunsaturated fatty acid consisting of a 20-atom long carbon chain with four double bonds and a carboxylic group at one end. Glutamate stimulates the production of arachidonic acid by activation of NMDA receptors or AMPA and metabotropic glutamate receptors, leading to an elevation in intracellular Ca^{2+}, which is necessary for activation of phospholipase A_2 and arachidonic acid release. Free arachidonic acid is produced by both neurones and astrocytes and may diffuse through the cell membrane to act as an intercellular messenger. Arachidonic acid may also be converted to various cyclooxgenase and lipoxygenase derivatives such as the prostaglandins and leukotrienes (reviewed by Attwell et al. *[58].*) Thus, the biological activities of arachidonic acid may be classified as to whether they are direct or are owing to the metabolic derivatives. Arachidonic acid is freely diffusible in the lipid membrane and thus may modulate the activity of a wide variety of membrane proteins either through direct interaction with the protein or via modification of the fluidity of the lipid membrane and thereby altering the energetics of conformational changes required for functional activity. Free arachidonic acid concentrations in extracellular spaces of the CNS are uncertain, but it is believed that concentrations up to 30 μM are generated under normal physiological conditions *(58)*. However, under ischaemic conditions, the increase in intracellular Ca^{2+} stimulates the activity of PLA_2 leading to significantly greater production of arachidonic acid and thus the effects of arachidonic acid may be more pronounced in these situations. It is difficult to precisely determine free arachidonic acid concentrations because arachidonic acid is freely diffusible in the lipid membrane, and at concentrations greater than 30 μM arachidonic acid forms

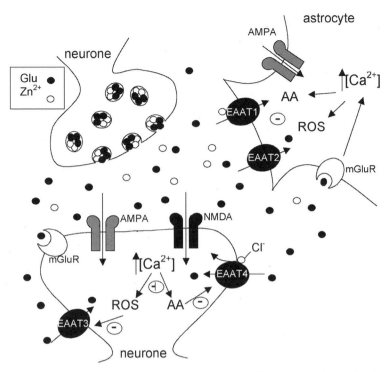

Fig. 4. Glutamate transporters are modulated allosterically by a number of endogenous factors including Zn^{2+} ions, arachidonic acid, and ROS. Zn^{2+} is stored in synaptic vesicles and may be co-released with glutamate into the synapse and bind to a variety of proteins in close proximity to the synapse, including glutamate transporters. Arachidonic acid, produced in response to elevations in glial and neuronal Ca^{2+} levels, can diffuse through membranes and modulate glutamate transporters EAAT1, EAAT2, and EAAT4. ROS produced in response to a variety of stimuli may also bind to and modulate the activity of glutamate transporters. See the text for more details on each of the endogenous allosteric modulators.

micelles. Therefore, effects of arachidonic acid at concentrations greater than 30–100 μM are likely to be at least partially caused by changes in membrane fluidity.

There are a number of reports using a variety of cell-expression systems concerning the functional consequences of arachidonic acid binding to glutamate transporters. In studies using the salamander retina Muller cells, brief exposure to arachidonic acid (10 μM) inhibits glutamate transport for prolonged periods, which suggests that arachidonic acid inhibits transport via a prolonged lipid-phase interaction with the transporters *(59)*. The predominant transporter subtype in these cells is the salamander homolog of EAAT1 and in a study using EAAT1 expressed in *X. laevis* oocytes similar levels of inhibition by arachidonic acid and rates of onset and reversal of inhibition were observed *(60)*. However, there have been conflicting reports on the effects of arachidonic acid on EAAT2 and the rat homolog GLT-1. Trotti et al. *(61)* reconstituted the purified GLT-1 in liposomes and in this preparation arachidonic acid (5 μM) inhibited transport via an aqueous-phase interaction *(61)*. However, in studies by Zerangue et al. *(60)* in which EAAT2 was expressed in *X. laevis* oocytes and also HEK-293 cells, arachidonic acid (100 μM) stimulated transport by decreasing the $K_{0.5}$ for glutamate. In unpublished studies from our laboratory, application of arachidonic acid to *X. laevis* oocytes expressing GLT1 or EAAT2 also caused stimulation, which is in contrast to that observed by Trotti et al. *(61)*, but consistent with that of Zerangue et al. *(60)*. These conflicting results suggest that the effects of arachidonic acid may be dependent on the cell in which the transporter is expressed, and/or the lipids used for reconstitution. With the recent greater appreciation of the roles of lipid rafts in determining functional responses of receptors *(62)*, it is reasonable to suggest that differences in the membrane environment surrounding transporter proteins may cause different functional responses to compounds such as arachidonic acid. Further work is required to isolate the precise molecular changes induced by arachidonic acid and under what conditions these changes may be expected to have an impact on transporter functions.

The interaction between arachidonic acid and EAAT4 results in a very novel change in transporter function. Arachidonic acid bound to EAAT4, in the presence of glutamate, activates an uncoupled proton conductance through the transporter *(63–65)*. The physiological role of this additional proton conductance is not clear, but it is interesting to note that a cyclooxygenase inhibitor, niflumic acid, can also stimulate a proton conductance through the transporter *(63)*. It is possible that arachidonic acid and niflumic acid bind to similar sites on EAAT4 or, at least induce similar conformational changes in the membrane protein, to activate the proton conductance. This raises the possibility that it may be possible to mimic pharmacologically the actions of arachidonic acid by compounds such as niflumic acid.

6.2. Zinc

Zinc modulates a number of proteins within excitatory synapses, including glutamate transporters *(66,67)* and glutamate receptors *(68,69)*. Zn^{2+} is stored in glutamatergic synaptic vesicles and is co-released with glutamate in various brain regions, such as the mossy fibers of the hippocampus, where it may directly regulate the various synaptic receptors, transporters, and channels *(70)*. Basal extracellular levels of Zn^{2+} in the brain are approx 10 nM *(70)* and upon stimulation, synaptic concentrations may rise to 118 μM *(66)*. Therefore, responses to Zn^{2+} within this range are expected to influence the dynamics of neurotransmission.

The effects of Zn^{2+} on glutamate transporters was first studied using Mueller and cone cells of the salamander retina *(66)*. The effects of Zn^{2+} were rapid and fully reversible, which suggests that Zn^{2+} binds directly with the transporters. Zn^{2+} acts as a noncompetitive inhibitor of glutamate transport with an IC_{50} in the low micromolar range *(66)*, which indicates that Zn^{2+} binds to a site on the transporter that is distinct from the binding sites for glutamate or the various co- and countertransported ions. In addition to inhibiting glutamate transport, Zn^{2+} differentially modulates the activity of the transporter-associated chloride channel. In Mueller cells, Zn^{2+} stimulates the chloride channel activity, but in the cones cells Zn^{2+} inhibits the chloride channel.

The actions of Zn^{2+} on the individual transporter subtypes expressed in *X. laevis* oocytes have also been investigated *(49,67)*. Application of Zn^{2+} to oocytes expressing EAAT1 rapidly reduces the current that is carried by glutamate, which fully recovers after washout of Zn^{2+}. Zn^{2+} does not appear to have any appreciable stimulatory or inhibitory effects on the chloride conductance of EAAT1. Zn^{2+} has no appreciable effects on EAAT2 and EAAT3, but does bind to EAAT4 and selectively inhibits the chloride conductance with little if any effect on the transport of glutamate *(49)*. Thus, Zn^{2+} appears to selectively manipulate transport and chloride channel functions of the different transporter subtypes. There also appears to be some subtle differences in the effects of Zn^{2+} on the human transporters expressed in oocytes compared with the salamander glutamate transporters. Nevertheless, the actions of Zn^{2+} on glutamate transporters demonstrate that is possible to allosterically and differentially regulate the transport conductance and the uncoupled chloride conductance of the transporters.

The Zn^{2+} binding sites on the glutamate transporters have been identified using site-directed mutagenesis. Mutation of either histidine 146 or histidine 156 to alanine in EAAT1 leads to a loss of Zn^{2+} sensitivity. The histidine corresponding to histidine 146 of EAAT1 is conserved between the other transporter subtypes, but the second histidine residue is conserved only in the EAAT4 subtype. Again mutation of either histidine residue in EAAT4 leads to loss of Zn^{2+} sensitivity, which confirms that the Zn^{2+} binding site is conserved between transporter subtypes. Furthermore, introduction of a histidine residue into EAAT2 at the site corresponding to the second histidine in EAAT1 leads to the formation of a Zn^{2+}-sensitive transporter. Thus, Zn^{2+} binds to very similar sites on EAAT1 and EAAT4 and these sites are distinct from the amino acid residues that form the glutamate and/or cation recognition sites. It is interesting to note that although the binding site for Zn^{2+} is conserved between EAAT1 and EAAT4, the functional effects mediated by Zn^{2+} bound to the transporters is quite different. In the case of EAAT1, Zn^{2+} inhibits transport with no effect on the chloride conductance, whereas for EAAT4, Zn^{2+} has no effect on the rate of glutamate transport but inhibits the chloride channel activity.

6.3. Free Radicals and Oxidants

Changes in oxygen supply to the brain can cause aberrant production of reactive oxygen species (ROS), which may have a variety of deleterious effects on cell metabolism. One such effect is a reduction or alteration in glutamate transporter function. Elevations in extracellular glutamate concentration will lead to increased glutamate receptor stimulation, Ca^{2+} influx and stimulation of nitric oxide synthase (NOS), phospholipase A_2, and xanthine oxidase, all of which may lead to the production of ROS. ROS have the capacity to oxidize many proteins, with sulfhydryl groups being particularly susceptible to oxidation. Cortical astrocytic cultures treated with xanthine, xanthine oxidase, or hydrogen peroxide have reduced glutamate transport activity *(71)* and the biological oxidants, peroxynitrate and peroxide, also inhibit glutamate uptake by selectively reducing the V_{max} *(72)*. The effects of thiol-oxidizing agents on transport activity of the recombinant glutamate transporters GLT-1, GLAST-1, and EAAC1 have also been investigated. Currents generated by glutamate transport can be enhanced and also inhibited by sequential treatment with the reducing agent ditriotrireitol (DTT) and the thiol-oxidizing agent 5,5′-dithio-bis(2-nitrobenzoic) acid (DTNB) *(73)*, suggesting that these glutamate transporters contain conserved sites in their structures that confer an SH-based redox regulatory mechanism. The redox interconversion of SH-groups on EAAC1 reduced the V_{max} of glutamate transport without affecting the Km or the uncoupled anion conductance *(74)*, suggesting that the two components of transport undergo independent modulation.

Amyotrophic lateral sclerosis (ALS) is a degenerative disorder of motor neurones. In 15–25% of cases, the genetic cause of the disease is a mutation of the enzyme Cu^+/Zn^{2+}

superoxide dismutase (SOD), which leads to elevated levels of ROS *(75)*. In patients with ALS, there is also a reduction in the levels of the glutamate transporter EAAT2, but the levels of EAAT3 are unaltered. Trotti et al. *(61)* were able to link these two effects by demonstrating that cells expressing the SOD mutants caused an increase in ROS and that this was directly responsible for the selective downregulation of the EAAT2 transporter *(76)*. Furthermore, it was the carboxy terminal region of the EAAT2 transporter that was particularly susceptible to damage by the increased levels of ROS. EAAT2 has a larger number of cysteine residues than the other transporters and it was argued that this property makes EAAT2 more susceptible to damage by the ROS.

7. OUTSTANDING ISSUES AND FUTURE DEVELOPMENTS

Although there have been a number advances in our understanding of how compounds interact with glutamate transporters to allosterically modulate the dynamics of transporter functions, a number of questions have arisen from these studies. Some of these questions include: To what extent do these processes influence transporter functions in vivo under physiological and pathological conditions, as may occur in conditions such as brain ischaemia or neurodegenerative disorders? Although the compounds may influence the dynamics of transporter functions, do they have significant impacts on the dynamics of synaptic transmission? Is it worthwhile to develop novel compounds that mimic the actions of some of these endogenous modulators as a way of manipulating extracellular glutamate concentrations to treat neurological disorders? At this stage, some of these issues are not well-understood, for a number of reasons. The first reason is that this field of research is still in its infancy, mainly because the molecular tools necessary for better understanding either do not exist or have only recently been identified. The cloning of the cDNAs encoding the transporters was a significant breakthrough, but further understanding of the molecular basis for transporter function will require detailed site-directed mutagenesis work or, ideally, the determination of high-resolution three-dimensional structures of the transporters. In particular, one of the unresolved issues about the structural basis for transporter functions is how transporters can function as both a transporter and an ion channel and what is the structural relationship between these two distinct functional properties. Of the endogenous allosteric modulators identified thus far, none of the compounds are highly selective for glutamate transporters over other proteins involved in synaptic transmission. However, with a detailed understanding of the structural basis for transporter functions and precise identification of the ligand binding sites for the various endogenous compounds, it should be possible to greatly enhance the development of compounds that selectively manipulate transporter functions. Once such compounds have been developed, or identified, some of the aforementioned issues raised concerning the physiological and pathological relevance of allosteric regulation of glutamate transporters can be better addressed. From this, it is hoped that alternate therapeutic drugs may be developed.

REFERENCES

1. Arriza, J. L., Eliasof, S., Kavanaugh, M. P., and Amara, S. G. (1997) Excitatory amino acid transporter 5, a retinal glutamate transporter coupled to a chloride conductance. *Proc. Natl. Acad. Sci. USA* **94,** 4155–4160.

2. Arriza, J. L., Fairman, W. A., Wadiche, J. I., Murdoch, G. H., Kavanaugh, M. P., and Amara, S. G. (1994) Functional comparisons of three glutamate transporter subtypes cloned from human motor cortex. *J. Neurosci.* **14,** 5559–5569.

3. Fairman, W. A., Vandenberg, R. J., Arriza, J. L., Kavanaugh, M. P., and Amara, S. G. (1995) An excitatory amino-acid transporter with properties of a ligand-gated chloride channel. *Nature* **375,** 599–603.

4. Storck, T., Schulte, S., Hofmann, K., and Stoffel, W. (1992) Structure, expression, and functional analysis of a Na(+)-dependent glutamate/aspartate transporter from rat brain. *Proc. Natl. Acad. Sci. USA* **89,** 10955–10959.

5. Tanaka, K. (1993) Cloning and expression of a glutamate transporter from mouse brain. *Neurosci. Lett.* **159,** 183–186.

6. Pines, G., Danbolt, N. C., Bjoras, M., Zhang, Y., Bendahan, A., Eide, L., et al. (1992) Cloning and expression of a rat brain L-glutamate transporter. [see comments]. [erratum appears in Nature 1992 Dec 24–31;360(6406):768]. *Nature* **360,** 464–467.

7. Kanai, Y. and Hediger, M. A. (1992) Primary structure and functional characterization of a high-affinity glutamate transporter. [see comments]. *Nature* **360,** 467–471.

8. Lehre, K. P. and Danbolt, N. C. (1998) The number of glutamate transporter subtype molecules at glutamatergic synapses: chemical and stereological quantification in young adult rat brain. *J. Neurosci.* **18,** 8751–8757.

9. Lehre, K. P., Davanger, S., and Danbolt, N. C. (1997) Localization of the glutamate transporter protein GLAST in rat retina. *Brain Res.* **744,** 129–137.

10. Yamada, K., Watanabe, M., Shibata, T., Tanaka, K., Wada, K., and Inoue, Y. (1996) EAAT4 is a post-synaptic glutamate transporter at Purkinje cell synapses. *Neuroreport* **7,** 2013–2017.

11. Furuta, A., Martin, L. J., Lin, C. L., Dykes-Hoberg, M., and Rothstein, J. D. (1997) Cellular and synaptic localization of the neuronal glutamate transporters excitatory amino acid transporter 3 and 4. *Neuroscience* **81,** 1031–1042.

12. Seal, R. P. and Amara, S. G. (1998) A reentrant loop domain in the glutamate carrier EAAT1 participates in substrate binding and translocation. *Neuron* **21,** 1487–1498.

13. Seal, R. P., Leighton, B. H., and Amara, S. G. (2000) A model for the topology of excitatory amino acid transporters determined by the extracellular accessibility of substituted cysteines. *Neuron* **25,** 695–706.

14. Grunewald, M., Bendahan, A., and Kanner, B. I. (1998) Biotinylation of single cysteine mutants of the glutamate transporter GLT-1 from rat brain reveals its unusual topology. *Neuron* **21,** 623–632.

15. Grunewald, M. and Kanner, B. I. (2000) The accessibility of a novel reentrant loop of the glutamate transporter GLT-1 is restricted by its substrate. *J. Biol. Chem.* **275,** 9684–9689.

16. Clements, J. D. (1996) Transmitter timecourse in the synaptic cleft: its role in central synaptic function. *Trends Neurosci.* **19,** 163–171.

17. Clements, J. D., Lester, R. A., Tong, G., Jahr, C. E., and Westbrook, G. L. (1992) The time course of glutamate in the synaptic cleft. *Science* **258,** 1498–1501.

18. Barbour, B., Brew, H., and Attwell, D. (1988) Electrogenic glutamate uptake in glial cells is activated by intracellular potassium. *Nature* **335,** 433–435.

19. Kanner, B. I. and Sharon, I. (1978) Active transport of L-glutamate by membrane vesicles isolated from rat brain. *Biochemistry (Mosc).* **17,** 3949–3953.

20. Stallcup, W. B., Bulloch, K., and Baetge, E. E. (1979) Coupled transport of glutamate, and sodium in a cerebellar nerve cell line. *J. Neurochem.* **32,** 57–65.

21. Zerangue, N. and Kavanaugh, M. P. (1996) Flux coupling in a neuronal glutamate transporter. *Nature* **383,** 634–637.

22. Levy, L. M., Warr, O., and Atwell, D. (1998) Stoichiometry of the glial glutamate transporter GLT-1 expressed inducibly in a chinese hamster ovary cell line selected for low endogenous Na$^+$-dependent glutamate uptake. *J. Neurosci.* **18,** 9620–9628.

23. Levy, L. M., Attwell, D., Hoover, F., Ash, J. F., Bjoras, M., and Danbolt, N. C. (1998) Inducible expression of the GLT-1 glutamate transporter in a CHO cell line selected for low endogenous glutamate uptake. [erratum appears in FEBS Lett 1998 May 1;427(1):152]. *FEBS Lett.* **422,** 339–342.

24. Brew, H. and Attwell, D. (1987) Electrogenic glutamate uptake is a major current carrier in the membrane of axolotl retinal glial cells. [erratum appears in Nature 1987 Aug 20–26;328(6132):742.]. *Nature* **327,** 707–709.

25. Billups, B., Rossi, D., and Attwell, D. (1996) Anion conductance behavior of the glutamate uptake carrier in salamander retinal glial cells. *J. Neurosci.* **16,** 6722–6731.

26. Wadiche, J. I., Amara, S. G., and Kavanaugh, M. P. (1995) Ion fluxes associated with excitatory amino acid transport. *Neuron* **15,** 721–728.

27. Grewer, C., Watzke, N., Wiessner, M., and Rauen, T. (2000) Glutamate translocation of the neuronal glutamate transporter EAAC1 occurs within milliseconds. *Proc. Natl. Acad. Sci. USA* **97,** 9706–9711.

28. Wadiche, J. I. and Kavanaugh, M. P. (1998) Macroscopic, and microscopic properties of a cloned glutamate transporter/chloride channel. *J. Neurosci.* **18,** 7650–7661.

29. Danbolt, N. C. (2001) Glutamate uptake. *Prog. Neurobiol.* **65,** 1–105.

30. Mager, S., Naeve, J., Quick, M., Labarca, C., Davidson, N., and Lester, H. A. (1993) Steady states, charge movements, and rates for a cloned GABA transporter expressed in Xenopus oocytes. *Neuron* **10,** 177–188.

31. Mager, S., Min, C., Henry, D. J., Chavkin, C., Hoffman, B. J., Davidson, N., and Lester, H. A. (1994) Conducting states of a mammalian serotonin transporter. *Neuron* **12,** 845–859.

32. Galli, A., Blakely, R. D., and DeFelice, L. J. (1996) Norepinephrine transporters have channel modes of conduction. *Proc. Natl. Acad. Sci. USA* **93,** 8671–8676.

33. Galli, A., Petersen, C. I., deBlaquiere, M., Blakely, R. D., and DeFelice, L. J. (1997) Drosophila serotonin transporters have voltage-dependent uptake coupled to a serotonin-gated ion channel. *J. Neurosci.* **17,** 3401–3411.

34. Galli, A., Jayanthi, L. D., Ramsey, I. S., Miller, J. W., Fremeau, R. T., Jr., and DeFelice, L. J. (1999) L-proline, and L-pipecolate induce enkephalin-sensitive currents in human embryonic kidney 293 cells transfected with the high-affinity mammalian brain L-proline transporter. *J. Neurosci.* **19,** 6290–6297.

35. Ingram, S. L. and Amara, S. G. (2000) Arachidonic acid stimulates a novel cocaine-sensitive cation conductance associated with the human dopamine transporter. *J. Neurosci.* **20,** 550–557.

36. Ingram, S. L., Prasad, B. M., and Amara, S. G. (2002) Dopamine transporter-mediated conductances increase excitability of midbrain dopamine neurons. *Nat. Neurosci.* **5,** 971–978.

37. Eliasof, S. and Jahr, C. E. (1996) Retinal glial cell glutamate transporter is coupled to an anionic conductance. *Proc. Natl. Acad. Sci. USA* **93,** 4153–4158.

38. Frenguelli, B. G., Blake, J. F., Brown, M. W., and Collingridge, G. L. (1991) Electrogenic uptake contributes a major component of the depolarizing action of L-glutamate in rat hippocampal slices. *Br. J. Pharmacol.* **102,** 355–362.

39. Grant, G. B. and Dowling, J. E. (1996) On bipolar cell responses in the teleost retina are generated by two distinct mechanisms. *J. Neurophysiol.* **76,** 3842–3849.

40. Sonders, M. S. and Amara, S. G. (1996) Channels in transporters. *Curr. Opin. Neurobiol.* **6,** 294–302.

41. Hebert, D. N. and Carruthers, A. (1992) Glucose transporter oligomeric structure determines transporter function. Reversible redox-dependent interconversions of tetrameric, and dimeric GLUT1. *J. Biol. Chem.* **267,** 23829–23838.

42. Jess, U., Betz, H., and Schloss, P. (1996) The membrane-bound rat serotonin transporter, SERT1, is an oligomeric protein. *FEBS Lett.* **394,** 44–46.

43. Scholze, P., Freissmuth, M., and Sitte, H. H. (2002) Mutations within an intramembrane leucine heptad repeat disrupt oligomer formation of the rat gaba transporter 1. *J. Biol. Chem.* **277,** 43682–43690.

44. Li, S., Mallory, M., Alford, M., Tanaka, S., and Masliah, E. (1997) Glutamate transporter alterations in Alzheimer disease are possibly associated with abnormal APP expression. *J. Neuropathol. Exp. Neurol.* **56,** 901–911.

45. Danbolt, N. C., Storm-Mathisen, J., and Kanner, B. I. (1992) An [Na+ + K+]coupled L-glutamate transporter purified from rat brain is located in glial cell processes. *Neuroscience* **51,** 295–310.

46. Haugeto, O., Ullensvang, K., Levy, L. M., Chaudhry, F. A., Honore, T., Nielsen, et al. (1996) Brain glutamate transporter proteins form homomultimers. *J. Biol. Chem.* **271,** 27715–27722.

47. Dehnes, Y., Chaudhry, F. A., Ullensvang, K., Lehre, K. P., Storm-Mathisen, J., and Danbolt, N. C. (1998) The glutamate transporter EAAT4 in rat cerebellar Purkinje cells: a glutamate-gated chloride channel concentrated near the synapse in parts of the dendritic membrane facing astroglia. *J. Neurosci.* **18,** 3606–3619.

48. Eskandari, S., Kreman, M., Kavanaugh, M. P., Wright, E. M., and Zampighi, G. A. (2000) Pentameric assembly of a neuronal glutamate transporter. *Proc. Natl. Acad. Sci. USA* **97,** 8641–8646.

49. Mitrovic, A. D., Plesko, F., and Vandenberg, R. J. (2001) $Zn(2^+)$ inhibits the anion conductance of the glutamate transporter EEAT4. *J. Biol. Chem.* **276,** 26071–26076.

50. Borre, L., Kavanaugh, M. P., and Kanner, B. I. (2002) Dynamic equilibrium between coupled, and uncoupled modes of a neuronal glutamate transporter. *J. Biol. Chem.* **277,** 13501–7.

51. Ryan, R. M. and Vandenberg, R. J. (2002) Distinct conformational states mediate the transport, and anion channel properties of the glutamate transporter EAAT-1. *J. Biol. Chem.* **277,** 13494–13500.

52. Billups, B. and Attwell, D. (1996) Modulation of non-vesicular glutamate release by pH. *Nature* **379,** 171–174.

53. Mort, D., Marcaggi, P., Grant, J., and Attwell, D. (2001) Effect of acute exposure to ammonia on glutamate transport in glial cells isolated from the salamander retina. *J. Neurophysiol.* **86,** 836–844.

54. Conradt, M. and Stoffel, W. (1997) Inhibition of the high-affinity brain glutamate transporter GLAST-1 via direct phosphorylation. *J. Neurochem.* **68,** 1244–1251.

55. Davis, K. E., Straff, D. J., Weinstein, E. A., Bannerman, P. G., Correale, D. M., Rothstein, J. D., and Robinson, M. B. (1998) Multiple signaling pathways regulate cell surface expression, and activity of the excitatory amino acid carrier 1 subtype of Glu transporter in C6 glioma. *J. Neurosci.* **18,** 2475–2485.

56. Lin, C. I., Orlov, I., Ruggiero, A. M., Dykes-Hoberg, M., Lee, A., Jackson, M., and Rothstein, J. D. (2001) Modulation of the neuronal glutamate transporter EAAC1 by the interacting protein GTRAP3-18. *Nature* **410,** 84–88.

57. Jackson, M., Song, W., Liu, M. Y., Jin, L., Dykes-Hoberg, M., Lin, C. I., et al. (2001) Modulation of the neuronal glutamate transporter EAAT4 by two interacting proteins. *Nature* **410,** 89–93.

58. Attwell, D., Barbour, B., and Szatkowski, M. (1993) Nonvesicular release of neurotransmitter. *Neuron* **11,** 401–407.

59. Barbour, B., Szatkowski, M., Ingledew, N., and Attwell, D. (1989) Arachidonic acid induces a prolonged inhibition of glutamate uptake into glial cells. *Nature* **342,** 918–920.

60. Zerangue, N., Arriza, J. L., Amara, S. G., and Kavanaugh, M. P. (1995) Differential modulation of human glutamate transporter subtypes by arachidonic acid. *J. Biol. Chem.* **270,** 6433–6435.

61. Trotti, D., Volterra, A., Lehre, K. P., Rossi, D., Gjesdal, O., Racagni, G., and Danbolt, N. C. (1995) Arachidonic acid inhibits a purified and reconstituted glutamate transporter directly from the water phase, and not via the phospholipid membrane. *J. Biol. Chem.* **270,** 9890–9895.

62. London, E. and Brown, D. A. (2000) Insolubility of lipids in triton X-100: physical origin, and relationship to sphingolipid/cholesterol membrane domains (rafts) Biochim. *Biophys. Acta* **1508,** 182–195.

63. Poulsen, M. V. and Vandenberg, R. J. (2001) Niflumic acid modulates uncoupled substrate-gated conductances in the human glutamate transporter EAAT4. *J. Physiol. (Lond).* **534,** 159–167.

64. Tzingounis, A. V., Lin, C. L., Rothstein, J. D., and Kavanaugh, M. P. (1998) Arachidonic acid activates a proton current in the rat glutamate transporter EAAT4. *J. Biol. Chem.* **273,** 17315–17317.

65. Fairman, W. A., Sonders, M. S., Murdoch, G. H., and Amara, S. G. (1998) Arachidonic acid elicits a substrate-gated proton current associated with the glutamate transporter EAAT4. *Nat. Neurosci.* **1,** 105–113.

66. Spiridon, M., Kamm, D., Billups, B., Mobbs, P., and Attwell, D. (1998) Modulation by zinc of the glutamate transporters in glial cells and cones isolated from the tiger salamander retina. *J. Physiol. (Lond).* **506,** 363–376.

67. Vandenberg, R. J., Mitrovic, A. D., and Johnston, G. A. (1998) Molecular basis for differential inhibition of glutamate transporter subtypes by zinc ions. *Mol. Pharmacol.* **54,** 189–196.

68. Westbrook, G. L. and Mayer, M. L. (1987) Micromolar concentrations of Zn2+ antagonize NMDA and GABA responses of hippocampal neurons. *Nature* **328,** 640–643.

69. Chen, N., Moshaver, A., and Raymond, L. A. (1997) Differential sensitivity of recombinant N-methyl-D-aspartate receptor subtypes to zinc inhibition. *Mol. Pharmacol.* **51,** 1015–1023.

70. Frederickson, C. J. (1989) Neurobiology of zinc and zinc-containing neurons. *Int. Rev. Neurobiol.* **31,** 145–238.

71. Volterra, A., Trotti, D., Tromba, C., Floridi, S., and Racagni, G. (1994) Glutamate uptake inhibition by oxygen free radicals in rat cortical astrocytes. *J. Neurosci.* **14,** 2924–2932.

72. Trotti, D., Rossi, D., Gjesdal, O., Levy, L. M., Racagni, G., Danbolt, N. C., and Volterra, A. (1996) Peroxynitrite inhibits glutamate transporter subtypes. *J. Biol. Chem.* **271,** 5976–5979.

73. Trotti, D., Rizzini, B. L., Rossi, D., Haugeto, O., Racagni, G., Danbolt, N. C., and Volterra, A. (1997) Neuronal and glial glutamate transporters possess an SH-based redox regulatory mechanism. *Eur. J. Neurosci.* **9,** 1236–1243.

74. Trotti, D., Nussberger, S., Volterra, A., and Hediger, M. A. (1997) Differential modulation of the uptake currents by redox interconversion of cysteine residues in the human neuronal glutamate transporter EAAC1. *Eur. J. Neurosci.* **9,** 2207–2212.

75. Rosen, D. R., Siddique, T., Patterson, D., Figlewicz, D. A., Sapp, P., Hentati, A., et al. (1993) Mutations in Cu/Zn superoxide dismutase gene are associated with familial amyotrophic lateral sclerosis. [see comments.] [erratum appears in Nature 1993 Jul 22;364(6435):362.]. *Nature* **362,** 59–62.

76. Trotti, D., Rolfs, A., Danbolt, N. C., Brown, R. H., Jr., and Hediger, M. A. (1999) SOD1 mutants linked to amyotrophic lateral sclerosis selectively inactivate a glial glutamate transporter. [erratum appears in Nat Neurosci 1999 Sep;2(9):848]. *Nat. Neurosci.* **2,** 427–433.

10

Characterization of the Substrate-Binding Site in GABA Transporters

Alan Sarup, Orla Miller Larsson, and Arne Schousboe

1. INTRODUCTION

γ-Aminobutyric acid (GABA) neurotransmission is characterized by the restricted expression of the GABA-synthesizing enzyme glutamate decarboxylase (GAD), the GABA receptors, and the GABA transporters in GABAergic synapses consisting of a presynaptic nerve ending, a postsynaptic entity, and surrounding astrocytes *(1)*. Additionally, the presynaptic nerve ending is characterized by the presence of the vesicles expressing the vesicular GABA transporter *(2,3)*. GABA catabolism, on the other hand, is not restricted to GABAergic synapses because the primary metabolic enzyme GABA transaminase is widely distributed not only in the central nervous system (CNS), but also in many other tissues, including the liver *(4,5)*. Among the different proteins involved in these processes, the receptors, transporters, and the GABA transaminase are all capable of recognizing, and binding the GABA molecule, each with a unique affinity and stringency regarding specificity *(6–8)*. Considering the high degree of flexibility of the GABA molecule, it is not surprising that GABA may be recognized by these different proteins in distinctly different conformations. Thus, the receptors (GABA$_A$) and the transporters have been known for several years to bind GABA in a more extended and somewhat folded conformation, respectively (*see* Krogsgaard-Larsen et al. *[9]*). Actually, in agreement with this, the GABA analogs of restricted conformation guvacine, nipecotic acid, THPO, and isoguvacine, isonipecotic acid, and THIP (Fig. 1) can be divided in two groups, each reflecting the GABA molecule in a folded and elongated conformation. The three folded analogs are specific ligands for GABA-transporters, whereas the three latter analogs specifically bind to the GABA$_A$ receptors with no affinity for the transporters. Hence, these GABA analogs have served as important lead structures in the design of new analogs interacting with either one or the other of the macromolecules constituting important functional components of the GABA synapses *(9)*. The present review is aimed at a further characterization of the molecular structures, which are recognized by the cloned GABA transporters as well as neurons and astrocytes expressing these transporters. For a number of years, it has been a challenge to explain on a rational pharmacological basis how these two cell types can

From: *Molecular Neuropharmacology: Strategies and Methods*
Edited by: A. Schousboe and H. Bräuner-Osborne © Humana Press Inc., Totowa, NJ

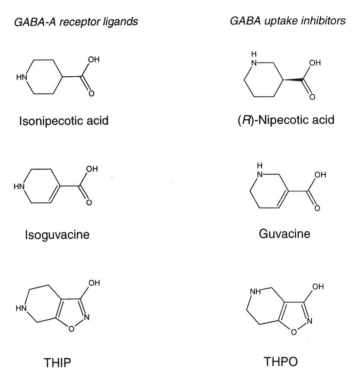

Fig. 1. Chemical structures of some common GABA mimitics of restricted comformation that bind to the substrate binding sites in the GABA$_A$ receptor complex and the GABA transporters, respectively.

exhibit completely different pharmacological properties regarding inhibitors of GABA transporters, considering the fact that at least the most abundant transporters are expressed in each of these cell types *(7,10)*. This aspect will be discussed extensively in this chapter.

2. CLONED TRANSPORTERS

The GABA transporters belong to the superfamily of Na$^+$- and Cl$^-$-dependent transporters for monoamine neurotransmitters and a number of neuroactive amino acids, including GABA, taurine, and glycine *(11–14)*. Additional homolog transporters in this superfamily of plasma membrane transporters have been cloned from various species *(see* refs. *15–17)*. Moreover, the members of this superfamily share a 12-transmembrane domain (TMD) topology with intracellular N- and C-terminals. *(16)*. It was, however, GABA transporter 1 (GAT-1) that was initially cloned from the rat brain *(11)*, and together with the cloned noradrenaline transporter *(18)*, provided the informational

Table 1
Common Nomenclature Used for Cloned GABA
Transporters From Human, Rat, and Mouse

Species	Name of homologous transporters			
Human	GAT-1	BGT-1	NC[a]	GAT-3
Rat	GAT-1	BGT-1	GAT-2	GAT-3
Mouse[b]	GAT1	GAT2	GAT3	GAT4

[a] NC, not cloned.

[b] Note that in the mouse GAT nomenclature, no hyphenation is used to separate the acronym from the number.

basis for the subsequent cloning series. Subsequent to the cloning of the rat GAT-1, highly homologous GABA transporters were cloned from different species, including humans *(19–26)*. One of these transporters has a higher affinity for the osmoregulator betaine (*N*-trimethyl-glycine) than for GABA and, hence, is termed the betaine transporter (BGT-1) *(24)*. However, in the mouse, it is called GABA transporter 2 (GAT2), creating some confusion in the nomenclature for GABA transporters among different species (Table 1). Thus, GAT2 and -3 have different meanings in the different species.

2.1. Neuronal and Astroglial Transporters

Studies of GABA transport in a variety of neural cell preparations, such as bulk-prepared cells, synaptosomes, cell lines, and primary cultures, have unequivocally demonstrated that GABA is transported with high affinity in a Na^+- and Cl^--dependent manner in both neurons and astrocytes (*see* Schousboe *[27]*). Very likely the uptake capacity is three- to fivefold higher in neurons than in astrocytes *(28)*. Quite clearly, in GABAergic nerve endings where the prevalent transporter is GAT1, the major function of the transporter is to allow recycling of GABA as a neurotransmitter by re-uptake of the released transmitter after its activation of its receptors. The exact functional role of this transport process may, however, be debated, because the inactivation of GABA as a neurotransmitter includes a receptor desensitization of the receptor, diffusion away from the receptor, binding to transporters, and finally internalization by the transmembrane transport. This final step provides the means for reutilization of GABA as a transmitter by its vesicular packaging and subsequent depolarization coupled release *(27)*. The functional role of astrocytic GABA transport, which is mediated primarily by GAT1 and to a lesser extent by the other transporters (Table 1) may be more difficult to define. However, it is clear from a number of studies (*see* Schousboe *[27]*) that this transport has a modulatory effect on the efficacy of GABAergic neurotransmission. Hence, inhibition of glial GABA uptake enhances the availability of GABA in the synaptic cleft and has an anti-seizure potential *(27,30)*. Studies of the expression of the different cloned transporters at the regional and cellular level have shown that marked differences exist with regard to the distribution of the individual transporters (*see* Gardea and Lopez-Colome *[30]* and Schousboe and Kanner *[10]*). However, the overall picture is that GAT1 is primarily neuronal, whereas GAT2-4 (mouse) may be primarily glial, although a clear selectivity in the distribution is probably not found. This lack of cellular specificity may be an important factor in understanding the enigmatic pharmacology of neuronal and glial GABA-transport (*see* below).

3. STRUCTURAL CLASSES OF GABA TRANSPORT INHIBITORS

As mentioned earlier, the cyclic structurally restricted GABA analogs guvacine, nipecotic acid, and THPO (31) have been important lead structures in the design of GABA analogs specifically recognizing the GABA carriers. In a number of cases, such analogs, which inhibit GABA uptake, are also substrates (32–34), but generally more lipophilic and bulky analogs are not transported by the carrier (35–37). Exemptions to this rule are the bicyclic isoxazoles THPO, THAO, and *exo*-THPO, which are not transported, despite their relatively high solubility in water (35–36,38). Table 2 provides a summary of GABA transport inhibitors with regard to inhibition kinetics and ability to act as a substrate for GAT1. It should be noted that among the GABA analogs that cannot be transported, differences exist with regard to the kinetic type of inhibition. Hence, some of the compounds exhibit competitive inhibition, whereas others are noncompetitive. The competitive inhibitors are likely to bind to the substrate (GABA) binding site, whereas the other types of inhibitors have a more complex mode of interaction with sites in addition to the GABA recognition site. Although it may be easy to understand why an inhibitor that does not bind to the substrate binding site cannot be transported, it may be more difficult to understand why an inhibitor that binds to this site cannot be transported. It does, however, indicate that binding to the substrate recognition site *per se* does not allow the transport cycle to be completed. Inspection of the molecular structures of the inhibitors that are competitive and transportable and those that are competitive and nontransportable does not *a priori* provide any clues as to the requirements with regard to molecular structure and ability to be transported. Thus, compounds such as the **N**-diphenylbutenyl (DPB) analog of nipecotic acid or guvacine as well as THPO and its DPB analogs are all competitive inhibitors, but are not transported. On the other hand, nipecotic acid is transported. In the latter case, the lipophilic side chain may explain why the compound becomes a nontransportable substrate, but in the case of THPO, which is structurally similar to nipecotic acid, it is not easily understood why it cannot be transported. However, molecular size may well be a determining factor and the size of nipecotic acid or guvacine may constitute the upper limit that is accepted by the transport machinery.

Similarly, a comparison of the molecular structures of competitive, nonsubstrate inhibitors and noncompetitive inhibitors may provide information about the requirements in molecular terms for binding to the substrate binding site. Again, it is not simple to provide the distinct molecular features that are required for binding. Thus, rather small changes in the lipophilic side chain of the derivatives of nipecotic acid may change an analog from a competitive to a noncompetitive inhibitor, i.e., from a compound capable of binding to the GABA recognition site to a compound not capable of binding to strictly to this site. It may still be an open question whether these types of analogs may have binding sites different from that of GABA. Studies of binding of tiagabine to GAT1 (39) have provided some clues to this issue. It may be concluded from these studies that tiagabine, which is not transported by GAT1, but which is a competitive inhibitor of GABA transport (Table 2), binds not only to the GABA transport site but also to a different site. In this context, it is interesting to note that binding of these lipophilic GABA analogs such as *N*-DPB-guvacine leads to a significant conformational change of the transporter. This is shown by the finding that the accessability of cystein-399 located in the intracellular loop between transmembrane domain 8 and 9 to the –SH reagent 2-amino-ethyl-methanethiosulfonate is strongly affected by binding of *N*-DPB-guvacine (40). Further insights into the significance of the

Table 2
Kinetic Characteristics of Inhibitors of GAT1- or Neuronal-Mediated GABA Uptake[a]

Inhibitor	Structure	$K_i(\mu M)$	Inhibition type	Substrate
(*R*)-nipecotic acid		70	Competitive	Yes
DPB-(*R*)-nipecotic acid		2.0[b]	Competitive	No
Tiagabine		0.11	Competitive	No
THPO		2000	Competitive	No
N-DPB-THPO		50	Competitive	ND
Exo-THPO		620	Competitive	ND
N-methyl-*exo*-THPO		275	Competitive	ND

(continues)

GABA-mimetic moiety and the structure of the "lipophilic" part of the GABA analogs used as inhibitors of GABA transport have been provided by a series of studies using derivatives of either nipecotic acid or THPO/*exo*-THPO. Interestingly, nipecotic acid inhibits GAT1, -3, and -4, but not GAT2 *(10)*. Although THPO and *exo*-THPO are less potent than nipecotic acid, these isoxazoles appear less discriminating between the four cloned mouse GABA transporters *(6,41)*. The DPB analogs of these GABA mimetic parent compounds are always about 100-fold more potent as inhibitors, but they generally reflect the subtype

Table 2
(Continued)

Inhibitor	Structure	$K_i(\mu M)$	Inhibition type	Substrate
4-Phenylbut-1-yl-*exo*-THPO		100	Noncompetitive	ND
N-DPB *N*-methyl-*exo*-THPO		4.2	Noncompetitive	ND

[a] Data adapted with permission from Larsson et al. *(34,35)*, Thomsen et al. *(55)*, Bolvig et al. *(41)*, and Sarup et al. *(6)*.
[b] Cultured neurons.
ND = not determined.

selectivity seen for the parent compounds. Hence, based on these observations, one would conclude that the "lipophilic" side chain plays no role with regard to interference with the substrate binding site other than quite dramatically enhancing the affinity. It is interesting that a comparison of the lipophilicity of the side chains and the corresponding affinity of the analogs with the GABA binding site leads to the conclusion that the lipophilicity and size of the side chain plays a significant role for the affinity (Table 3). Larger and more lipophilic *N*-substitutions lead to significant increases in affinity (lower IC_{50} values) regardless of the GABA-mimetic parent compound.

Electronegative atoms in the side chain of the diaryl analogs of nipecotic acid and guvacine did not dramatically alter the nanomolar potency for these compounds as inhibitors of GABA uptake in rat brain synaptosomes *(42)*. However, it cannot be excluded that these compounds may exhibit significant activities on GAT2 to 4, although GAT1-mediated transport is likely to represent the majority of the transmembrane GABA flux in the synaptosomal preperation. In agreement with these observations, the (*R*)-1-(4,4-bis (3-methyl-2-thienyl)-3-butenyl) and the *para*-substituted trifluromethyl-biphenyl-methoxyethyl analog of nipecotic acid and guvacine, Tiagabine and CI966, respectively, are selective potent inhibitors of GAT1-mediated GABA uptake *(43)*. Interestingly, the *para*-methoxy-substituted triarylnipecotic acid analog, SNAP-5114, exhibits a moderate selectivity towards GAT4-mediated GABA uptake *(44)*. The properties of this compound indicate that modifications in the distal part of the lipophillic side chain of the aforementioned analogs could provide novel pharmacological properties for inhibition of GABA uptake exhibited by others transporters than GAT1. However, the complexity of the inhibition kinetics exhibited by some of the GABA inhibitors may constitute a major challenge in developing pharmacophore models for inhibition of the individual GABA

Table 3
Correlation Between Lipophilicity and Size of the Side Chain on Inhibition of GAT1- or Neuronal-Mediated GABA Uptake[a]

Parent compound	Side Chain	IC$_{50}$ (μM)
Nipecotic acid	–	20
		0.2[b]
		0.8
THPO	–	1300
		30
Exo-THPO	–	1000
		450
		320
		>500
		550
		190
		7
		6

[a] Data from Larsson et al. *(37)*, Bolvig et al. *(41)*, and Sarup et al. *(6)*.
[b] Cultured neurons.

transporters *(6)*. Moreover, the recent demonstration of homooligomers of GAT1 makes such models even more complicated *(45)*. Interestingly, mutations in an intramembrane nonpolar heptad repeat of leucine residues disrupts oligomer formation *(46)*. It is likely that lipophilic GABA uptake inhibitors may interfere with such complex formation.

It has been demonstrated that it is not only the lipophilicity and bulkyness of the ring structures (DPB, thiophene rings, etc.) that may influence the affinity for the transporter. It is clear that the nature of the spacer or linker between these groups and the nitrogen atom in the GABA-mimetic part of the inhibitor play an important role *(42,47)*. Thus, the affinity depends on the presence of a double bond (butenyl vs butyl), the presence of an oxygen atom (vinyl ether), or a nitrogen atom plus an oxygen atom (oxime). It appears that the double bond is significant, but it can be replaced by an oxygen atom (vinyl ether). Moreover, generally analogs containing the vinyl ether or the oxime linker show similar affinities for the transporter *(42)*.

Common to the linkers described earlier is the electronegative property conferred by the double bond or the ether and oxime bonds. Therefore, it has been concluded that this property of the linker plays an important role for the binding to the substrate recognition site of the GABA transporter in synaptosomes *(42)*, which is likely to reflect primarily GAT1. This is compatible with the conclusions from a number of mutagenesis studies in which it was demonstrated that the positively charged amino acid residues tryptophane-68 and -222 and arginine-69 are essential for binding and transport of GABA and sodium *(48,49)*. Additionally, tyrosine-140 has been shown to play a significant role in GABA binding because this is essentially abolished by mutating the tyrosine to a phenylalanine *(50)*. This suggests that the hydroxyl group of tyrosine may participate in hydrogen bonding *via* water molecules in analogy with ligand binding in the AMPA receptor subunits GluR1 and GluR4 *(51)*. Recently, it has been demonstrated that mutation of glycine-63 in the TMD1, to serine or cysteine, resulted in a defective GABA transport and transport currents *(52)*. It is tempting to speculate whether glycine-63 could be a part of the substrate binding pocket in GAT1. Along this line, it is worth noting that the conserved aspartate-98 in the bioamine transporters is important for binding of their endogenous ligands *(53)*. The aspartate-98 is conserved in the bioamine transporters and is positioned analogous to the GAT1 glycine-63; hence, a similar role could be ascribed to the glycine.

It should also be mentioned that a longer linker containing a nitrogen atom has been shown to lead to lower affinities for the aryl lipophilic analogs of the above mentioned GABA mimetics *(47)*. In addition, modifications of one of the two aryl groups in *N*-Ω asymmetrically substituted nipecotic acid analogs (oxime and vinylether linkers) seem to have a moderate effect on the potency of inhibition of synaptosomal GABA uptake *(54)*. Novel sets of lipophilic molecules that do not contain the nipecotic acid moiety, have been shown to have higher affinity for GAT2 and GAT4 than for GAT1 and GAT3 *(55)*. This, together with the previous demonstration that SNAP-5114 has higher affinity for GAT4 than for the other cloned transporters *(44)*, shows that lipophilic, bulky molecules that do not contain a strict GABA mimetic moiety can be used to distinguish between the different cloned GABA transporters.

4. PHARMACOLOGY OF NEURONAL AND ASTROGLIAL GABA UPTAKE

In the past, neuronal and astroglia GABA transport were distinguished using β-alanine as an inhibitor of glial GABA uptake and diaminobutyric acid as inhibitor of neuronal GABA uptake *(56)*. However, subsequent kinetic analyses of these GABA analogs in primary cultures of astrocytes and GABAergic cerebral cortical neurons *(57,58)* have revealed questions about the selectivity of these compounds. In particular, it was surprising that β-alanine was found not to be a competitive inhibitor/substrate in either cell preparation, whereas it was a perfect substrate of the taurine carrier expressed in both cell types *(58)*.

During the last two decades, a large number of GABA analogs based on *N*-substitutions of the GABA mimetics nipecotic acid, guvacine, THPO, THAO, and *exo*-THPO

(Fig. 1) using side chains of increasing size and lipophilicity have been investigated as inhibitors of neuronal and astroglial GABA uptake. The results of some of these studies have been summarized in Table 4. Among the parent GABA-mimetic compounds, only the bicyclic isoxazoles and in particular *exo*-THPO exhibit somewhat higher affinity for astroglial GABA transport than for its neuronal counter part. Moreover, it is interesting that this marginal glial selectivity disappears completely in the DPB analogs of all of these compounds while at the same time the potency as inhibitors increases by a factor of 10 or more. Interestingly *N*-methyl-*exo*-THPO and *N*-acetyloxyethyl-*exo*-THPO turned out to be among the most highly glial selective GABA transport inhibitors reported to date *(6,59–60)*. In this context, it is puzzling that the closely related *N*-ethyl- and *N*-2-hydroxyethyl-exo-THPO analogs show no cell selectivity at all. Thus, one cannot at this stage point to any molecular, structural background to rationally explain the structure-activity difference between the neuronal and astroglial GABA transport. The fact that both of the aforementioned molecules specifically inhibit GAT1 and exhibit no affinity for the other cloned GABA transporters *(59,60)* makes the situation more puzzling. Thus, in order to explain this ability of certain GAT1 inhibitors to preferentially inhibit astroglial GABA uptake, it may be hypothesized that a molecular interaction among different GABA transporters expressed in these cells may influence the pharmacological characteristics of GAT1 in a thus far not well-understood manner *(60)*. Alternatively, GABA-transporters not yet cloned and with unknown pharmacological characteristics may exist *(10)*. Attempting to solve this puzzle will be a challenge for future studies.

5. GABA TRANSPORT INHIBITORS AS ANTICONVULSANTS

From a mechanistic point of view based on the basic mode of action of GABAergic synapses in which GABA appears to be both recycled as a neurotransmitter and produced by *de novo* synthesis from glutamate catalyzed by glutamate decarboxylase *(61)*, it was hypothesized that inhibition of GABA uptake might lead to an increased availability of GABA in the synaptic cleft *(62–65)*. Hence, it was shown that the prodrug of nipecotic acid, its pivoloylic acid ester, had anticonvulsant activity in audiogenic seizure prone DBA mice after intraperitoneal administration leading to production of nipecotic acid intracerebrally *via* hydrolysis catalyzed by unspecific esterase activity *(64)*. It was also shown that THPO administered in chicks with a less well-developed blood-brain barrier (BBB) that allowed penetration of THPO into the brain protected against sound-induced seizures *(66)*. At the same time, it was demonstrated that the lipophilic DPB analogs of nipecotic acid and guvacine capable of penetrating the BBB were highly potent and efficient anticonvulsant compounds *(67,68)*. This subsequently led to the development of tiagabine, a potent inhibitor of GAT1, as an antiepileptic drug *(69–71)*. Based on the notion that uptake of GABA into astroglial cells would prevent GABA from being directly re-utilized as a neurotransmitter *(62,72)*, a number of the glial-selective GABA uptake inhibitors have been assessed for anticonvulsant activity in animal models *(59,73–75)*. It was recently demonstrated that a better correlation exists between anticonvulsant activity and inhibition of astroglial GABA uptake than between anticonvulsant activity and inhibition of neuronal GABA transport *(60)*. Based on this concept, it seems attractive to focus future studies on gaining a better understanding of the molecular pharmacological properties of astroglial GABA transporters. This includes an elu-

Table 4
IC$_{50}$ Values of a Series of *N*-Substituted Analogs of the GABA Mimetics Shown in Fig. 1 and *exo*-THPO analogs (Tables 2 and 3)[a]

GABA mimetic	N-substitution −R$_1$	−R$_2$	IC$_{50}$ (μM) Neurons	Astrocytes
GABA			8[b]	32[b]
(**R**)-Nipecotic acid			12	16
N-DPB-Nipecotic acid	[structure]		1.3	2.0
Tiagabine	[structure]		0.4	0.2
Guvacine			32	29
N-DPB-guvacine	[structure]		4.9	4.2
THPO			487	258
N-methyl-*exo*-THPO	[structure]		38	26
***Exo*-THPO**			800	200
N-methyl-*exo*-THPO	[CH$_3$]		405	48
N, N-methyl-*exo*-THPO	[CH$_3$]	[CH$_3$]	>1000	>1000
N-ethyl-*exo*-THPO	[CH$_3$]		390	301
N-2-hydroxyethyl-*exo*-THPO	[OH]		300	200
N-acetyloxyethyl-*exo*-THPO	[structure]		200	18
N-allyl-*exo*-THPO	[structure]		220	73
N-4-phenylbutyl-*exo*-THPO	[structure]		100	15
N-DPB-*exo*-THPO	[structure]		1.4	0.6
N-DPB-*N*-methyl-*exo*-THPO	[structure]	[CH$_3$]	5	2

[a] Data from Larsson et al. *(35,36)*, Falch et al. *(59)*, White et al. *(60)*, and Sarup et al. *(6)*.
[b] K$_m$ value.

cidation of the molecular identity of the transporters expressed, as well as a characterization of the mechanisms regulating their expression *(27)*.

6. CONCLUDING REMARKS

The GABA transporters constitute important entities in the central nervous system (CNS), where they control and terminate the pivotal inhibitory GABAergic neurotransmission. Since the early 1970s, major successful steps have been undertaken to develop and synthesize inhibitors of GABA uptake. However, in the 10–15 yr before GATs were cloned, the general view was that GABA uptake systems existed in both neurons and astrocytes. Moreover, these GABA uptake systems differ in their pharmacology, and even today some of these differences are not fully understood. With the advent of cloning of GATs, the complexity of GABA uptake became more clear, and the development of new drug candidates selectively targeting GABA transport was rapidly accelerated. Because neuronal/GAT1-mediated uptake may be quantitatively the most important in most brain regions, is it not surprising that focus was on this transporter. Today, potent and selective GAT1 inhibitors are available. The structural requirements for ligands acting as inhibitors of GABA uptake are not straightforward. The GABA mimetic part, the linker region, and the lipophilic aromatic substituents clearly have profound effects on selectivity and potency as inhibitors of these carriers, with the effects essentially restricted to GAT1. These effects can be explained partly by interactions with identified charged amino acid residues in the GAT1 transporter. Surprisingly, only a minor effect has been identified for GAT2 to 4, in spite of the high amino acid sequence homology among the GATs. However, development of pharmacophore models for these particular inhibitors is a complex task and is still at an early stage. Although the substrate binding site has not been fully characterized, important information on the biophysics of the transport process may be helpful in the context of understanding ligand transporter interactions. Recent evidence has revealed aspects of the higher organization of the transporters (i.e., oligomerization, trafficking, regulation). Finally, additional reports underscoring the functional importance of astroglial GABA uptake have recently been published. Thus it is now important that steps be taken towards development of inhibitors that selectively target GAT2 to-4 or astroglial GABA uptake in general. This research may provide essential new insights into the physiological functions and potentially therapeutical uses of the GABA transporter subtypes.

ACKNOWLEDGMENTS

The expert secretarial assistance of Ms. Hanne Danø is greatly appreciated. This work has been supported by grants from the Danish MRC (52-00-1011) and the Lundbeck Foundation.

REFERENCES

1. Waagepetersen, H. S., Sonnewald, U., Schousboe, A. (1999) The GABA paradox: multiple roles as metabolite, neurotransmitter, and neurodifferentiative agent. *J. Neurochem.* **73**, 1335–1342.
2. Sagne, C., El Mestikawy, S., Isambert, M. F., Hamon, M., Henry, J. P., Giros, B., and Gasnier, B. (1997) Cloning of a functional vesicular GABA and glycine transporter by screening of genome databases. *FEBS Lett.* **417**, 177–183.

3. McIntire, S. L., Reimer, R. J., Schuske, K., Edwards, R. H., and Jørgensen, E. M. (1997) Identification and characterization of the vesicular GABA transporter. *Nature* **389,** 870–876.
4. Schousboe, A., Wu, J. Y., and Roberts, E. (1973) Purification and characterization of the 4-aminobutyrate-2,ketoglutarate transaminase from mouse brain. *Biochemistry* **12,** 2868–2873.
5. Schousboe, A., Saito, K., and Wu, J.-Y. (1980) Characterization and cellular and subcellular localization of GABA-transaminase. *Brain Res. Bull.* **5,** 71–76.
6. Sarup, A., Larsson, O. M., Bolvig, T., Frølund, B., Krogsgaard-Larsen, P., and Schousboe, A. (2003) Effects of 3-hydroxy-4-amino-4,5,6,7-tetrahydro-1,2-benzisoxazol (*exo*-THPO) and its N-substituted analogs on GABA transport in cultured neurons and astrocytes and by the four cloned mouse GABA transporters. *Neurochem. Int.* **43,** 445–451.
7. Sarup, A., Larsson, O. M., and Schousboe, A. (2003) GABA transporters and GABA-transaminase as drug targets, in *Current Drug Targets: CNS & Neurological Disorders,* **2,** 269–277.
8. Frølund, B., Jørgensen, A. T., Liljefors, T., Mortensen, M., and Krogsgaard-Larsen, P. (2003) Characterization of the GABA-A receptor recognition site through ligand design and pharmacophore modelling. *Published in this volume.*
9. Krogsgaard-Larsen, P., Frølund, B., and Frydenvang, K. (2000) GABA uptake inhibitors. Design, molecular pharmacology and therapeutic aspects. *Curr. Pharm. Des.* **6,** 1193–1209.
10. Schousboe, A. and Kanner, B. I. (2002) GABA transporters: functional and pharmacological properties, in *Glutamate and GABA Receptors and Transporters: Structure, Function and Pharmacology* (Egebjerg, J., Schousboe, A., and Krogsgaard-Larsen, P., eds.), Francis & Taylor, London, pp. 337–349.
11. Guastella, J., Nelson, N., Nelson, H., Czyzyk, L., Keynan, S., Miedel, M., et al. (1990) Cloning and expression of a rat brain GABA transporter. *Science* **249,** 1303–1306.
12. Liu, Q. R., Nelson, N., Mandiyan, S., Lopéz-Corcuera, B., and Nelson, N. (1992) Cloning and expression of a cDNA encoding the transporter of taurine and beta-alanine in mouse brain. *FEBS Lett.* **305,** 110–114.
13. Liu, Q. R., Lopéz-Corcuera, B., Mandiyan, S., Nelson, H., and Nelson, N. (1993) Cloning and expression of a spinal cord- and brain-specific glycine transporter with novel structural features. *J. Biol. Chem.* **268,** 22802–22808.
14. Smith, K. E., Borden, L. A., Wang, C. H., Hartig, P. R., Branchek, T. A., and Weinshank, R. L. (1992) Cloning and expression of a high affinity taurine transporter from rat brain. *Mol. Pharmacol.* **42,** 563–569.
15. Amara, S. G. and Kuhar, M. J. (1993) Neurotransmitter transporters: recent progress. *Annu. Rev. Neurosci.* **16,** 73–93.
16. Nelson, N. (1998) The family of Na$^+$/Cl$^-$ neurotransmitter transporters. *J. Neurochem.* **71,** 1785–1803.
17. Torres, G. E., Gainetdinov, R. R., and Caron, M. G. (2003) Plasma membrane monoamine transporters: structure, regulation and function. *Nat. Rev. Neurosci.* **4,** 13–25.
18. Pacholczyk, T., Blakely, R. D., and Amara, S. G. (1991) Expression cloning of a cocaine- and antidepressant-sensitive human noradrenaline transporter. *Nature* **350,** 350–354.
19. Nelson, H., Mandiyan, S., and Nelson, N. (1990) Cloning of the human brain GABA transporter. *FEBS Lett.* **269,** 181–184.
20. Lopéz-Corcuera, B., Liu, Q. R., Mandiyan, S., Nelson, H., and Nelson, N. (1992) Cloning and expression of a cDNA encoding the transporter of taurine and beta-alanine in mouse brain. *J. Biol. Chem.* **267,** 17491–17493.
21. Borden, L. A., Smith, K. E., Hartig, P. R., Branchek, T. A., and Weinshank, R. L. (1992) Molecular heterogeneity of the gamma-aminobutyric acid (GABA) transport system. *Receptors Channels* **3,** 129–146.
22. Borden, L. A., Smith, K. E., Hartig, P. R., Branchek, T. A., and Weinshank, R. L. (1992) Cloning of two novel high affinity GABA transporters from rat brain. *J. Biol. Chem.* **267,** 21098–21104.

23. Borden, L. A., Dhar, T. G., Smith, K. E., Branchek, T. A., Gluchowski, C., and Weinshank, R. L. (1994) Cloning of the human homologue of the GABA transporter GAT-3 and identification of a novel inhibitor with selectivity for this site. *Receptors Chann.* **2,** 207–213.

24. Borden, L. A., Smith, K. E., Gustafson, E. L., Branchek, T. A., and Weinshank, R. L. (1995) Cloning and expression of a betaine/GABA transporter from human brain. *J. Neurochem.* **64,** 977–984.

25. Liu, Q. R., Lopéz-Corcuera, B., Nelson, H., Mandiyan, S., and Nelson, N. (1992) Cloning and expression of a cDNA encoding the transporter of taurine and beta-alanine in mouse brain. *Proc. Natl. Acad. Sci. USA* **89,** 12145–12149.

26. Liu, Q. R., Lopéz-Corcuera, B., Mandiyan, S., Nelson, H., and Nelson, N. (1993) Molecular characterization of four pharmacologically distinct gamma-aminobutyric acid transporters in mouse brain. *J. Biol. Chem.* **268,** 2106–2112.

27. Schousboe, A. (2003) Role of astrocytes in maintenance and modulation of glutamatergic and GABAergic neurotransmission. *Neurochem. Res.* **28,** 347–352.

28. Hertz, L. and Schousboe, A. (1987) Primary cultures of GABAergic, and glutamatergic neurons as model systems to study neurotransmitter functions. I. Differentiated cells, in *Model Systems of Development and Aging of the Nervous System* (Vernadakis, A., Privat, A., Lauder, J. M., Timiras, P. S., and Giacobini, E., eds.), M. Nijhoff Publishing Company, Boston, MA, pp. 19–31.

29. Schousboe, A. (2000) Pharmacological and functional characterization of astrocytic GABA transport: a short review. *Neurochem Res.* **25,** 1241–1244.

30. Gadea, A. and Lopez-Colome, A. M. (2001) Glial transporters for glutamate, glycine and GABA: II. GABA transporters. *J. Neurosci. Res.* **63,** 461–468.

31. Krogsgaard-Larsen, P. and Johnston, G. A. R. (1975) Inhibition of GABA uptake in rat brain slices by nipecotic acid, various isoxazoles and related compounds. *J. Neurochem.* **25,** 797–802.

32. Johnston, G. A. R., Krogsgaard-Larsen, P., and Stephanson, A. L. (1975) Betel nut constituents as inhibitors of gamma-aminobutyric acid uptake. *Nature.* **258,** 627–628.

33. Johnston, G. A. R., Stephanson, A. L., and Twitchin, B. (1976) Uptake and release of nipecotic acid by rat brain slices. *J. Neurochem.* **26,** 83–87.

34. Larsson, O. M., Krogsgaard-Larsen, P., and Schousboe, A. (1980) High-affinity uptake of (RS)-nipecotic acid in astrocytes cultured from mouse brain. Comparison with GABA transport. *J. Neurochem.* **34,** 970–977.

35. Larsson, O. M., Krogsgaard-Larsen, P., and Schousboe, A. (1985) Characterization of the uptake of GABA, nipecotic acid and cis-4-OH-nipecotic acid in cultured neurons and astrocytes. *Neurochem. Int.* **7,** 853–860.

36. Larsson, O. M., Falch, E., Krogsgaard-Larsen, P., and Schousboe, A. (1988) Kinetic characterization of inhibition of γ-aminobutyric acid uptake into cultured neurons and astrocytes by 4,4-diphenyl-3-butenyl derivatives of nipecotic acid and guvacine. *J. Neurochem.* **50,** 818–823.

37. Larsson, O. M., Falch, E., Schousboe, A., and Krogsgaard-Larsen, P. (1991) GABA uptake inhibitors: Kinetics and molecular pharmacology, in *Presynaptic Receptors, and Neuronal Transporters* (Langer, S. Z., Galzin, A. M., and Costentin, J., eds.), Pergamon Press, Oxford, UK, pp. 197–200.

38. Juhász, G., Kékesi, K. A., Nyitrai, G., Dobolyi, A., Krogsgaard-Larsen, P., and Schousboe, A. (1997) Differential effects of nipecotic acid, and 4,5,6,7-tetrahydroisoxazolo[4,5-c]pyridin-3-ol on extracellular gamma-aminobutyrate levels in rat thalamus. *Eur. J. Pharmacol.* **331,** 139–144.

39. Mager, S., Kleinberger-Doron, N., Keshet, G. I., Davidson, N., Kanner, B. I., and Lester, H. A. (1996) Ion binding and permeation at the GABA transporter GAT1. *J Neurosci.* **16,** 5405–5414.

40. Golovanevsky, V. and Kanner, B. I. (1999) The reactivity of the gamma-aminobutyric acid transporter GAT-1 toward sulfhydryl reagents is conformationally sensitive. Identification of a major target residue. *J. Biol. Chem.* **274,** 23020–23026.

41. Bolvig, T., Larsson, O. M., Pickering, D., Nelson, N., Falch, E., Krogsgaard-Larsen, P., and Schousboe, A. (1999) Action of bicyclic isoxazole GABA analogues on GABA transporters and its relation to anticonvulsant activity. *Eur. J. Pharmacol.* **375,** 367–374.

42. Knutsen, L. J., Andersen, K. E., Lau, J., et al. (1999) Synthesis of novel GABA uptake inhibitors. 3. Diaryloxime and diarylvinyl ether derivatives of nipecotic acid and guvacine as anticonvulsant agents. *J. Med. Chem.* **42,** 3447–3462.

43. Borden, L. A., Dhar, T. G., Smith, K. E., Weinshank, R. L., Branchek, T. A., and Gluchowski, C. (1994) Tiagabine, SK&F 89976-A, CI-966, and NNC-711 are selective for the cloned GABA transporter GAT-1. *Eur. J. Pharmacol.* **269,** 219–224.

44. Dhar, T. G., Borden, L. A., Tyagarajan, S., Smith, K. E., Branchek, T. A., Weinshank, R. L., and Gluchowski, C. (1994) Design, synthesis and evaluation of substituted triarylnipecotic acid derivatives as GABA uptake inhibitors: identification of a ligand with moderate affinity and selectivity for the cloned human GABA transporter GAT-3. *J. Med. Chem.* **37,** 2334–2342.

45. Schmid, J. A., Scholze, P., Kudlacek, O., Freissmuth, M., Singer, E. A., and Sitte, H. H. (2001) Oligomerization of the human serotonin transporter and of the rat GABA transporter 1 visualized by fluorescence resonance energy transfer microscopy in living cells. *J. Biol. Chem.* **276,** 3805–3810.

46. Scholze, P., Freissmuth, M., and Sitte, H. H. (2002) Mutations within an intramembrane leucine heptad repeat disrupt oligomer formation of the rat GABA transporter 1. *J. Biol. Chem.* **277,** 43682–43690.

47. Andersen, K. E., Sørensen, J. L., Huusfeldt, P. O., Knutsen, L. J., Lau, J., Lundt, B. F., et al. (1999) Synthesis of novel GABA uptake inhibitors. 4. Bioisosteric transformation and successive optimization of known GABA uptake inhibitors leading to a series of potent anticonvulsant drug candidates. *J. Med. Chem.* **42,** 4281–4291.

48. Kleinberger-Doron, N. and Kanner, B. I. (1994) Identification of tryptophan residues critical for the function and targeting of the gamma-aminobutyric acid transporter (subtype A). *J. Biol. Chem.* **269,** 3063–3067.

49. Pantanowitz, S., Bendahan, A., and Kanner, B. I. (1993) Only one of the charged amino acids located in the transmembrane alpha-helices of the gamma-aminobutyric acid transporter (subtype A) is essential for its activity. *J. Biol. Chem.* **268,** 3222–3225.

50. Bismuth, Y., Kavanaugh, M. P., and Kanner, B. I. (1997) Tyrosine 140 of the gamma-aminobutyric acid transporter GAT-1 plays a critical role in neurotransmitter recognition. *J. Biol. Chem.* **272,** 16096–16102.

51. Banke, T. G., Greenwood, J. R., Christensen, J. K., Liljefors, T., Traynelis, S. F., Schousboe, A., and Pickering, D. S. (2001) Identification of amino acid residues in GluR1 responsible for ligand binding and desensitization. *J. Neurosci.* **21,** 3052–3062.

52. Kanner, B. I. (2003) Transmembrane domain I of the gamma-aminobutyric acid transporter GAT-1 plays a crucial role in the transition between cation leak and transport modes. *J. Biol. Chem.* **278,** 3705–3712.

53. Barker, E. L., Moore, K. R., Rakhshan, F., and Blakely, R. D. (1999) Transmembrane domain I contributes to the permeation pathway for serotonin and ions in the serotonin transporter. *J. Neurosci.* **19,** 4705–4717.

54. Andersen, K. E., Lau, J., Lundt, B. F., Petersen, H., Huusfeldt, P. O., Suzdak, P. D., and Swedberg, M. D. (2001) Synthesis of novel GABA uptake inhibitors. Part 6: preparation and evaluation of N-Omega asymmetrically substituted nipecotic acid derivatives. *Bioorg. Med. Chem.* **9,** 2773–2785.

55. Thomsen, C., Sørensen, P. O., and Egebjerg, J. (1997) 1-(3-(9H-carbazol-9-yl)-1- propyl)-4-(2-methoxyphenyl)-4-piperidinol, a novel subtype selective inhibitor of the mouse type II GABA-transporter. *Br. J. Pharmacol.* **120,** 983–985.

56. Iversen, L. L. and Kelly, J. S. (1975) Uptake and metabolism of gamma-aminobutyric acid by neurones and glial cells. Biochem. *Pharmacol.* **24,** 933–938.
57. Larsson, O. M., Johnston, G. A. R., and Schousboe, A. (1983) Differences in uptake kinetics of cis-3-aminocyclohexane carboxylic acid into neurons and astrocytes in primary cultures. *Brain Res.* **260,** 279–285.
58. Larsson, O. M., Griffiths, R., Allen, I. C., and Schousboe, A. (1986) Mutual inhibition kinetic analysis of gamma-aminobutyric acid, taurine and beta-alanine high-affinity transport into neurons, and astrocytes: evidence for similarity between the taurine and beta-alanine carriers in both cell types. *J. Neurochem.* **47,** 426–432.
59. Falch, E., Perregaard, J., Frølund, B., et al. (1999) Selective inhibitors of glial GABA uptake: synthesis, absolute stereochemistry, and pharmacology of the enantiomers of 3-hydroxy-4-amino-4,5,6,7-tetrahydro-1,2-benzisoxazole (*exo*-THPO) and analogues. *J. Med. Chem.* **42,** 5402–5414.
60. White, H. S., Sarup, A., Bolvig, T., et al. (2002) Correlation between anticonvulsant activity and inhibitory action on glial gamma-aminobutyric acid uptake of the highly selective mouse gamma-aminobutyric acid transporter 1 inhibitor 3-hydroxy-4-amino-4,5,6,7-tetrahydro-1,2-benzisoxazole and its N-alkylated analogs. *J. Pharmacol. Exp. Ther.* **302,** 636–644.
61. Roberts, E. (1974) Disinhibition as an organizing principle in the nervous system. The role of gamma-aminobutyric acid. *Adv. Neurol.* **5,** 127–143.
62. Schousboe, A. (1979) Effects of GABA-analogues on the high-affinity uptake of GABA in astrocytes in primary cultures. *Adv. Exp. Med. Biol.* **123,** 219–237.
63. Wood, J. D., Schousboe, A., and Krogsgaard-Larsen, P. (1980) In vivo changes in the GABA content of nerve endings (synaptosomes) induced by inhibitors of GABA uptake. *Neuropharmacology* **19,** 1149–1152.
64. Meldrum, B. S., Croucher, M. J., and Krogsgaard-Larsen, P. (1981) GABA-uptake inhibitors as anticonvulsant agents, in *Problems in GABA Research: From Brain to Bacteria* (Okada, Y. and Roberts, E., eds.) Excerpta Medica, Amsterdam, pp. 182–191.
65. Krogsgaard-Larsen, P., Falch, E., Larsson, O. M., and Schousboe, A. (1987) GABA uptake inhibitors: relevance to antiepileptic drug research. *Epilepsy Res.* **1,** 77–93.
66. Wood, J. D., Johnson, D. D., Krogsgaard-Larsen, P., and Schousboe, A. (1983) Anticonvulsant activity of the glial-selective GABA uptake inhibitor, THPO. *Neuropharmacology* **22,** 139–142.
67. Ali, F. E., Bondinell, W. E., Dandridge, P. A., et al. (1985) Orally active and potent inhibitors of gamma-aminobutyric acid uptake. *J. Med. Chem.* **28,** 653–660.
68. Yunger, L. M., Fowler, P. J., Zarevics, P., and Setler, P. E. (1984) Novel inhibitors of gamma-aminobutyric acid (GABA) uptake: anticonvulsant actions in rats and mice. *J. Pharmacol. Exp. Ther.* **228,** 109–115.
69. Bræstrup, C., Nielsen, E. B., Sonnewald, U., Knutsen, L. J., Andersen, K. E., Jansen, J. A., et al. (1990) (R)-N-[4,4-bis(3-methyl-2-thienyl)but-3-en-1-yl]nipecotic acid binds with high affinity to the brain gamma-aminobutyric acid uptake carrier. *J. Neurochem.* **54,** 639–647.
70. Richens, A., Chadwick, D. W., Duncan, J. S., et al. (1995) Adjunctive treatment of partial seizures with tiagabine: a placebo-controlled trial. *Epilepsy Res.* **21,** 37–42.
71. Ben-Menachem, E., Söderfelt, B., Hamberger, A., Hedner, T., and Persson, L. I. (1995) Seizure frequency and CSF parameters in a double-blind placebo controlled trial of gabapentin in patients with intractable complex partial seizures. *Epilepsy Res.* **21,** 231–236.
72. Schousboe, A., Larsson, O. M., Wood, J. D., and Krogsgaard-Larsen, P. (1983) Transport and metabolism of gamma-aminobutyric acid in neurons and glia: implications for epilepsy. *Epilepsia.* **24,** 531–538.
73. Gonsalves, S. F., Twitchell, B., Harbaugh, R. E., Krogsgaard-Larsen, P., and Schousboe, A. (1989) Anticonvulsant activity of intracerebroventricularly administered glial GABA uptake inhibitors and other GABAmimetics in chemical seizure models. *Epilepsy Res.* **4,** 34–41.

74. Gonsalves, S. F., Twitchell, B., Harbaugh, R. E., Krogsgaard-Larsen, P., and Schousboe, A. (1989) Anticonvulsant activity of the glial GABA uptake inhibitor, THAO, in chemical seizures. *Eur. J. Pharmacol.* **168,** 265–268.

75. White, H. S., Hunt, J., Wolf, H. H., Swinyard, E. A., Falch, E., Krogsgaard-Larsen, P., and Schousboe, A. (1993) Anticonvulsant activity of the gamma-aminobutyric acid uptake inhibitor N-4,4-diphenyl-3-butenyl-4,5,6,7-tetrahydroisoxazolo[4,5-*c*]pyridin-3-ol. *Eur. J. Pharmacol.* **236,** 147–149.

11

Insights From Endogenous and Engineered Zn²⁺ Binding Sites in Monoamine Transporters

Claus Juul Loland and Ulrik Gether

1. INTRODUCTION

The availability in the synaptic cleft of dopamine, serotonin, and norepinephrine, referred to as the monoamines, is tightly regulated by specific transport proteins that mediate rapid uptake into the presynaptic nerve terminals utilizing the Na^+ gradient across the plasma membrane as the driving force *(1–3)*. Three distinct monoamine transporters have been identified: the dopamine transporter (DAT) (Fig. 1), the norepinephrine transporter (NET), and the serotonin transporter (SERT) *(1–3)*. These transporters belong to the family of Na^+/Cl^--coupled transporters that also include transporters for other neurotransmitters such as GABA (γ-aminobutyric acid) and glycine *(1–3)*. The homology among the Na^+/Cl^--dependent neurotransmitter transporters is striking with, e.g., 67% sequence identity between DAT and NET, 49% between DAT and SERT, and 45% between DAT and the (GABA transporter-1 (GAT-1) *(2,4)*. Recently, it has become clear that homologs of Na^+/Cl^--dependent transporters also exist in prokaryotes. BLAST searches of newly sequenced bacterial genomes have revealed the existence of genes in many bacteria and archae (~50) encoding proteins with up to 25% sequence identity with the human DAT (see also Chapter 12). The function of these putative transporters is unknown, except that one transporter from the thermophilic bacteria *Symbiobacterium Thermophillum* very recently has been identified as a highly selective tryptophan transporter *(5)*.

The monoamine transporters are targets for the action of several drugs. This includes both the most commonly used antidepressants, which are selectively blocking the function of the serotonin transporter (selective serotonin reuptake inhibitors [SSRIs]); and widely abused psychostimulants, such as cocaine, amphetamine, and "ecstasy" *(2,3)*. Not surprisingly, these transporters have therefore been the focus of intensive research. During recent years, these research efforts have improved our understanding of their role in drug-abuse mechanisms, and progress has been made in our understanding of how the transporters operate at the molecular level *(2)*. A high-resolution structure is, however, still not available for a Na^+/Cl^--dependent transporter or for any other transport protein utilizing a transmembrane ion gradient as the driving force for the trans-

From: *Molecular Neuropharmacology: Strategies and Methods*
Edited by: A. Schousboe and H. Bräuner-Osborne © Humana Press Inc., Totowa, NJ

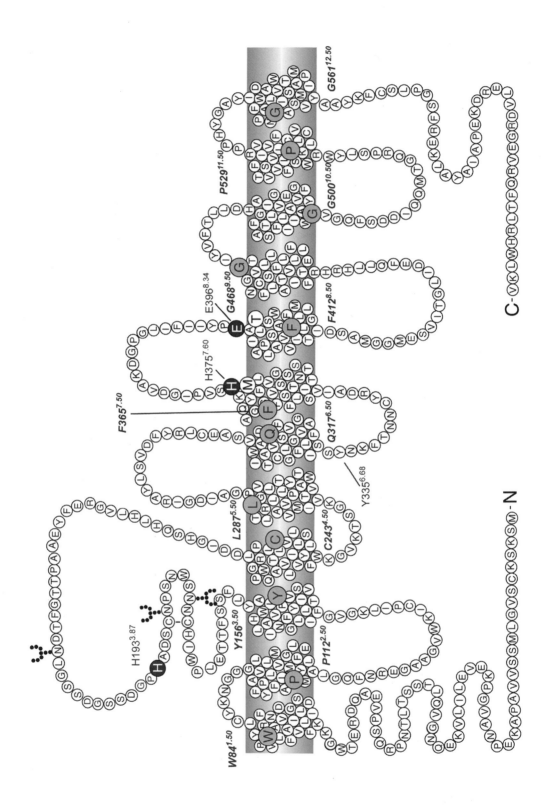

Fig. 1. Schematic representation of the human dopamine transporter. The large second extracellular loop contains three sites for *N*-linked glycosylation and two cysteines that are conserved among all biogenic amine transporters and believed to form a disulfide bridge *(88–90)*. The three residues coordinating Zn^{2+} binding to the wild-type hDAT are highlighted (enlarged black circles) *(50,51)*. The most conserved residue in each transmembrane segment is also highlighted (gray circles). To generate a generic numbering scheme allowing direct comparison of positions between the individual members of the transporter family, these residues have been assigned arbitrarily by the number of the helix and the number 50. Other residues are then numbered according to their position relative to this conserved residue. For example, 1.55 indicates a residue in transmembrane segment 1 (TM1) five residues carboxyterminal to the most conserved residue in this TM (Trp1.50). Throughout the chapter all residues are indicated by both their actual amino acid number in the transporter and in superscript by this generic numbering scheme (see also Chapter 12).

port process (secondary active transporters). Notably, low-resolution, three-dimensional structures have now been published for the bacterial Na^+/H^+-antiporter NhaA *(6)* and of the bacterial oxalate transporter OxlT based on cryo-electron microscopy of two-dimensional crystals *(7)*. These structures have provided the first glimpses of the higher structural organization of secondary active transport proteins, but to what degree these very different proteins may resemble that of the Na^+/Cl^--dependent transporters to which they share no sequence homology remains elusive.

In the absence of a high-resolution structure, a broad spectrum of noncrystallographic strategies have proven highly useful to obtain insight into structure/function relationships in the family of Na^+/Cl^--dependent neurotransmitter transporters (for detailed review, *see* ref. *[2]* and Chapter 12). The use of immunochemical techniques *(8,9)*, as well as site-selective labeling with membrane permeable and nonpermeable cysteine/amine reactive biotinylation reagents *(10,11)*, have provided support for the originally proposed topology shown in Fig. 1. This topology is characterized by 12 transmembrane segments, a large second extracellular loop, and an intracellular location of both the N- and C-termini (Fig. 1). Furthermore, construction of chimeric transporters *(12–15)*, point mutational analysis *(see,* e.g., refs. *[16–22])*, photo-affinity labeling *(23)*, and biophysical techniques *(24)* have provided information about residues and domains critical for the translocation process and for substrate and/or blocker recognition. The application of the substituted cysteine accessibility method (SCAM) has been used to predict secondary structure relationships and identify conformationally sensitive positions in the transporter molecules *(11,25,26–28)*. Moreover, the application of fluorescence resonance energy tranfer (FRET) *(29)* and intermolecular cross-linking strategies *(30)* have suggested that Na^+/Cl^--dependent neurotransmitter transporters exist in the membrane as an oligomeric structure although the functional significance of this still needs to be clarified.

In this chapter, it will be described how we have utilized endogenous and engineered Zn^{2+} binding sites to explore the structure and molecular function of Na^+/Cl^--dependent neurotransmitter transporters. The work has not only allowed the definition of the first structural constraints in the tertiary structure of this class of transporters, but also provided new insight into both conformational changes accompanying substrate translocation and mechanisms governing conformational isomerization in the translocation cycle. In the chapter, we will also review the theoretical and practical basis for application of "Zn^{2+} site engineering" and the results obtained from application of the technique to several other membrane proteins.

2. INSIGHTS FROM NATURALLY OCCURRING ZN^{2+} BINDING SITES IN SOLUBLE PROTEINS

Zn^{2+} binding sites are found in numerous soluble proteins, including particular enzymes and DNA binding transcription factors *(31–34)*. In enzymes, Zn^{2+} may have a direct catalytic role (catalytic sites) or Zn^{2+} may stabilize the active site structure of the enzyme (structural sites) *(34)*. In DNA binding transcription factors, Zn^{2+} serves a purely structural role, stabilizing the geometry of the so-called Zn^{2+} finger motif *(31,33)*. Another class of sites is co-catalytic sites, where two or three metal ions are situated in immediate proximity and in which two of the metals are bridged by a side-chain moiety of a single amino acid residue *(34)*. Only few of these sites contain only Zn^{2+} and more often contain Zn^{2+} in combination with Cu, Fe, or Mg *(34)*. Interestingly, Zn^{2+} binding sites may also occur at the interface between two protein molecules (protein interface sites) serving either a catalytic or a structural role *(34)*.

The availability of the three-dimensional structure of now approx 200 soluble Zn^{2+} binding proteins has provided profound insight into the structural basis for the interaction of Zn^{2+} with polypeptides and defined the strict structural constraints for coordination of the zinc(II) ion *(33,34)*. As for other metal ions, the residues that can act as ligands for Zn^{2+} are those that contain electron-donating atoms (S, O, or N) or have amino acid side chain with ionizable groups *(35)*. Although this includes serine, tyrosine, arginine, and lysine, the strongest interactions are with the imidazole side chain of histidines, the sulfhydryl side chain of cysteines, and the carboxylate side chains of glutamates and aspartates *(35)*. The predominant coordination geometry found among Zn^{2+}-binding sites in soluble proteins is tetrahedral *(33,34)*. For catalytic sites, this will involve the side chains of three residues plus a water molecule, with histidine being the most abundant ligand and only occasional involvement of cysteines, aspartates, and glutamates *(33,34)*. Importantly, the water molecule present in the catalytic site plays a key role for the catalytic process. Ionization/polarization of the water molecule can for example provide hydroxide ions at neutral pH, or alternatively Lewis acid catalysis by the catalytic Zn^{2+} ion may occur by displacement of water or by expansion of the coordination sphere *(34)*. In contrast to the catalytic sites, the structural sites are usually buried and have no bound water molecule, thus involving the side chains of four residues, with cysteines being most common closely followed by histidine and only rarely aspartate or glutamate *(33,34)*. More infrequently, the Zn^{2+} binding sites may have five or six coordinates involving a variable number of water molecules *(34)*. The sites having five coordinates adopt most often a trigonal bipyramidal geometry and less often a square-based pyramidal geometry, whereas sites with six ligands are found to adopt octahedral geometry *(34)*. The average distance between the coordinating ligand and Zn^{2+} is largely independent of the coordination geometry and is in the range of 2.0–2.3 Å *(33)*. Accordingly, two residues involved in coordinating the same Zn^{2+}(II) ion must be in close proximity in the tertiary structure of the protein, with the α-carbons being approx 13 Å apart *(33)*.

The crystal structures of Zn^{2+} binding proteins not only revealed well-defined tertiary structure constraints to accommodate Zn^{2+} binding, but also well-defined con-

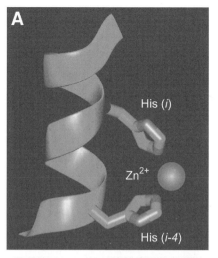

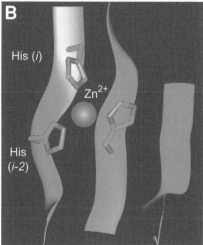

Fig. 2. Structural constraints for binding of Zn²⁺ between two histidine residues in an α-helix or β-strand. **(A)** Coordination of Zn²⁺ between two histidines present in an α-helix requires that the two histidines are positioned as i and i4 with i assuming the *gauche*+ rotamer and i4 the *trans* rotamer. The figure shows ¹⁴⁰His and ¹⁴⁴His in the endogenous Zn²⁺ binding site of the elastase of *Pseudomonas aeruginosa* (1.5 Å resolution) *(91)*. **(B)** Coordination of Zn²⁺ between two histidines present in a β-strand requires that the two histidines are positioned as i and i2 with i assuming the *gauche*+ rotamer and i2 the *trans* rotamer. The figure shows ⁹⁴His and ⁹⁶His in the endogenous Zn²⁺ binding site of carbonic anhydrase II (1.54 Å resolution) *(92)*. The histidine in the adjacent strand (¹¹⁹His) forms the third coordinate in the binding site. Reproduced with permission from ref. *(36)*.

straints at the secondary structure level. For example, coordination of Zn²⁺ between two histidines that are located in the same α-helix requires that the two histidines are positioned as i and i4 with i assuming the *gauche*+ rotamer and i4 the *trans* rotamer *(33,36)* (Fig. 2). If the two residues instead are positioned as either i and i3 or i and i5, binding of Zn²⁺ will involve a substantial distortion of the helix *(33,36)*. A similar strict pattern was observed for pairs of Zn²⁺-binding histidines in β-strands. Two histidines within a β-strand must be positioned as i and i2 to bind Zn^{2+} without distortion of the strand *(33,36)* (Fig. 2). As described later in this chapter, this information may allow useful inferences about not only tertiary structure relationships, but also secondary structure-relationships from engineered Zn^{2+} binding sites.

3. ENDOGENOUS ZN^{2+} BINDING SITES IN MEMBRANE PROTEINS

During recent years, evidence has been obtained indicating the presence of Zn^{2+} binding sites also in many membrane proteins. This includes: ligand-gated ion channels such as the $GABA_A$ receptor *(37,38)*, the glycine receptor *(39,40)*, and the NMDA receptors *(41–45)*; G protein coupled receptors such as the melanocortin receptor-1 and 4 *(46)*, the neurokinin-3 receptor *(47)*, and the $β_2$ adrenergic receptor *(48,49)*; and neurotransmitter transporters such as the DAT *(50,51)* and the glutamate transporter EAAT-1 *(52)*. The three-dimensional structures of these proteins have not been solved; thus, the presence of endogenous Zn^{2+} binding sites has been discovered indirectly by the ability of low-micromolar concentrations of Zn^{2+} to modulate the function of these proteins and by subsequent identification of Zn^{2+} coordinating residues using site-directed mutagenesis.

The physiological role of the Zn^{2+} binding sites identified in membrane proteins remains unclear. The extracellular free Zn^{2+} concentration is generally kept at a low level in the brain (estimated to be in the range from 1 pM–10 nM *(31)*; however, in glutaminergic nerve terminals, especially at the mossy-fiber synapses in the hippocampus, Zn^{2+} is found in very high concentration in the synaptic vesicles (3–30 m*M*) *(53)*. Upon the arrival of an action potential at the terminal, Zn^{2+} has been shown to be co-released with glutamate. Based on studies using hippocampal slice preparations, this has been estimated to result in free Zn^{2+} concentration that transiently may reach 10–30 μ*M* *(54–56)*. Older studies have even estimated that the concentration may reach 100–300 μ*M* upon stimulation of glutaminergic neurons *(57)*. It is, therefore, likely that Zn^{2+} in sufficiently high concentrations at least transiently may be present in the synapses to modulate the function of receptors and transporters containing endogenous Zn^{2+} binding sites. Final evidence for this will have to await the application of, for example, transgenic strategies involving homologous substitution with Zn^{2+}-insensitive mutants (knockins).

4. ENGINEERING OF ZN^{2+} BINDING SITES: BACKGROUND

Owing to the well-defined structural constraints, Zn^{2+} binding sites have been artificially introduced into proteins in many cases *(58)*. Zn^{2+} is particularly well-suited for this purpose. As a divalent cation, it has a completely filled *d* shell with 10 *d* electrons *(58)*. This electron configuration has three important consequences. First, because of the filled *d* shell, Zn^{2+} has no ligand field stabilization energy when coordinated by ligands in any geometry, in contrast to ions with partially filled *d* shells, which can favor certain arrangements of ligands over others. Second, divalent zinc is—in contrast to, e.g., Cu^{2+}—not redox active; neither the potential oxidized form, Zn^{3+}, nor the potential reduced form, Zn^+, is accessible under physiologic conditions. Third, Zn^{2+} is relatively labile in kinetic terms, meaning that it undergoes ligand-exchange reactions relatively rapidly *(58)*.

Zn^{2+} binding sites have been artificially introduced into proteins to stabilize them against denaturation or proteolytic degradation and into enzymes for enhanced regulation. They have also been engineered into proteins to assist their purification, crystallization, and X-ray crystal structure determination *(58)*. Moreover, Zn^{2+} binding sites have been artificially introduced into receptor and transporter proteins with the purpose of defining structural constraints in the secondary, tertiary, and quaternary structure of these proteins *(36,51,59–67)*. The simple principle of this "Zn^{2+} site engineering" approach is based on the ability of properly oriented Zn^{2+} coordinating residues to chelate Zn^{2+} in low micromolar concentrations. As Zn^{2+} ligands, histidines should be preferred over cysteines, aspartates, and glutamates. Even though cysteines can bind Zn^{2+} with high affinity, they have several disadvantages. The presence of a free sulfhydryl might interfere with protein folding owing to spontaneous formation of disulfide bridges between a pair of inserted cysteines. Also, the oxidation state of cysteine may be difficult to control in the presence of metal ions. Histidines should be preferred over aspartates and glutamates simply because of the higher affinity of Zn^{2+} for the imidazol side chain.

To achieve Zn^{2+} binding and accordingly generate a successful Zn^{2+}-binding site between mutationally inserted histidines, several criteria must be satisfied. First, the distances between the α-carbons of the substituted side chains should be less than 13 Å to allow the corresponding imidazole side chain to bind Zn^{2+}. Correspondingly, it should be possible for the coordinating nitrogen of the imidazol rings to be ~4 Å apart to allow coordination by Zn^{2+} *(33)*. Second, the insertion/substitution of a preexisting residue with histidine should be structurally and functionally tolerated by the protein of interest. In this context, it is important to note that cysteines in some cases are preferable because cysteines substitutions often are better tolerated than histidine substitutions. Third, the residues substituted must be exposed to the aqueous environment to allow Zn^{2+} binding. Fourth, binding of Zn^{2+} to the engineered binding site should be measurable. In membrane proteins that are not easily purified in large quantities to allow direct structural analyses, the only readily applicable way of detecting Zn^{2+} binding is indirectly via a putative effect of Zn^{2+} on the function of the protein. Such effects can include, for example, inhibition or stimulation of radioligand binding, receptor G protein coupling *(59,68)*, or transport activity *(50,52)*.

It is of interest to note that—at least theoretically—it should be possible to predict the number of Zn^{2+} coordinating residues in a given Zn^{2+} binding site depending on observed affinity for Zn^{2+} *(69)*. A bidentate site would be predicted to have an affinity constant in the range of 2–200 μ*M*, a tridentate site would be predicted to have an affinity constant in the range of 36–3700 n*M,* whereas a tetrahedral Zn^{2+}-binding site will lie between 2 fM and 25 n*M (70,69)*. Obviously, the ranges are quite wide because the Zn^{2+} affinity is critically dependent on a wide variety of factors besides the ability of the coordinating sides chains to chelate Zn^{2+}.

Metal ion binding sites have been engineered in several polytopic membrane proteins. Kaback and coworkers designed a series of metal ion binding sites in the Lac permease, a proton coupled lactose transporter from *Escherichia coli (71–73)*. Binding of the metal ion (Mn^{2+}) was determined directly on purified protein by the use of electron paramagnetic resonance spectroscopy and this data provided new insight into packing of the 12 transmembrane helices present in this transport protein *(71–73)*. Elling et al. introduced the use of Zn^{2+} site engineering in G protein-coupled receptors by substituting an antagonist binding site in the NK-1 (substance P) receptor with a tridentate histidine Zn^{2+} binding site *(59)*. Subsequent engineering of a series of bis-His Zn^{2+} binding sites in the NK-1 receptor defined a series of distance constraints in the tertiary structure of receptor *(62)*. Most importantly, these data provided information about the organization of the seven-helix bundle in GPCRs *(62)*. The structural constraints defined by the engineered sites could, for example, only be interpreted in the context of a counterclockwise orientation (seen from the extracellular side) of the seven helices *(62)*. Similarly, as described in the next section, our application of Zn^{2+} site engineering to Na^+/Cl^--dependent neurotransmitter transporters has enabled definition of important secondary and tertiary relationships in this class of proteins. Evidently, the structural information achieved from these studies is rather limited as compared with a high-resolution crystal structure. However, it is important to note that the application of Zn^{2+} site engineering has provided (and can provide) interesting structural information long before a high-resolution structure has been (is) available. It can also be used, for example, to develop "evolutionary fingerprints" for deducing the degree of structural conservation across a protein family, and it may address specific structural questions that might be relevant even if high-resolution structural information is available *(59,62,63,65,66,68,74)*.

It is also important to emphasize that the generation of Zn^{2+} binding sites can not only provide structural insights, but also offer novel insights into protein function. The substitution of a nonpeptide antagonist binding site in the NK-1 (substance P receptor) not only provided new structural insight, but also supported a concept of "allosteric competitive" antagonism, in which antagonists and agonists mutually exclude each other's binding to the receptor without necessarily sharing an overlapping binding site by stabilizing distinct inactive and active conformations, respectively *(59)*. Indirectly, Zn^{2+} binding sites may also convey information about conformational changes important for protein function. In the G protein-coupled light receptor rhodopsin, Zn^{2+} has been shown to inhibit receptor activation by binding to an engineered bis-His Zn^{2+} binding site between the cytoplasmic extensions of transmembrane segment (TM)-3 and -6 *(75)*. A conceivable explanation for this observation is that Zn^{2+} by binding to this site constrains motions between the two domains critical for the receptor activation process *(75)*. This conclusion has been supported by studies employing spectroscopic techniques *(76,77)* and the substituted cystein accessibility method *(78)*. Also, as described in detail in the next section, we have in the DAT and GAT-1 engineered inhibitory Zn^{2+} binding sites between histidines and/or cysteines in TM7 and -8 *(36,51,65)*. By binding to these sites, Zn^{2+} acts as a noncompetitive inhibitor of substrate translocation representing strong, though still indirect, evidence that relative movements between these domains are critical for this process.

Zn^{2+} binding sites may also allow insight into the structure of specific functional states of a given protein. In the β_2 adrenergic receptor, it has been demonstrated that coordination of Zn^{2+} or Cu^{2+} between a cysteine inserted in TM7 and the aspartate in TM3, which bind the positively charge amine of the agonists and antagonists for this receptor, leads to receptor activation *(68)*. The observation defined an important distance constraint in an active conformation of the receptor. This is of obvious interest given that the only available high-resolution structure of a GPCR is the inactive dark state of rhodopsin. The information could potentially be useful in structure-based drug-discovery processes aimed at developing agonist ligands. In the same receptor, it has recently been found that binding of Zn^{2+} to a naturally occurring binding site on the intracellular side of the membrane leads to allosteric enhancement receptor activation *(49)*. The site presumably involves Glu225 at the cytoplasmic side of TM5 and His-269 plus Cys265 at the cytoplasmic side of TM6 and, thus, the data provide additional support for a conformational rearrangement in this part of the receptor during activation and G protein coupling *(49)*. In the metabotropic glutamate receptor-1 (mGluR1), a Zn^{2+} binding site has recently been generated in the large extracellular aminoterminal domain (ATD) *(79)*. At this site, Zn^{2+} acts as a noncompetitive antagonist, presumably by preventing formation of the "closed" active conformation of the ligand binding site that is contained within the ATD *(79)*. Finally, it is noteworthy that discrete mutations in, for example, the DAT have been observed to elicit dramatic alterations in the Zn^{2+} effect at this protein *(66)*. Because such alterations can be a direct consequence of, for example, changes in the distribution between different functional states, such mutations may—together with Zn^{2+}—represent valuable tools for deciphering the molecular basis for the biological function of a given protein (*see* next section).

5. AN ENDOGENOUS ZN²⁺ BINDING IN THE DOPAMINE TRANSPORTER

By investigating the susceptibility of the unmodified human dopamine transporter (hDAT) and the homologous human norepinephrine transporter (hNET) to Zn^{2+}, we obtained evidence that the hDAT, but not the hNET, contains an endogenous high-affinity Zn^{2+} binding site *(50)*. As illustrated in Fig. 3, it was found that Zn^{2+} in micromolar concentrations is a potent inhibitor of [³H]dopamine uptake in hDAT but not in hNET *(50)*. The inhibition in hDAT was biphasic with an IC_{50} value for the high-affinity phase of approx 1 μM and > 1000 μM for the low-affinity phase *(50)*. In the NET, only low-affinity inhibition was observed. The low-affinity inhibition by Zn^{2+} at millimolar concentrations most likely represents nonspecific toxic effects of Zn^{2+}; however, the high-affinity inhibition in hDAT conceivably reflected interaction of Zn^{2+} with a specific site within the hDAT *(50)*. The presence of an endogenous Zn^{2+} binding site in the hDAT was further supported by the observation that Zn^{2+} in micromolar concentrations potentiated binding of the cocaine analog [³H]WIN 35 428 to the hDAT, but not the hNET *(50)* (Fig. 3).

Saturation uptake experiments in the presence and absence of micromolar concentrations of Zn^{2+} showed that the high-affinity inhibition of uptake in the hDAT was owing to a decrease in the V_{MAX} for [³H]dopamine uptake with no change in K_M, consistent with a noncompetitive mechanism of action *(50)*. It was also observed that dopamine inhibits binding of the cocaine analog [³H]WIN 35 428 with the same potency in the absence and presence of Zn^{2+}, altogether supporting that dopamine can bind to the Zn^{2+} occupied transporter but that translocation is blocked *(50)*. The most likely explanation is that Zn^{2+} is restricting movements but not blocking the translocation cycle completely, causing the transporter to translocate at a lower efficacy. Another possibility that cannot be excluded is that in the presence of Zn^{2+}, the translocation process occurs at an unchanged rate, but dopamine is less likely to be released to the intracellular environment. Thus, Zn^{2+} increases the chance that the transporter reorients from its putative inward-facing conformation to its outward-facing conformation with the dopamine still bound.

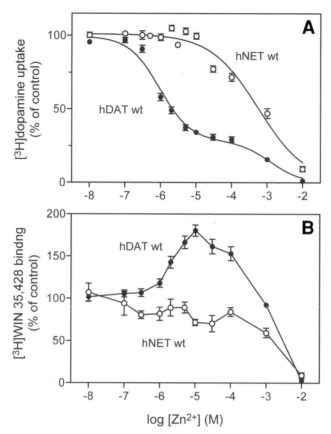

Fig. 3. Evidence for an endogenous Zn^{2+}-binding site in the DAT but not in the NET. **(A)** Zn^{2+} inhibition of [^{3}H]dopamine uptake in COS-7 cells transiently expressing hDAT (filled circles) or hNET (open circles). **(B)** Effect of Zn^{2+} on binding of the cocaine analog [^{3}H]WIN 35,428 to hDAT (filled circles) and hNET (open circles). Values are percent of control ([^{3}H]dopamine uptake or [^{3}H] WIN 35,428 binding in the absence of Zn^{2+}) expressed as means ± S.E. of 3–5 experiments performed in triplicate.

The structural basis for the high-affinity interaction with hDAT was investigated by application of a systematic mutagenesis approach aimed at identifying Zn^{2+} coordinating residues within the transporter molecule. Because the NET, in contrast to hDAT, was insensitive to Zn^{2+}, at least one coordinating residue could be expected to be nonconserved between the hDAT and the hNET. Moreover, because Zn^{2+} is unlikely to penetrate the plasmamembrane, it would be expected that the coordinating residues would be on the extracellular face of the transporter molecule. Indeed, a systematic knockout of all nonconserved histidine residues at the extracellular side of the DAT identified a histidine (His193[3.87])[1] in the second extracellular loop (Fig. 1) that, upon mutation to lysine, elim-

---

[1] All residues are indicated both by their actual amino acid number in the transporter and in superscript by the generic numbering scheme described in the chapter by Goldberg et al. According to the scheme, the most conserved residue in each transmembrane segment has been given the number 50, and each residue is numbered according to its position relative to this conserved residue. For example, 1.55 indicates a residue in TM1 five residue carboxyterminal to the most conserved residue in this TM (Trp1.50).

inated high-affinity Zn^{2+} susceptibility *(50)*. The involvement of His193$^{3.87}$ was further supported by the fact that if a histidine were inserted into the corresponding position in the NET (NET-K189H) hDAT-like Zn^{2+} susceptibility was transferred to this transporter *(50)*. This observation also gave the important information that the remaining coordinate(s) must be found among residues conserved between hDAT and hNET. In full agreement with this prediction, subsequent systematic mutation of all the conserved extracellular histidines identified an additional coordinating residue (His375$^{7.60}$ at the top of TM7, Fig. 1) that, upon mutation into an alanine, revealed the same dramatic decrease in Zn^{2+} susceptibility as observed for the H193K mutation *(50)*.

Because the vast majority of naturally occurring Zn^{2+} binding sites contains three or four coordinating residues and because the apparent Zn^{2+} affinity for the hDAT (~1 μM) would be most consistent with a tridentate binding site (though in the low-affinity range; *see* previous section), we predicted that at least one additional residue was part of the hDAT Zn^{2+} binding site. Mutation of multiple cysteines assumed to be accessible from the extracellular side did not indicate participation of these residues in Zn^{2+} binding (L. Norregaard, J. Ferrer, J. Javitch, and U. Gether, unpublished observation); however, mutation of all conserved extracellular aspartates and glutamates identified a glutamate as the third residue in the Zn^{2+} binding site located in the top of TM8 at position 396$^{8.34}$ *(51)* (Fig. 1). The involvement of both His375$^{7.60}$ and Glu396$^{8.34}$ was additionally supported by mutation of the corresponding residues in the Zn^{2+}-sensitive NET-K189H mutant (His372 and Glu393, respectively), resulting in eliminated high-affinity Zn^{2+} inhibition of [³H]dopamine uptake *(51)*. Moreover, it was found that a histidine could fully substitute for the glutamate (hDAT-E396H) with respect to Zn^{2+} inhibition of [³H]dopamine uptake *(51)*. This minimized the possibility that the loss of Zn^{2+} sensitivity upon mutating Glu396$^{8.34}$ to glutamine in hDAT is owing to an indirect structural effect caused by the removal of a negative charge, rather than the disruption of a direct interaction with Zn^{2+}.

The identification of the three residues involved in Zn^{2+} binding to the endogenous Zn^{2+} binding site imposes an important set of distance constraints in the tertiary structure of the DAT. According to the hydrophobicity plot, His375$^{7.60}$ and Glu396$^{8.34}$ are predicted to be located right at the extracellular end of TM7 and TM8, respectively (Fig. 1). They are separated with a large loop of 20 residues, allowing TM7 and TM8 to be rather far apart in the tertiary structure. However their common participation in the Zn^{2+} site outlines the close association between the two transmembrane segments. Moreover, the participation of His193$^{3.87}$ in binding of Zn^{2+} to the DAT outline the association of the outer portion of TM7 and TM8 with the large extracellular loop connecting TM3 and TM4.

6. STRUCTURAL INSIGHT FROM ENGINEERED ZN²⁺ BINDING SITES IN NA⁺/CL⁻-DEPENDENT NEUROTRANSMITTER TRANSPORTERS

The identification of the coordinates in the endogenous Zn^{2+} binding site in the DAT suggested that His375$^{7.60}$ at the top of TM7 is facing Glu396$^{8.34}$ at the top of TM8. To further explore this spatial proximity between TM7 and TM8, a series of engineered Zn^{2+} binding sites was established *(51,36)*. If Glu396$^{8.34}$ is situated in an α-helical environment, it could be expected that His375$^{7.60}$ also is close to the residue situated one helical turn from Glu396$^{8.34}$, i.e. the residues in the i4 or i3 positions, while the i2 position would be expected to be located on the other side of the helix. Accordingly, three mutant transporters were generated in which $His193^{3.87}$ and $Glu396^{8.34}$ were removed, $His375^{7.60}$ preserved, and cysteine residues inserted in position i2, i3, and i4 relative to position 396 *(51)*. The insertion of cysteines was chosen in favor of the stronger binding histidines because the residues are smaller and therefore likely better tolerated in the transmembrane domains, where space can be limited. Indeed, Zn^{2+} could inhibit [^{3}H]DA uptake when a cysteine were inserted in position $400^{8.38}$ (i4) displaying an IC_{50} value of 24 μM as compared to 660 μM for the background mutant (hDAT-H193K-E396Q) *(51)*. By probing cysteine residues in TM8 against the $His375^{7.60}$ in TM7, the results strongly support that the top of TM8 is an α-helical configuration.

In another set of experiments, the secondary structure at the external end of TM7 was probed by engineering sites with two coordinating residues within the same hypothetical helix. It was attempted specifically to take advantage of the finding that coordination of Zn^{2+} between two histidines located in an α-helix requires that the two histidines are positioned as i and i4 position, with i assuming the *gauche*+ rotamer and i4 the *trans* rotamer (Fig. 2) *(36)*. First, $His193^{3.87}$ in ECL 2 was substituted with a lysine (the corresponding residue in the Zn^{2+}-insensitive hNET is a lysine). In the background of this mutation (H193K), we introduced a histidine inserted in the i4 position relative to $His375^{7.60}$ (H193K-M371H). In this mutant, Zn^{2+} was a potent inhibitor of [^{3}H]dopamine uptake, indicating coordination of Zn^{2+} between $His375^{7.60}$, $Glu396^{8.34}$, and M371H and thus supporting an α-helical configuration of TM7 *(36)*. This was further corroborated by the observation that no increase in the apparent Zn^{2+} affinity was found upon introduction of at the i2, i3, and i5 positions relative to $His375^{7.60}$. Interestingly, a different pattern was observed when histidines were introduced at positions i2 V377H) i3 (P378H), and i4(I379H) and i5 (G380H) relative to $His375^{7.60}$. Thus, a marked increase in the apparent Zn^{2+} affinity was observed by introducing a histidine not only at the i4 position from $His375^{7.60}$ but also at the i2 and i3 positions *(36)*. These data are inconsistent with an α-helical configuration between residue 375 and 379 and indicate the absence of a well-defined secondary structure. Thus, the data suggest an approximate boundary between the end of the TM7 helix and the succeeding loop around position 375 *(36)*.

Altogether, the identification of the coordinating residues in the endogenous hDAT Zn^{2+} binding site followed by the engineering artificial sites have defined an important series of structural constraints in this transporter. This includes not only a series of proximity relationships in the tertiary structure, but also secondary structure relationships. The data also provided information about the orientation of TM7 relative to TM8. A model of the "TM7/8 microdomain" that incorporates all these structural constraints is shown in Fig. 4 *(36)*. The model is an important example of how structural inferences derived from a series of Zn^{2+} binding sites can provide sufficient information for at least an initial structural mapping of a selected protein domain.

The fact that the endogenous Zn^{2+}-binding site in hDAT could be transferred to the homologous hNET *(50)* suggests the possibility that Zn^{2+} site engineering can be used as an "evolutionary fingerprint," meaning that if a Zn^{2+} binding site can be transferred from one protein to an other by mutation of the corresponding residues, it is an indication of that the two proteins adopt a similar tertiary structure even if the primary amino acid

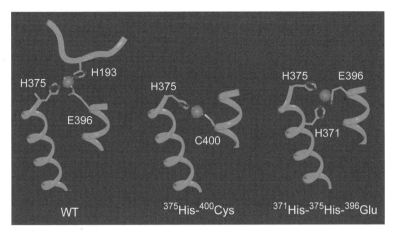

Fig. 4. Molecular Model of the TM7/8 microdomain. The model incorporates the constraints defined by data achieved in ref. **(50,51,36)**. Three different Zn²⁺ binding sites were included in the modeling procedure. Left panel, the endogenous Zn²⁺ binding site in hDAT ([193]His-[375]His-[396]Glu) *(50,51,36)*. Middle panel, the engineered site [375]His-[400]Cys **(50,51,36)**. Right panel, the engineered site [371]His-[375]His-[396]Glu *(36,50,51)*. The three Zn²⁺ binding sites were all modeled assuming a tetrahedral coordination geometry. Fragments [347]Ile-[375]His (TM7) and [396]Glu-[401]Leu (TM8) were modeled as two ideal anti-parallel α-helices with a predicted crossing angle of 161.3° between the axes and a distance between the α-carbon of [375]His and [396]Glu of 10.4 Å *(36)*. This relative backbone orientation of TM7 and TM8 satisfies all three Zn²⁺ binding sites through side-chain rotations alone without changes in the position of the two backbones. Reproduced with permission from ref. *(36)*.

sequence is different. This hypothesis was further supported by the observation that some of the established Zn²⁺ sites in the DAT also could be engineered into the GABA transporter, GAT-1 *(65)*. Introduction of a histidine in position 349[7.60] (position 375 in hDAT) together with either a histidine in position 370[8.34] (position 396 in hDAT) or a cysteine in position 374[8.38] (position 400 in hDAT) resulted in bidentate Zn²⁺ binding sites where Zn²⁺ binding lead to potent inhibition of [³H]GABA uptake *(65)*.

Electrophysiological analysis of the mutant transporters upon expression in *Xenopus laevis* oocytes showed that the inhibition of uptake was accompanied by a corresponding inhibition of the substrate-induced current *(65)*. Interestingly, analysis of the uncoupled Li+ conductance found in the GAT-1 showed that Zn²⁺ strongly inhibited the leak conductance in the T349H-E370H mutant, whereas no inhibition by Zn²⁺ was observed in T349H-Q374C *(65)*. This differential effect provides strong evidence that the leak conductance represents a unique operational mode of the transporter involving conformational changes distinct from those associated with substrate translocation *(65)*. Moreover, it reflects yet another example of how engineering of Zn²⁺ binding site can be helpful for the dissection of not only structural relationships but also specific functional mechanisms in membrane proteins.

7. THE COMPLEXITY OF ZN²⁺ ACTION AT HDAT: OPPOSITE EFFECT ON UPTAKE AND RELEASE

A further complexity of the action of Zn²⁺ at the hDAT was revealed when we investigated the effect of Zn²⁺ on reverse transport. Notably, the monoamine transporters are like other Na$^+$ coupled transporters, capable not only of Na$^+$-dependent uptake but also of reverse transport of their substrate(s) *(80)*. This reverse transport occurs either in response to addition of external substrate, upon disruption of the Na$^+$-gradient or by membrane depolarization *(81)*. The exact mechanism by which DAT mediates this efflux of substrate is poorly understood. According to the facilitate-exchange diffusion model of transporter function, inward and outward transport is predicted to be stoichiometrically linked and strictly coupled *(82)*. Accordingly, any inhibition of uptake would result in a concomitant inhibition of efflux. It was therefore particularly interesting to observe that in contrast to substrate uptake where Zn^{2+} acts as an inhibitor, Zn^{2+} enhances efflux of substrate; when cells are preloaded with [^{3}H]MPP$^+$ and then challenged with amphetamine, the addition of Zn^{2+} causes a substantial increase in [^{3}H]MPP$^+$ efflux *(83)* (Fig. 5). Similarly, Zn^{2+} enhanced efflux upon depolarization as well as augmented efflux in response to amphetamine in striatal slices *(83)*. The augmentation of efflux mediated by Zn^{2+} was importantly found to be a result of Zn^{2+} binding to the same endogenous Zn^{2+} binding site as that mediating inhibition of uptake *(83)*. Also, it is important to note that the enhancement of release was observed selectively for the endogenous Zn^{2+} binding site and could not be mimicked by an engineered site, such as that involving His375$^{7.60}$, I377H, and Glu396$^{8.34}$ *(83)*. Taken together, the data not only disputes the facilitated exchange-diffusion model, but also shows the first example of an allosteric modulator acting at hDAT that differentially can modulate inward and outward transport. This might be of interest also in an in vivo context. Reverse transport of dopamine via the DAT has recently been suggested to be responsible for nonexocytotic, Ca^{2+}-independent release of dopamine in the substantia nigra (SN) upon excitation of glutaminergic neurons projecting from the subthalamic nucleus *(84)*. This release contributes to the autoinhibitory effects mediated by the dopamine D$_2$-receptors to regulate overstimulatory inputs from the subthalamic nucleus *(84)*. Because Zn^{2+} in many brain regions is stored in synaptic vesicles and co-released together with glutamate leading transiently to a free Zn^{2+} concentration in the higher micromolar range *(53)*, it is conceivable that Zn^{2+} may modulate DA release in SN. Thus, our observations may not merely reveal a biochemical peculiarity, but could be physiologically relevant.

8. CONFORMATIONAL ISOMERIZATION IN THE TRANSPORT CYCLE: CONVERSION OF THE INHIBITORY ZN^{2+} SWITCH IN HDAT TO AN ACTIVATING ZN^{2+} SWITCH

It is a general assumption that Na$^+$/Cl$^-$-dependent transporters operate by an alternating access mechanism, where the transporter interchanges between an "outward-facing" conformation, in which the substrate binding site is accessible to the extracellular medium, and an "inward facing" conformation, in which the binding site is accessible to the intracellular environment *(85)*. The translocation process is energetically coupled to the transmembrane Na$^+$-gradient and it is believed that the initial event in the translocation cycle is binding of sodium ions to the transporter *(85)*. The model predicts that the transporter, in the presence of Na$^+$ but in the absence of substrate, primarily resides in the outward-facing conformation, ready to bind extracellular substrate. Binding of substrate initiates translocation, causing transition of the transporter to the inward-facing conformation followed by release of substrate and Na$^+$ to the intracellular environment.

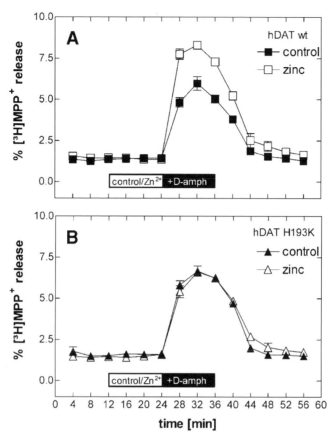

Fig. 5. Zn²⁺ enhances amphetamine induced release of [³H]MPP+. HEK-293 cells stably expressing the hDAT were preloaded with [³H]MPP+ and superfused upon reaching a stable baseline (basal efflux: mean of the three fractions before drug addition; hDAT wt: **(A)** basal efflux: $0.247 \pm 0.004\%$ ' min⁻¹., i.e., 245.6 ± 6.7 dpm' min⁻¹, $n = 60$ observations of randomly chosen experiments performed on different days; hDAT-H193K: **(B)** basal [³H]MPP+ efflux: $0.433 \pm 0.08\%$' min⁻¹, i.e., 181.2 ± 7.1 dpm' min⁻¹, $n = 47$. The experiment was started with the collection of 4-min fractions. After three fractions (12 min) of basal efflux, cells were exposed to Zn²⁺ (10 µM), or left at control conditions as indicated. After six fractions (from 24 min and onward), amphetamine (10 µM) was added to all superfusion channels. After nine fractions (from 36 min and onward), all channels were switched back to control conditions. Data are presented as fractional efflux, i.e., each fraction is expressed as the percentage of radioactivity present in the cells at the beginning of that fraction. Symbols represent means ± S.E. of 6–12 observations (one observation equals one superfusion chamber; all experiments were performed in triplicate). Reproduced with permission from ref. *(83)*.

A prerequisite for this model is the existence of both external and internal "gates," i.e., protein domains that are capable of occluding access to the substrate binding site from the extracellular and intracellular environment, respectively. Little is known about such domains in this family of transporters. The molecular mechanisms governing the cooperative function of the putative gating domains also remain unknown. It could be predicted, however, that stabilization of the transporter in the outward-facing conformation in the absence of substrate but in the presence of Na⁺ requires a network of con-

straining intramolecular interactions—possibly in the gating domains themselves—
which is released upon substrate binding to the transporter and thus controls the conformational equilibrium of the translocation cycle. It follows that mutation of residues
that are part of this stabilizing network of intramolecular interactions may cause spontaneous changes in the distribution between different conformational states in the
translocation cycle. In principle, such mutations would be analogous to constitutively
activating mutations in G protein-coupled receptors *(86,87)*. In these mutants, it is
believed that the agonist-independent shift in the equilibrium between inactive and
active states of the receptor is owing to the release of constraining intramolecular interactions important for maintaining the receptor preferentially in its inactive state in the
absence of agonist *(86,87)*.

Interestingly, we have recently identified a mutation of a tyrosine in the third intracellular loop of the hDAT that causes a major alteration in the conformational equilibrium of the transport cycle, and thus as such is comparable to mutants on G
protein-coupled receptors causing constitutive isomerization of the receptor to the
active state *(66)*. Most importantly, this conclusion is based on the observation that
mutation of the tyrosine completely reverts the effect of Zn^{2+} at the endogenous Zn^{2+}
binding site in the hDAT *(50,51)* from potent inhibition of transport to potent stimulation of transport (Fig. 6). In the absence of Zn^{2+}, transport capacity is reduced to less
than 1% of that observed for the wild-type, however, the presence of Zn^{2+} in only
micromolar concentrations causes a close to 30-fold increase in uptake *(66)*. Moreover,
it is found that the apparent affinities for cocaine and several other inhibitors are substantially decreased, whereas the apparent affinities for substrates are markedly
increased *(66)*. Notably, the decrease in apparent cocaine affinity was around 150-fold
and thus to date the most dramatic alteration in cocaine affinity reported upon mutation
of a single residue in the monoamine transporters *(66)*.

It is our interpretation that the mutation of Tyr335[6.68] to an alanine leads to disruption of intramolecular interactions critical for stabilizing the transporter in a conformation in which extracellular substrate can bind and initiate transport. As a consequence,
the transporter accumulates in the inward-facing conformation and/or putative intermediate states between the inward- and outward-facing conformations. A changed conformational equilibrium is strongly supported by the substantial changes in apparent
affinities for substrate and it can explain the reduced transport capacity of the Y335A
mutant because only a small fraction of transporters would reside in the outward-facing
resting conformation and thus be available for transport. Because Zn^{2+} likely stabilizes
the outward-facing conformation of the transporter, it is conceivable that Zn^{2+} is capable of reestablishing a conformational equilibrium similar to the wild-type DAT, allowing substrate binding and translocation to occur *(66)*. In this context, it is important to
note that occupancy of the endogenous high-affinity Zn^{2+} binding site in the wild-type
does not result in full inhibition of uptake, but only in maximum of approx 75% inhibition. Thus, the Zn^{2+}-occupied transporter is still is capable of translocating substrate,
albeit with reduced efficiency.

The mutation of Tyr335[6.68] introduces a new paradigm in the family of Na^+/Cl^--
dependent transporters, i.e., mutation causing spontaneous changes in the distribution
between individual functional states in the translocation cycle. Importantly, by systematic mutations of conserved intracellular residues in the hDAT, we have now identified

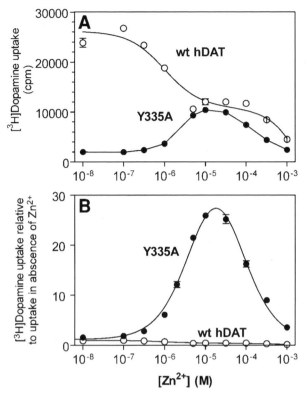

Fig. 6. Conversion of an inhibitory Zn^{2+} switch to a stimulatory Zn^{2+} switch by mutation of Ty335[6.68]. **(A)** The effect of Zn^{2+} on [³H]dopamine uptake in COS-7 cells transiently expressing hDAT-Y335A (filled circles) or WT hDAT (open circles). Each data point is expressed as mean ± S.E. in count per minutes (CPM) of triplicate determinations from a representative experiment. **(B)** Zn^{2+} effect on [³H]dopamine uptake in COS-7 cells transiently expressing hDAT-Y335A (filled circles) or WT hDAT (open circles). Values are percent of control ([³H]dopamine uptake in the absence of Zn^{2+}) expressed as means ± S.E. of 5 (hDAT-Y335A) and 3 (hDAT) experiments performed in triplicate. Reproduced with permission from ref. *(66)*.

three additional residues situated in the second, third, and fourth intracellular loop that upon mutation display a similar phenotype as the Y335A mutation (Loland and Gether, in preparation). It important to emphasize that corresponding constitutively activating mutations in GPCRs have proven to be extremely valuable tools for gaining insight into the molecular function of this family of membrane proteins *(87)*. It is also our expectation that the present tyrosine mutation could serve as an important tool in future studies of neurotransmitter transporters. Assuming that Zn^{2+} is capable of switching the Y335A mutant between discrete steps in the translocation cycle, the mutant may be used, for example, to map structural differences between such distinct steps. This could be done by employing the substituted cysteine accessibility method (SCAM), thus enabling study of the changes in the accessibility pattern for cysteine substituted residues in absence and presence of Zn^{2+}. A similar approach has been applied previously to characterize structural differences between inactive and active conformations in GPCRs using constitutively activated receptor mutants *(78)*.

ACKNOWLEDGMENTS

Dr. Søren Rasmussen is thanked for helpful comments on the manuscript.

REFERENCES

1. Blakely, R. D. and Bauman, A. L. (2000) Biogenic amine transporters: regulation in flux. *Curr. Opin. Neurobiol.* **10,** 328–336.
2. Norregaard, L. and Gether, U. (2001) The monoamine neurotransmitter transporters: structure, conformational changes and molecular gating. *Curr. Opin. Drug Discov. Dev.* **4,** 591–601.
3. Torres, G. E., Gainetdinov, R. R., and Caron, M. G. (2003) Plasma membrane monoamine transporters: structure, regulation and function. *Nat. Rev. Neurosci.* **4,** 13–25.
4. Miller, J. W., Kleven, D. T., Domin, B. A., and Fremau Jr., R. T. (1997) Cloned sodium- (and chloride-) dependent high affinity transporters for GABA, glycine, proline, betaine, taurine, and creatine, in *Neurotransmitter Transporters: Structure, Function, and Regulation* (Reith, M. E. A., ed.). Humana Press, Totowa, NJ, pp. 101–150.
5. Androutsellis-Theotokis, A., Ueda, K., Beppu, T., Goldberg, N. R., Das, S., Javitch, J. A., and Rudnick, G. (2003) Characterization of a functional bacterial homologue of sodium-dependent neurotransmitter transporters. *J. Biol. Chem.*
6. Williams, K. A. (2000) Three-dimensional structure of the ion-coupled transport protein NhaA. *Nature* **403,** 112–115.
7. Heymann, J. A., Sarker, R., Hirai, T., Shi, D., Milne, J. L., Maloney, P. C., and Subramaniam, S. (2001) Projection structure and molecular architecture of OxlT, a bacterial membrane transporter. *Embo. J.* **20,** 4408–4413.
8. Melikian, H. E., McDonald, J. K., Gu, H., Rudnick, G., Moore, K. R., and Blakely, R. D. (1994) Human norepinephrine transporter. Biosynthetic studies using a site-directed polyclonal antibody. *J. Biol. Chem.* **269,** 12290–12297.
9. Bruss, M., Hammermann, R., Brimijoin, S., and Bonisch, H. (1995) Antipeptide antibodies confirm the topology of the human norepinephrine transporter. *J. Biol. Chem.* **270,** 9197–9201.
10. Chen, J. G., Liu-Chen, S., and Rudnick, G. (1998) Determination of external loop topology in the serotonin transporter by site-directed chemical labeling [In Process Citation]. *J. Biol. Chem.* **273,** 12675–12681.
11. Ferrer, J. and Javitch, J. A. (1998) Cocaine alters the accessibility of endogenous cysteines in putative extracellular and extracellular loops of the human dopamine transporter. *Proc. Natl. Acad. Sci. USA* **95,** 9238–9243.
12. Barker, E. L., Kimmel, H. L., and Blakely, R. D. (1994) Chimeric human and rat serotonin transporters reveal domains involved in recognition of transporter ligands. *Mol. Pharmacol.* **46,** 799–807.
13. Buck, K. J. and Amara, S. G. (1994) Chimeric dopamine-norepinephrine transporters delineate structural domains influencing selectivity for catecholamines and 1methyl-4-phenylpyridinium. *Proc. Natl. Acad. Sci. USA* **91,** 12584–12588.
14. Giros, B., Wang, Y. M., Suter, S., McLeskey, S. B., Pifl, C., and Caron, M. G. (1994) Delineation of discrete domains for substrate, cocaine, and tricyclic antidepressant interactions using chimeric dopamine-norepinephrine transporters. *J. Biol. Chem.* **269,** 15985–15988.
15. Buck, K. J. and Amara, S. G. (1995) Structural domains of catecholamine transporter chimeras involved in selective inhibition by antidepressants and psychomotor stimulants. *Mol. Pharmacol.* **48,** 1030–1037.
16. Kitayama, S., Shimada, S., Xu, H., Markham, L., Donovan, D. M., and Uhl, G. R. (1992) Dopamine transporter site-directed mutations differentially alter substrate transport and cocaine binding. *Proc. Natl. Acad. Sci. USA* **89,** 7782–7785.
17. Barker, E. L. and Blakely, R. D. (1996) Identification of a single amino acid, phenylalanine 586, that is responsible for high affinity interactions of tricyclic antidepressants with the human serotonin transporter. *Mol. Pharmacol.* **50,** 957–965.

18. Bismuth, Y., Kavanaugh, M. P., and Kanner, B. I. (1997) Tyrosine 140 of the gamma-aminobutyric acid transporter GAT-1 plays a critical role in neurotransmitter recognition. *J. Biol. Chem.* **272,** 16096–16102.
19. Barker, E. L. and Blakely, R. D. (1998) Structural determinants of neurotransmitter transport using cross-species chimeras: studies on serotonin transporter. *Methods Enzymol* **296,** 475–498.
20. Barker, E. L., Moore, K. R., Rakhshan, F., and Blakely, R. D. (1999) Transmembrane domain I contributes to the permeation pathway for serotonin and ions in the serotonin transporter [In Process Citation]. *J. Neurosci.* **19,** 4705–4717.
21. Adkins, E. M., Barker, E. L., and Blakely, R. D. (2001) Interactions of tryptamine derivatives with serotonin transporter species variants implicate transmembrane domain I in substrate recognition. *Mol. Pharmacol.* **59,** 514–523.
22. Chen, N., Vaughan, R. A., and Reith, M. E. (2001) The role of conserved tryptophan and acidic residues in the human dopamine transporter as characterized by site-directed mutagenesis. *J. Neurochem.* **77,** 1116–1127.
23. Vaughan, R. A. (1995) Photoaffinity-labeled ligand binding domains on dopamine transporters identified by peptide mapping. *Mol. Pharmacol.* **47,** 956–964.
24. Rasmussen, S. G., Carroll, F. I., Maresch, M. J., Jensen, A. D., Tate, C. G., and Gether, U. (2001) Biophysical characterization of the cocaine binding pocket in the serotonin transporter using a fluorescent cocaine analogue as a molecular reporter. *J. Biol. Chem.* **276,** 4717–4723.
25. Chen, J. G., Sachpatzidis, A., and Rudnick, G. (1997) The third transmembrane domain of the serotonin transporter contains residues associated with substrate and cocaine binding. *J. Biol. Chem.* **272,** 28321–28327.
26. Chen, N., Ferrer, J. V., Javitch, J. A., and Justice, J. B., Jr. (2000) Transport-dependent accessibility of a cytoplasmic loop cysteine in the human dopamine transporter. *J. Biol. Chem.* **275,** 1608–1614.
27. Androutsellis-Theotokis, A., Ghassemi, F., and Rudnick, G. (2001) A conformationally sensitive residue on the cytoplasmic surface of serotonin transporter. *J. Biol. Chem.* **9,** 9.
28. Androutsellis-Theotokis, A. and Rudnick, G. (2002) Accessibility and conformational coupling in serotonin transporter predicted internal domains. *J. Neurosci.* **22,** 8370–8378.
29. Schmid, J. A., Scholze, P., Kudlacek, O., Freissmuth, M., Singer, E. A., and Sitte, H. H. (2000) Oligomerization of the human serotonin transporter and of the rat GABA transporter 1 visualized by fluorescence resonance energy transfer microscopy in living cells. *J. Biol. Chem.* **8,** 8.
30. Hastrup, H., Karlin, A., and Javitch, J. A. (2001) Symmetrical dimer of the human dopamine transporter revealed by cross-linking Cys-306 at the extracellular end of the sixth transmembrane segment. *Proc. Natl. Acad. Sci. USA* **98,** 10055–10060.
31. Frederickson, C. J. (1989) Neurobiology of zinc and zinc-containing neurons. *Int. Rev. Neurobiol.* **31,** 145–238.
32. Vallee, B. L. and Falchuk, K. H. (1993) The biochemical basis of zinc physiology. *Physiol. Rev.* **73,** 79–118.
33. Alberts, I. L., Nadassy, K., and Wodak, S. J. (1998) Analysis of zinc binding sites in protein crystal structures. *Protein. Sci.* **7,** 1700–1716.
34. Auld, D. S. (2001) Zinc coordination sphere in biochemical zinc sites. *Biometals* **14,** 271–313.
35. Schwabe, J. W. and Klug, A. (1994) Zinc mining for protein domains [news; comment]. *Nat. Struct. Biol.* **1,** 345–349.
36. Norregaard, L., Visiers, I., Loland, C. J., Ballesteros, J., Weinstein, H., and Gether, U. (2000) Structural probing of a microdomain in the dopamine transporter by engineering of artificial Zn(²⁺) binding sites. *Biochemistry* **39,** 15836–15846.

37. Draguhn, A., Verdorn, T. A., Ewert, M., Seeburg, P. H., and Sakmann, B. (1990) Functional and molecular distinction between recombinant rat GABAA receptor subtypes by Zn^{2+}. *Neuron* **5**, 781–788.

38. Horenstein, J. and Akabas, M. H. (1998) Location of a high affinity Zn^{2+} binding site in the channel of alpha1beta1 gamma-aminobutyric acidA receptors. *Mol. Pharmacol.* **53**, 870–877.

39. Bloomenthal, A. B., Goldwater, E., Pritchett, D. B., and Harrison, N. L. (1994) Biphasic modulation of the strychnine-sensitive glycine receptor by Zn^{2+}. *Mol. Pharmacol.* **46**, 1156–1159.

40. Laube, B., Kuhse, J., Rundstrom, N., Kirsch, J., Schmieden, V., and Betz, H. (1995) Modulation by zinc ions of native rat and recombinant human inhibitory glycine receptors. *J. Physiol. (Lond.)* **483**, 613–619.

41. Peters, S., Koh, J., and Choi, D. W. (1987) Zinc selectively blocks the action of *N*-methyl-D-aspartate on cortical neurons. *Science* **236**, 589–593.

42. Westbrook, G. L. and Mayer, M. L. (1987) Micromolar concentrations of Zn^{2+} antagonize NMDA and GABA responses of hippocampal neurons. *Nature* **328**, 640–643.

43. Hollmann, M., Boulter, J., Maron, C., Beasley, L., Sullivan, J., Pecht, G., and Heinemann, S. (1993) Zinc potentiates agonist-induced currents at certain splice variants of the NMDA receptor. *Neuron* **10**, 943–954.

44. Choi, Y. B. and Lipton, S. A. (1999) Identification and mechanism of action of two histidine residues underlying high-affinity Zn^{2+} inhibition of the NMDA receptor. *Neuron* **23**, 171–180.

45. Fayyazuddin, A., Villarroel, A., Le Goff, A., Lerma, J., and Neyton, J. (2000) Four residues of the extracellular N-terminal domain of the NR2A subunit control high-affinity Zn^{2+} binding to NMDA receptors. *Neuron* **25**, 683–694.

46. Holst, B., Elling, C. E., and Schwartz, T. W. (2002) Metal ion-mediated agonism and agonist enhancement in melanocortin MC1 and MC4 receptors. *J. Biol. Chem.* **277**, 47662–47670.

47. Rosenkilde, M. M., Lucibello, M., Holst, B., and Schwartz, T. W. (1998) Natural agonist enhancing bis-His zinc-site in transmembrane segment V of the tachykinin NK3 receptor. *FEBS Lett.* **439**, 35–40.

48. Swaminath, G., Steenhuis, J., Kobilka, B., and Lee, T. W. (2002) Allosteric modulation of beta2-adrenergic receptor by Zn($2+$) *Mol. Pharmacol.* **61**, 65–72.

49. Swaminath, G., Lee, T. W., and Kobilka, B. (2003) Identification of an allosteric binding site for Zn^{2+} on the beta2 adrenergic receptor. *J. Biol. Chem.* **278**, 352–356.

50. Norregaard, L., Frederiksen, D., Nielsen, E. O., and Gether, U. (1998) Delineation of an endogenous zinc-binding site in the human dopamine transporter. *Embo. J.* **17**, 4266–4273.

51. Loland, C. J., Norregaard, L., and Gether, U. (1999) Defining proximity relationships in the tertiary structure of the dopamine transporter. Identification of a conserved glutamic acid as a third coordinate in the endogenous Zn($2+$)-binding site. *J. Biol. Chem.* **274**, 36928–36934.

52. Vandenberg, R. J., Mitrovic, A. D., and Johnston, G. A. (1998) Molecular basis for differential inhibition of glutamate transporter subtypes by zinc ions. *Mol. Pharmacol.* **54**, 189–196.

53. Frederickson, C. J. and Bush, A. I. (2001) Synaptically released zinc: physiological functions and pathological effects. *Biometals* **14**, 353–366.

54. Vogt, K., Mellor, J., Tong, G., and Nicoll, R. (2000) The actions of synaptically released zinc at hippocampal mossy fiber synapses. *Neuron* **26**, 187–196.

55. Li, Y., Hough, C. J., Frederickson, C. J., and Sarvey, J. M. (2001) Induction of mossy fiber → Ca3 long-term potentiation requires translocation of synaptically released Zn^{2+}. *J. Neurosci.* **21**, 8015–8025.

56. Li, Y., Hough, C. J., Suh, S. W., Sarvey, J. M., and Frederickson, C. J. (2001) Rapid translocation of Zn($2+$) from presynaptic terminals into postsynaptic hippocampal neurons after physiological stimulation. *J. Neurophysiol.* **86**, 2597–2604.

57. Assaf, S. Y. and Chung, S. H. (1984) Release of endogenous Zn²⁺ from brain tissue during activity. *Nature* **308,** 734–736.

58. Berg, J. M. and Shi, Y. (1996) The galvanization of biology: a growing appreciation for the roles of zinc. *Science* **271,** 1081–1085.

59. Elling, C. E., Nielsen, S. M., and Schwartz, T. W. (1995) Conversion of antagonist-binding site to metal-ion site in the tachykinin NK-1 receptor. *Nature* **374,** 74–77.

60. Voss, J., Hubbell, W. L., and Kaback, H. R. (1995) Distance determination in proteins using designed metal ion binding sites and site-directed spin labeling: application to the lactose permease of Escherichia coli. *Proc. Natl. Acad. Sci. USA* **92,** 12300–12303.

61. Voss, J., Salwinski, L., Kaback, H. R., and Hubbell, W. L. (1995) A method for distance determination in proteins using a designed metal ion binding site and site-directed spin labeling: evaluation with T4 lysozyme. *Proc. Natl. Acad. Sci. USA* **92,** 12295–12299.

62. Elling, C. E. and Schwartz, T. W. (1996) Connectivity and orientation of the seven helical bundle in the tachykinin NK-1 receptor probed by zinc site engineering. *Embo. J.* **15,** 6213–6219.

63. Thirstrup, K., Elling, C. E., Hjorth, S. A., and Schwartz, T. W. (1996) Construction of a high affinity zinc switch in the kappa-opioid receptor. *J. Biol. Chem.* **271,** 7875–7878.

64. Elling, C. E., Thirstrup, K., Nielsen, S. M., Hjorth, S. A., and Schwartz, T. W. (1997) Engineering of metal-ion sites as distance constraints in structural and functional analysis of 7TM receptors. *Fold Des.* **2,** S76–80.

65. MacAulay, N., Bendahan, A., Loland, C. J., Zeuthen, T., Kanner, B. I., and Gether, U. (2001) Engineered Zn{super2}+ switches in the GABA transporter-1: differential effects on GABA uptake and currents. *J. Biol. Chem.* **29,** 29.

66. Loland, C. J., Norregaard, L., Litman, T., and Gether, U. (2002) Generation of an activating Zn(2+) switch in the dopamine transporter: mutation of an intracellular tyrosine constitutively alters the conformational equilibrium of the transport cycle. *Proc. Natl. Acad. Sci. USA* **99,** 1683–1688.

67. Norgaard-Nielsen, K., Norregaard, L., Hastrup, H., Javitch, J. A., and Gether, U. (2002) Zn(2+) site engineering at the oligomeric interface of the dopamine transporter. *FEBS Lett.* **524,** 87–91.

68. Elling, C. E., Thirstrup, K., Holst, B., and Schwartz, T. W. (1999) Conversion of agonist site to metal-ion chelator site in the beta(2)- adrenergic receptor. *Proc. Natl. Acad. Sci. USA* **96,** 12322–12327.

69. Regan, L. (1995) Protein design: novel metal-binding sites. *Trends Biochem. Sci.* **20,** 280–285.

70. Ippolito, J. A., Baird, T. T., Jr., McGee, S. A., Christianson, D. W., and Fierke, C. A. (1995) Structure-assisted redesign of a protein-zinc-binding site with femtomolar affinity. *Proc. Natl. Acad. Sci. USA* **92,** 5017–5021.

71. He, M. M., Voss, J., Hubbell, W. L., and Kaback, H. R. (1995) Use of designed metal-binding sites to study helix proximity in the lactose permease of Escherichia coli. 2. Proximity of helix IX (Arg302) with helix X (His322 and Glu325). *Biochemistry* **34,** 15667–15670.

72. He, M. M., Voss, J., Hubbell, W. L., and Kaback, H. R. (1995) Use of designed metal-binding sites to study helix proximity in the lactose permease of Escherichia coli. 1. Proximity of helix VII (Asp237 and Asp240) with helices X (Lys319) and XI (Lys358). *Biochemistry* **34,** 15661–15666.

73. Jung, K., Voss, J., He, M., Hubbell, W. L., and Kaback, H. R. (1995) Engineering a metal binding site within a polytopic membrane protein, the lactose permease of Escherichia coli. *Biochemistry* **34,** 6272–6277.

74. Holst, B., Elling, C. E., and Schwartz, T. W. (2000) Partial agonism through a zinc-Ion switch constructed between transmembrane domains III and VII in the tachykinin NK(1) receptor. *Mol. Pharmacol.* **58,** 263–270.

75. Sheikh, S. P., Zvyaga, T. A., Lichtarge, O., Sakmar, T. P., and Bourne, H. R. (1996) Rhodopsin activation blocked by metal-ion-binding sites linking transmembrane helices C and F. *Nature* **383,** 347–350.

76. Farrens, D. L., Altenbach, C., Yang, K., Hubbell, W. L., and Khorana, H. G. (1996) Requirement of rigid-body motion of transmembrane helices for light activation of rhodopsin. *Science* **274,** 768–770.

77. Jensen, A. D., Guarnieri, F., Rasmussen, S. G., Asmar, F., Ballesteros, J. A., and Gether, U. (2001) Agonist-induced conformational changes at the cytoplasmic side of TM6 in the {beta}2 adrenergic receptor mapped by site-selective fluorescent labeling. *J. Biol. Chem.* **276,** 9279–9290.

78. Ballesteros, J. A., Jensen, A. D., Liapakis, G., Rasmussen, S. G., Shi, L., Gether, U., and Javitch, J. A. (2001) Activation of the beta 2-adrenergic receptor involves disruption of an ionic lock between the cytoplasmic ends of transmembrane segments 3 and 6. *J. Biol. Chem.* **276,** 29171–29177.

79. Jensen, A. A., Sheppard, P. O., Jensen, L. B., O'Hara, P. J., and Brauner-Osborne, H. (2001) Construction of a high affinity zinc binding site in the metabotropic glutamate receptor mGluR1: noncompetitive antagonism originating from the amino-terminal domain of a family C G protein-coupled receptor. *J. Biol. Chem.* **276,** 10110–10118.

80. Levi, G. and Raiteri, M. (1993) Carrier-mediated release of neurotransmitters. *Trends Neurosci.* **16,** 415–419.

81. Sitte, H. H., Huck, S., Reither, H., Boehm, S., Singer, E. A., and Pifl, C. (1998) Carrier-mediated release, transport rates, and charge transfer induced by amphetamine, tyramine, and dopamine in mammalian cells transfected with the human dopamine transporter. *J. Neurochem.* **71,** 1289–1297.

82. Fischer, J. F. and Cho, A. K. (1979) Chemical release of dopamine from striatal homogenates: evidence for an exchange diffusion model. *J. Pharmacol. Exp. Ther.* **208,** 203–209.

83. Scholze, P., Norregaard, L., Singer, E. A., Freissmuth, M., Gether, U., and Sitte, H. H. (2002) The role of zinc ions in reverse transport mediated by monoamine transporters. *J. Biol. Chem.* **8,** 8.

84. Falkenburger, B. H., Barstow, K. L., and Mintz, I. M. (2001) Dendrodendritic inhibition through reversal of dopamine transport. *Science* **293,** 2465–2470.

85. Rudnick, G. (1997) Mechanisms of biogenic amine neurotransporters, in *Neurotransmitter Transporters: Structure, Function, and Regulation* 1st ed. (Reith, M. E. A., ed.), Humana Press, Totowa, NJ, pp. 73–100.

86. Lefkowitz, R. J., Cotecchia, S., Samama, P., and Costa, T. (1993) Constitutive activity of receptors coupled to guanine nucelotide regulatory proteins. *Trends Pharmacol. Sci.* **14,** 303–307.

87. Gether, U. (2000) Uncovering molecular mechanisms involved in activation of G protein-coupled receptors. *Endocr. Rev.* **21,** 90–113.

88. Wang, J. B., Moriwaki, A., and Uhl, G. R. (1995) Dopamine transporter cysteine mutants: second extracellular loop cysteines are required for transporter expression. *J. Neurochem.* **64,** 1416–1419.

89. Chen, J. G., Liu-Chen, S., and Rudnick, G. (1997) External cysteine residues in the serotonin transporter. *Biochemistry* **36,** 1479–1486.

90. Sur, C., Schloss, P., and Betz, H. (1997) The rat serotonin transporter: identification of cysteine residues important for substrate transport. *Biochem. Biophys. Res. Commun.* **241,** 68–72.

91. Thayer, M. M., Flaherty, K. M., and McKay, D. B. (1991) Three-dimensional structure of the elastase of Pseudomonas aeruginosa at 1.5-A resolution. *J. Biol. Chem.* **266,** 2864–2871.

92. Hakansson, K., Carlsson, M., Svensson, L. A., and Liljas, A. (1992) Structure of native and apo carbonic anhydrase II and structure of some of its anion-ligand complexes. *J. Mol. Biol.* **227,** 1192–1204.

12

A Structural Context for Studying Neurotransmitter Transporter Function

Naomi R. Goldberg, Thijs Beuming, Harel Weinstein, and Jonathan A. Javitch

1. INTRODUCTION

The Na$^+$/Cl$^-$-dependent neurotransmitter transporters (NTs) constitute a family of homologous membrane proteins responsible for the reuptake from the synaptic cleft of neurotransmitters, including dopamine, serotonin (5-HT), norepinephrine, γ-aminobutyric acid (GABA), and glycine, as well as other small molecules, such as proline, creatine, betaine, and taurine. These transporters couple the movement of sodium down its electrochemical gradient to the translocation of substrate across the plasma membrane. The biogenic amine transporters, including the dopamine transporter (DAT), norepinephrine transporter (NET), and the serotonin transporter (SERT), are the molecular targets for psychostimulant drugs, including cocaine and amphetamine. These transporters are also targets for many antidepressants.

The absence of a high-resolution structure for any member of this group limits current understanding of the relationship between the structure and function of these transporters. Insights into structure–function properties of the transporters have emerged from analysis of the effects of site-directed mutations and chimeras *(1,2)*. This chapter reviews what is known about the topology, structure, and molecular basis of binding and transport in the family of NTs. To aid in the comparative analysis, we introduce here a novel generic numbering scheme to facilitate the comparison of aligned positions within the sequences of the entire NT family.

2. GENERIC NUMBERING SCHEME

We have developed a common residue numbering scheme that facilitates comparison of the sequences of different NTs. The numbering scheme is informative of the relative position of each amino acid, the amino acid present at that position, and the actual amino acid number in a particular transporter. Each index number starts with the number of the transmembrane segment (TM), e.g., 1 for TM1, and is followed by a number indicating the position relative to a reference residue that is the most conserved position in that TM. That reference residue is arbitrarily assigned the number 50. For exam-

From: *Molecular Neuropharmacology: Strategies and Methods*
Edited by: A. Schousboe and H. Bräuner-Osborne © Humana Press Inc., Totowa, NJ

ple, as our reference residue in TM1 we chose, in the manner described below, a tryptophan whose index number would be 1.50, i.e., $Trp_{1.50}$. When referring to a particular transporter, the index number can be preceded by the number of the residue in the particular sequence. In the human dopamine transporter (hDAT), the tryptophan in TM1 is at position 84, so this residue will be referred to as $Trp84_{1.50}$. A serine residue located five amino acids C-terminal to $Trp84_{1.50}$ will be $Ser89_{1.55}$. A similar numbering scheme has been developed for G protein-coupled receptors *(3)*.

The conserved residues identified as 50 were chosen from a sequence alignment analysis. There are more than 100 residues among a set of 51 mammalian sodium-dependent NTs of known function that are 100% conserved (alignment not shown). To decide which residue is the most appropriate reference residue for each of the TMs, we searched for additional sequences with similarity to the mammalian NTs, and investigated the decrease in conservation for alignments of increasing size.

Blasting the Swissprot and TrEMBL databases with several query NT sequences yielded 246 hits with e-values less than 0.001. (No fragments were included in the alignment.) These included 102 mammalian proteins (of which 33 are orphans), 36 drosophila proteins (of which 32 are orphans), 16 proteins from *Caenochobditis* elegans (of which 14 are orphans), 57 bacterial proteins[1], 12 archaeal proteins, and 23 proteins from other species. These sequences were ranked in terms of their similarity to the original set of 51 mammalian transporters. Each of the 246 sequences was aligned individually against each of the 51 mammalian transporters, and the average similarity (% identity) was taken as a measure of the similarity of that sequence to the group of mammalian transporters.

Multiple alignments were generated with gradually decreasing average sequence identity. The alignments included all sequences with average identity higher then 30% (69 sequences, including all mammalian NTs), 20% (108 sequences), 15% (135 sequences), and 10% (197 sequences), respectively. Based on these alignments, we have chosen the most appropriate "reference residue" in each TM based on the extent of conservation in as large an alignment as possible, with the additional criterion of it being located within the putative transmembrane segment (see below). The 12 amino acids chosen as reference residues are shown in Table 1 and are indicated by blue circles in Figs. 1 and 2.

All reference residues are 100% conserved in the mammalian NTs, and all are highly conserved in the sequences that are assumed to share a similar structural framework (i.e., identity > 20%). (Although $Cys_{4.50}$ is 86% conserved in the 108 sequences with greater than 20% sequence identity, threonine is present in another 9% of these sequences, for a total conservation of 95%.) Most reference residues have a high degree of conservation in all sequences, except for $Cys/Thr_{4.50}$, which is only conserved among all sequences with identity > 15% (but which is nonetheless the most conserved residue in TM4 in the larger alignments). For the larger alignments (≤15% average identity), a number of sequences contain insertions and deletions in TM9 and TM10, in particular at positions $Gly_{9.50}$ and $Gly_{10.50}$.

---

[1] TnaT, a member of this family from *Symbiobacterium thermophilum,* has recently been discovered to be a sodium-dependent tryptophan transporter (Androutsellis-Theotokis, A., Goldberg, N. R., Ueda, K., et al. (2003) Characterization of a functional bacterial homologue of sodium-dependent neurotransmitter transporters. *J. Biol. Chem.* **278,** 12703–12709).

Table 1
Neurotransmitter Transporter Index Positions

TM	Index number	Reference residue in hDAT	Conservation in all 108 sequences with >20% identity (%)	Conservation in all 246 sequences (%)
1	1.50	W84	99	87
2	2.50	P112	99	88
3	3.50	Y156	97	88
4	4.50	C243	86	60
5	5.50	L287	100	88
6	6.50	Q317	100	88
7	7.50	F365	98	84
8	8.50	F412	100	85
9	9.50	G468	99	91
10	10.50	G500	99	74
11	11.50	P529	100	81
12	12.50	G561	97	68

3. SECONDARY STRUCTURE PREDICTION OF NA⁺/CL⁻-COUPLED NEUROTRANSMITTER TRANSPORTERS

To develop a structural context for the management of structure–function data, the secondary structure of this transporter family was predicted, based on approaches and algorithms that have served in the study of other families of unknown structure (e.g., the GPCRs *[3,4]*). The methods have been collected in a suite of programs named ProperTM (http://icb.med.cornell.edu/services/propertm/start), which allows for user-driven sequential applications of various algorithms that have been applied broadly and have been validated repeatedly by subsequent determined structures (for reviews, refs. *see [3,4]*). In brief, these methods include the calculation of properties (hydrophobicity, volume, conservation) associated with positions in a multiple sequence alignment, and the prediction of secondary structure and protein-lipid interfaces from this information. The methods in ProperTM have been designed to analyze membrane proteins, but some of the applications (e.g., the calculation of a conservation index) should be useful for the analysis of nonmembrane proteins as well. The representations of the structural features in the annotated helical nets in Fig. 1 display information obtained with this suite of programs, calculated from the alignment of NT sequences. The structural characteristics of the transporters summarized in the figures include the following.

3.1. Hydrophobicity

The parsing of the transporter sequences into the TM domains shown in Fig. 1A represents the consensus result of three different methods. Average hydrophobicity was calculated with ProperTM using different window sizes and the Kyte and Doolittle scale *(7)*. TMHMM, a hidden Markov model-based approach *(8)*, and PHDHTM, a profile-based neural network method *(9)*, were then utilized to refine the predictions.

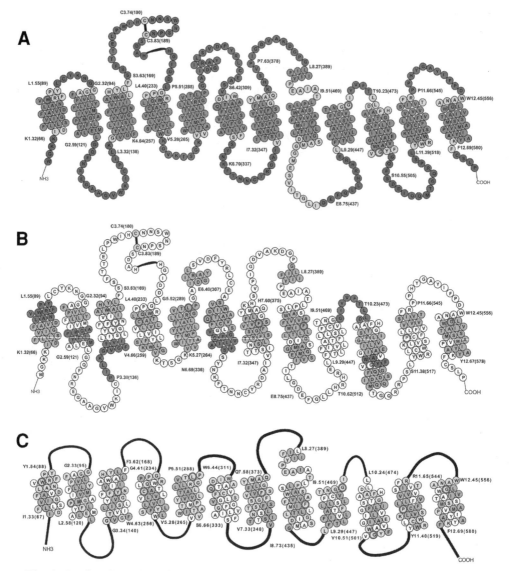

Fig. 1. Predicted structural properties of neurotransmitter transporters. The helical nets represent the human dopamine transporter, with the single letter code of each amino acid within a circle. The most conserved residues in the 12 transmembrane segments (index number TM#.50–*see* text) are identified by thick blue lines surrounding the circles. The index numbers are shown for selected residues, with human DAT residue numbers in parentheses. In (A) and (B), cysteines in the second extracellular loop thought to form a disulfide bond are shown in yellow. When not shown, residues in the N- and C-termini and in loops are indicated by thick black lines. **(A)** Predicted transmembrane topology. Based on hydrophobicity analysis, cyan residues are predicted to be located within the lipid bilayer, and red residues are predicted to be solvent exposed outside the lipid boundaries. Regions that are not well-defined by the prediction methods are shown in pink. The assignment of residues to the transmembrane segment is based on this hydrophobicity analysis as well as on predictions of secondary structure, lipid accessibility and the summarized experimental data, all of which are discussed below and in the text. **(B)** Secondary structure propensity based on the spatial periodicity of residue properties. Predicted α-helical segments are shown in orange; sequential residues with periodicity consistent with β-strand are shown in green. Segments shown in white did not exhibit an identifiable periodicity. **(C)** Probability of lipid exposure. Positions shown in purple are predicted to lie on the helix-lipid interface, positions in yellow are predicted to face the protein interior, and white regions are undefined. (see text).

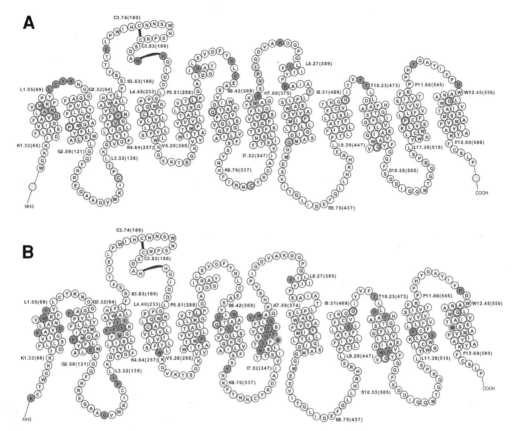

Fig. 2. Summary of experimental findings from the neurotransmitter transporter family. As in Fig. 1, the helical nets represent the human dopamine transporter, with the single letter code of each amino acid within a circle and with the index positions shown in blue. Residues not shown in the N- and C-termini are indicated by thick black lines. (**A**) Cysteine accessibility and zinc sites. Endogenous cysteines or substituted cysteines in DAT (or at the aligned positions in other NTs) are shown in orange if they were found to be accessible to impermeant sulfhydryl reagents when applied extracellularly. All accessible residues in the TMs are shown in yellow, and residues accessible to impermeant reagents only in membranes or after permeabilization are shown in light yellow. Residues that are protected from reaction by the presence of substrate and/or inhibitor are circled in magenta. Endogenous zinc sites in DAT are shown in bright green, whereas engineered zinc sites are shown in light green. (**B**) Mutations affecting substrate and/or inhibitor recognition. As described in the text, mutations in DAT, or in aligned positions in homologous NTs, that decreased the apparent affinity for substrate or the binding affinity for inhibitors are shown in orange. Also shown in orange are the random mutations in TM7 that decreased or abolished transport and a number of other mutations that affected the function of the transporters (*see* text). Residues shown in orange that were predicted to face lipid in (**C**) are shown surrounded by gold circles.

3.2. Secondary-Structure Propensity

A central element in the prediction of secondary structure is the periodicity of sequence conservation, which has proven to be a good indicator in a number of membrane proteins *(5)*. The periodicity is quantified by Fourier transform (FT) analysis. A property profile is calculated over a window size N to produce a power spectrum P(ω); if the sequence contained within the window N adopts α-helical conformation, a peak in the power spectrum should appear around 105°, the angle between adjacent side chains in an α-helix viewed down its axis. For β strands, the peak should appear between 166° and 180°. The α-helical and β-strand periodicities calculated from P(ω) characterize the relative extent of periodicity in the α-helical/β-strand region compared to the entire spectrum. These are shown in Fig. 1B.

3.3. Conservation and Lipid-Facing Probability

An approach to the calculation of conservation has been proposed on the basis of the polytope-method and the information content *(4,6)*. Briefly, the degree of conservation at each position in the alignment is determined by the number of different amino acids, the probability of finding a particular residue replaced by another, and the frequency of appearance of each type of residue. These factors are integrated to calculate the conservation index (CI).

The prediction of lipid-facing orientations of the TM segments was based on the analysis of properties of individual residues in the context of the alignment. Thus, lipid-facing residues were considered to have a low conservation index, high average hydrophobicity (hdp), and a low hydrophobicity standard deviation (σ_{hdp}). These factors were integrated in ProperTM to calculate the probability that an amino acid lies on a protein-lipid interface. This method was tested on the known structure of the photo-reaction center and a multiple sequence alignment of 12 photo-reaction center sequences (Ballesteros et al., unpublished) and was found to be very discriminating when used with an appropriate threshold value. Lipid-facing probabilities are shown in Fig. 1C. Note, however, that the strong dependence of the prediction of lipid-facing probability on the variability-criterion may misidentify inward-facing residues involved in subtype-selective functions as pointing outwards toward lipid. One possible example is the region in TM3 surrounding the conserved $Tyr_{3.50}$, in which nonconserved hydrophobic residues (e.g., $Val_{3.46}$) have been shown to be involved in ligand binding (*see* below).

4. TOPOLOGY

Various biochemical approaches have been used to support the proposed topology of the NTs. For example, comparing the effects of permeant and impermeant sulfhydryl reagents in cells and membranes enables inferences about the topology of transporters. Methanethiosulfonate (MTS) reagents comprise a group of sulfhydryl-reactive compounds that are highly selective for water-accessible ionized cysteines *(10,11)*. MTS ethylammonium (MTSEA), a weak base, is membrane permeant, as it can cross the membrane in the unprotonated state *(12)*. MTS ethyltrimethylammonium (MTSET), a charged quaternary ammonium, and MTS ethylsulfonate (MTSES), ionized at neutral pH, are both relatively membrane-impermeant *(11,12)*. Thus, in membranes, the MTS reagents are expected to have similar access to cysteines on the extracellular and cytoplasmic surfaces, whereas, in cells, water-accessible cysteines on the cytoplasmic surface should react with MTSEA at a significantly greater rate than with MTSET or MTSES. Studies of a mutant DAT transporter, "X5C," in which five endogenous cysteines were simultaneously replaced by other residues, as well as mutants in which each cysteine was restored one at a time within the X5C background, showed that $Cys90_{1.56}$, $Cys135_{3.29}$, $Cys306_{6.39}$, and $Cys342_{6.75}$ reacted with MTS reagents, as assessed by changes in binding of the cocaine analog, 2-β-carbomethoxy-3-β-(4-fluorophenyl) tropane (CFT) *(13)*. Moreover, the rates of reaction of $Cys90_{1.56}$ and $Cys306_{6.39}$ with MTSEA, MTSES, and MTSET were similar in cells and membranes, whereas $Cys135_{3.29}$ and $Cys342_{6.75}$ reacted much more rapidly with MTSEA in membranes than in cells. Thus, these studies support the proposed topology, predicting extracellular locations of $Cys90_{1.56}$ and $Cys306_{6.39}$, and cytoplasmic locations of $Cys135_{3.29}$ and $Cys342_{6.75}$ (see Fig. 2A). The aligned residue in the first extracellular loop (EL1) of SERT, $Cys109_{1.56}$, is also accessible to external MTSET in a cation-dependent manner *(14)*, and $Cys74_{1.56}$ in the GABA transporter (GAT) is accessible to external MTSET as well *(15)*.

Two highly conserved cysteines in EL2 are likely to form a disulfide bond. In SERT, mutation of one EL2 cysteine, $Cys200_{3.74}$, increased sensitivity to MTSET, suggesting increased extracellular accessibility of a previously unreactive cysteine residue *(14)*. The sensitivity to impermeant MTS reagents and decreased cell-surface delivery of $C200_{3.74}S$ were both reversed in the $C200_{3.74}S$-$C209_{3.83}S$ double mutant. These data support the existence of a disulfide bond between $Cys200_{3.74}$ and $Cys209_{3.83}$ and suggest that this disulfide is a structurally conserved feature among the transporter family. Consistent with this view, when EL2 $Cys180_{3.74}$ and $Cys189_{3.83}$ in DAT were replaced with alanine, transport activity was abolished *(16)*. Immunofluorescence studies of the DAT mutants indicated that the transporter was trapped in intracellular compartments and was unable to traffic to the plasma membrane. However, the absence of these cysteines and therefore of a disulfide bond in the bacterial and archaeal transporters, suggests that this disulfide is not essential to the sodium-dependent transport function of these proteins, but rather that the disulfide is important for biosynthesis in the eukaryotic members of the family.

Cysteine mutations in GAT verified an intracellular position for $Cys399_{8.63}$ in the fourth intracellular loop (IL4) between TM8 and TM9 *(17)*, and an extracellular location for positions $73_{1.55}$–$76_{1.58}$ in EL1 *(15)*, as cysteines at these positions conferred sensitivity to the sulfhydryl reagents, MTSEA and *N*-ethylmaleimide (for $Cys399_{8.63}$) and MTSET (residues $73_{1.55}$–$76_{1.58}$).

The studies described earlier, adaptations of the substituted cysteine accessibility method (SCAM) *(11)*, rely upon a functional readout for the chemical modification of a cysteine. If an effect is observed from the modification with a sulfhydryl reagent in a cysteine mutant, but not in the background construct, the substituted cysteine is inferred to be water-accessible. This is usually a reasonable inference, but can be wrong in isolated cases. Mutation of residues in TM7 of SERT have been shown to increase the accessibility of $Cys109_{1.56}$ (*see* below) and thereby to confer sensitivity to sulfhydryl reagents indirectly *(18)*. Thus, it is important to consider the potential effects of mutation on the reactivity of endogenous cysteines. Only a Cys-less background can completely avoid this potential complexity, but this has not been possible to achieve for any of the eukaryotic transporters in this family. A second and more common problem is that of a false-negative determination of reactivity based on the lack of a functional effect of reaction with a given cysteine. Just as many positions tolerate mutation to another residue, chemical modification may also be tolerated with limited functional impact. Therefore, methods have been developed to assess for reaction using direct biochemical readouts of reaction that do not rely upon altered function. For example, reaction of a biotinylated sulfhydryl or lysine reagent can be detected via precipitation with avidin agarose and subsequent detection via immunoblotting.

These methods have been used systematically to explore the topology of SERT. Cells expressing a mutant SERT in which all four endogenous extracellular lysines were mutated (eK-less) showed dramatically less labeling with the impermeant sulfo-NHS-SS-biotin reagent, as compared to wild-type *(19)*. When mutants containing only one of the original lysines in EL2, EL3, or EL4, were treated with sulfo-NHS-SS-biotin, they showed a substantial increase in labeling over that of the eK-less background, suggesting that each of these positions was accessible from the extracellular milieu. Similarly, MTSEA-biotin labeled wild-type SERT, but not a mutant in which $Cys109_{1.56}$, the sole endogenous cysteine predicted to reside in an extracellular loop, was replaced with alanine. Lysines and cysteines inserted into the remaining putative extracellular loops, caused the transporter to be more heavily labeled than the respective backgrounds, eK-less or $C109_{1.56}A$, suggesting accessibility of each of these loops to the extracellular milieu (*see* Fig. 2A). A similar analysis was conducted to examine cysteine insertion mutants in putative intracellular loops of SERT. In support of the predicted topology, cysteine residues in each predicted cytoplasmic domain, including the five predicted ILs and the N- and C- termini, reacted with MTS reagents upon permeabilization with digitonin, or in membranes, but not in intact cells *(20)*.

Also consistent with this labeling in SERT, antibodies raised against EL2 and EL4 of NET showed reactivity with intact COS cells expressing NET, as assessed by immunofluorescence, whereas antibodies against the amino and carboxyl terminal regions reacted only after cells were permeabilized *(21)*.

These studies, summarized in Fig. 2A, argue strongly in favor of the proposed common topology of the NT proteins and, thereby, against alternative topologies proposed for GAT and the glycine transporter (GLYT) *(22–24)*, which were based on glycosylation site scanning and fusions to reporter sequences. The use of small impermeant reagents, such as MTSET, in deducing topology is more attractive methodologically, as it allows detection of residues in small loops that may not be accessible to macromolecular probes, such as proteases or glycosylation machinery.

5. SECONDARY, TERTIARY, AND QUATERNARY STRUCTURE

5.1. Probing Secondary Structure

Mutagenesis studies of the neurotransmitter transporters have shed some light on the secondary, tertiary, and quaternary structure of various members of the family. By mutating to cysteine, one at a time, 20 residues in TM3 of SERT, Chen et al., identified three positions, $Ile172_{3.46}$, $Tyr176_{3.50}$, and $Ile179_{3.53}$, which, when substituted by cysteine, rendered SERT functionally sensitive to the sulfhydryl reagent MTSET *(25)*. Moreover, mutating $Ile172_{3.46}$ or $Tyr176_{3.50}$ to cysteine disrupted binding of 5-HT and cocaine to SERT. The spacing of the three residues suggested that they form a patch on an α-helix, thereby supporting the notion that TM3 is helical, with one side facing a binding pocket for substrate and inhibitor (*see* Fig. 2A).

Random mutagenesis studies on TM7 of rat SERT identified six amino acid residues as functionally significant by their sensitivity to nonconservative mutations *(26)*. Four of the six residues, $Asn368_{7.38}$, $Gly376_{7.46}$, $Phe380_{7.50}$, and $Gly384_{7.54}$, fall on a stripe that runs at an angle down one side of the predicted α-helix (Fig. 2B). An additional residue within the stripe, $Ser372_{7.42}$, was shown to be functionally significant, as mutation affected sodium dependence, although transport activity was retained despite a nonconservative substitution. The pattern of sensitivity to mutation formed by these five residues suggests an α-helical structure of TM7. The helical pattern breaks down at the carboxyl end of the helix, where three consecutive residues ($Gly384_{7.54}$, $Tyr385_{7.55}$, and $Met386_{7.56}$) were sensitive to mutation. This was interpreted as a boundary between the end of the α-helix and beginning of the hydrophilic loop, although the alternative explanation, that mutations at these positions of a continuing α-helix are not functionally tolerated is more consistent with our secondary structure predictions (Fig. 1A,B) and with the Zn^{2+} site data discussed below.

Engineering of metal binding sites in DAT has provided some insight into secondary and tertiary structure. An endogenous Zn^{2+} binding site has been identified *(27,28)*, involving $His193_{3.87}$ (in EL2), $His375_{7.60}$ (at the extracellular end of TM7), and $Glu396_{8.34}$ (at the extracellular end of TM8). By interacting at this site, Zn^{2+} acts as a potent noncompetitive blocker of dopamine uptake. In a mutant in which $His193_{3.87}$ and $Glu396_{8.34}$ were removed but $His375_{7.60}$ was maintained, addition of another cysteine at position $400_{8.38}$, i+4 from $Glu396_{8.34}$, produced a new Zn^{2+} binding site, which resulted in Zn^{2+} inhibiting dopamine uptake with a 30-fold increase in affinity *(28)*. In contrast, much smaller increases in affinity were observed with cysteines added at the (i-2) and (i-3) positions. Thus, the ability to engineer an artificial bidentate Zn^{2+} binding site between $His375_{7.60}$ and position $400_{8.38}$, but not $398_{8.36}$ or $399_{8.37}$, suggests the existence of an α-helical configuration in this region at the extracellular end of TM8, as well as the proximity of TM7 and TM8, which enables the formation of a geometrically defined Zn^{2+} binding pocket (*see* below).

In addition, Zn^{2+} was shown to inhibit dopamine uptake in a mutant containing an engineered tridentate zinc site, in which the i-4 site from $His375_{7.60}$, $Met371_{7.56}$, was replaced with histidine, whereas the introduction of histidines at the i-2, i-3, and i-5 position did not increase Zn^{2+} affinity *(29)*. In contrast, histidines at positions i+2, i+3, and i+4 all resulted in potent inhibition of dopamine uptake by Zn^{2+}. The incorporation of these data in a model of secondary structure provides evidence for an α-helical configuration of the extracellular portion of TM7, as well as the absence of well-defined secondary structure between positions $375_{7.60}$ and $379_{7.64}$ (Fig. 1B), thereby suggesting an approximate boundary between the C-terminal end of the helix and the beginning of EL4 *(29)*.

5.2. Inferences About Tertiary Structure

The restrictive geometry of Zn^{2+} binding sites also enables inferences regarding spatial proximity relations among various elements of secondary structure in the transporter. Such evidence for spatial proximity relationships has been inferred based both on the delineation of an endogenous Zn^{2+} binding site in DAT and the engineering of artificial binding sites. High-resolution structures of zinc-binding proteins show that the average distance between the Zn^{2+} ion and the coordinating atom of the zinc-binding residue averages 2 Å *(30)*. Because $His193_{3.87}$, $His375_{7.60}$, and $Glu396_{8.34}$ were shown to form three major coordinates for zinc binding, suggesting that these three residues are in close proximity to each other, they offered an important distance constraint in the tertiary structure of DAT involving part of EL2 near TM7 and the TM8 segment *(27,28)*. Moreover, as described earlier, the engineering of artificial Zn^{2+}-binding sites, the bidentate $His375_{7.60}$–$Cys400_{8.38}$ *(28)*, and the tridentate $His371_{7.56}$–$His375_{7.60}$–$Glu396_{8.34}$ *(29)*, further supports the close association between TM7 and TM8, and suggests that $His375_{7.60}$ must face TM8 and be positioned between $Glu396_{8.34}$ and $Thr400_{8.38}$. The functional transfer of these artificial Zn^{2+}-binding sites between TM7 and TM8 to rat GAT (rGAT) further supports the close proximity between TM7 and TM8 in the tertiary structure of GAT, and also suggests that the structural organization of the TM7/8 microdomain is evolutionarily conserved among NTs *(31)*.

5.3. The Identification of Quaternary Structure

A variety of experimental methods have been directed toward understanding the quaternary structure of the neurotransmitter transporters. Co-immunoprecipitation of differentially epitope-tagged forms of SERT in detergent suggested that SERT assembles into homo-oligomeric units *(32)*. In addition, in cells in which a sensitive and a resistant SERT construct were co-expressed in different ratios, the effect of modification with MTSEA on uptake lagged behind the fraction of sensitive SERT. This implies that SERT is in a complex and that it is necessary to inhibit more than one SERT to disrupt uptake, consistent with a dimeric form of SERT, or possibly with association of dimers into a higher order complex. Further evidence of oligomerization in vivo and in vitro was provided by fluorescence resonance energy transfer (FRET) microscopy of CFP- and YFP-labeled SERT and GAT *(33)*. No alteration in FRET was detected when transporter was co-incubated with substrates or blockers, suggesting that the homo-oligomeric state is not altered by these compounds.

In two radiation-inactivation studies, DAT was inferred to be a dimer *(34)* and a tetramer *(35)*. In more recent work, treatment of DAT-expressing HEK293 cells with cysteine-reactive cross-linking reagents and subsequent co-immunoprecipitation of differentially tagged DAT molecules, demonstrated that DAT in the plasma membrane can be cross-linked into a dimer *(36)*. $Cys306_{6.39}$, at the extracellular end of TM6, in each of the two DATs was shown to be the cross-linked residue. Interestingly, the motif GVXXGVXXA, which promotes dimerization in model systems *(37,38)*, is found at the intracellular end of TM6 in DAT, and is partially conserved in many other NTs. When either glycine in this motif was mutated, DAT expression and function were impaired. This suggested that the intracellular end of TM6, like the extracellular end, is part of a dimerization interface. Engineering of a Zn^{2+} binding site comprised of $Cys306_{6.39}$ and a histidine at position $310_{6.43}$ at the predicted dimeric interface resulted in Zn^{2+} inhibition of dopamine transport *(39)*. These results suggest that conformational changes necessary for substrate translocation may occur at the oligomeric interface between two DAT molecules.

An additional symmetrical interface involving residues in TM4 of DAT has recently been identified by cysteine cross-linking (Hastrup and Javitch, submitted). Reintroduction of the endogenous $Cys243_{4.50}$ and $Cys306_{6.39}$ into the Cys-depleted DAT background construct led to cross-linking into dimer, trimer, and tetramer, suggesting that the TM4 and TM6 interfaces are distinct and that DAT is likely to form a tetramer in the plasma membrane. Notably, cocaine-like DAT inhibitors protected against cross-linking of $Cys243_{4.50}$, suggesting that ligand-related helix movement may occur at the TM4 interface.

A requirement for dimeric or higher order complexes of DAT for proper trafficking to the plasma membrane has been inferred from a study that identified dominant-negative mutants of DAT *(40)*. Two nonfunctional mutants, $Y335_{6.68}A$ and $D79_{1.45}G$, were trafficked to the cell surface and inhibited wild-type DAT uptake activity. Because coexpression of the mutant DAT did not affect the surface expression of the wild-type DAT, the resulting inhibition presumably resulted from interactions between wild-type and mutant transporter. Also, trafficking-defective amino and carboxyl truncation mutants inhibited wild-type function by formation of oligomeric complexes that could not traffic to the cell-surface, as shown by immunofluorescence microscopy. This dominant-negative effect was disrupted by mutation of a leucine-repeat motif ($Leu99_{2.37}$, $Met 106_{2.44}$, $Leu113_{2.51}$, and $Leu120_{2.58}$) in TM2. Mutants in the repeat, which were inactive owing to a trafficking defect, did not exert a dominant-negative effect on wild-type, and the authors inferred that this resulted from an inability of the TM2 mutants to dimerize with wild-type DAT. The relationship between a dimeric TM2 interface and the TM4 and TM6 interfaces is not yet known, but recent experiments have shown that several substituted cysteines in TM2 of DAT can be cross-linked in a Cys-depleted DAT background (Sen et al., in preparation).

Mutation of leucines in a GAT-TM2 leucine heptad repeat also interfered with oligomerization, as assessed by a loss or elimination of FRET, and resulted in intracellular retention of GAT mutants, with the exception of $L97_{2.51}A$, which did reach the cell surface but with a considerable loss of energy transfer *(41)*. Furthermore, intermolecular FRET was observed during the maturation of wild-type GAT in the endoplasmic reticulum (ER). These findings support the proposal that oligomerization of GAT during biosynthesis is important for export from the ER, and that mutation of the leucine heptad repeat in TM2 impairs oligomerization.

Although there is abundant evidence supporting oligomerization of the NTs, it is possible that not all members of the family share an identical quaternary structure. Blue native gel electrophoresis, coupled with selective surface-labeling methods, was used to infer that glycine transporters, GLYT1 and GLYT2, exist as monomers in the plasma membrane of *Xenopus* oocytes *(42)*. Treatment with the cross-linker glutaraldehyde also did not induce oligomerization of GLYT. Although the absence of cross-linking by this nonspecific lysine reactive reagent does not prove the absence of oligomeric structure in GLYT, because endogenous lysines in a dimer may not be appropriately positioned to cross-link, it nonetheless raises the possibility that organization into higher-order oligomeric complexes may not be a general feature of the NTs.

6. SUBSTRATE AND INHIBITOR BINDING SITES

Delineation of the binding sites for substrates and inhibitors in the NT family remains a major hurdle to progress in this field, in spite of intensive studies aimed at resolving the question of which domains or amino acids constitute the binding sites. As summarized in this section, general regions have been inferred from studies with chimeras or from photoaffinity labeling analyses, whereas individual residues have been suggested to have a role in binding based on site-directed mutagenesis studies. It is often difficult, however, to determine whether the functional effect of a site-directed mutation is direct or indirect, and therefore, interpretation of mutagenesis studies must be approached with caution.

6.1. Affinity Labeling

Functional domains of DAT have been studied using photoaffinity ligands. Epitope-specific immunoprecipitation of proteolytic fragments of labeled DAT showed that [125I]1-[2-(diphenylmethoxy)-ethyl]-4-[2-(4-azido-3-iodophenyl) ethyl] piperazine ([125I]DEEP) was incorporated into a fragment containing TM1 and TM2. In contrast, [125I]-3β-(p-chlorophenyl)tropane-2β-carboxylic acid, 4′-azido-3′-iodophenylethyl ester ([125I]RTI 82), a cocaine analog, was incorporated into a fragment containing TM4-TM7 *(43,44)*. Despite having a tropane-based structure (like RTI-82), [125I]GA 2–34, a potent benztropine analog, was also found to label TM1-TM2 *(45)*, as did [125I]DEEP. More recently, a piperidine-based photoaffinity label [125I]4-[2-(diphenyl-methoxy)ethyl]-1-[(4-azido-3-iodophenyl)methyl]-piperidine ([125I]AD-96-129) was shown to label both TM1-TM2 and TM4-TM7; to a lesser extent this was found with DEEP as well *(46)*. These results suggest that these two regions might be in proximity in the tertiary structure of the transporter, but it is also possible that the greater flexibility of the latter ligands allows them to access different regions within the binding site.

These data, together with extensive evidence for divergent structure activity relationships between the two tropane-based classes of dopamine uptake inhibitors (the 3-aryl tropane compounds, such as RTI 82 and cocaine, vs benztropine analogs such as GA 2–34) (reviewed in ref. *[47]*), support the hypothesis that distinct structural classes of dopamine uptake inhibitors may interact with different binding domains in DAT. Thus, different stereoselectivities and patterns of tolerated tropane ring substitutions raise the possibility that the 3-aryl tropanes and benztropines may bind in distinctive orientations.

6.2. Chimeric Transporters

Construction of chimeras between transporters from different species or between two transporters within this family has been useful in localizing general areas that may confer a specific effect on function. For example, analysis of chimeras between human and bovine DATs revealed that CFT binding requires human DAT sequences in the regions encompassing TM3 and TM6-8 *(48)*. Cross-species chimeras between rat and human SERT were also helpful in identifying a region between $Ser532_{10.65}$ and position 12.65 as being responsible for species selectivity of the tricyclic antidepressant (TCA) imipramine *(49,50)*. Exchanging the carboxyl tails of SERT and NET or of rat and human SERTs did not alter antidepressant recognition, suggesting that the C-terminus is not involved in ligand recognition *(50,51)*. Additionally, analysis of binding and uptake in DAT mutants with multiple substitutions in transmembrane polar residues was interpreted as supporting the contribution of polar residues in TM4 and TM11 to cocaine recognition, as well as to the involvement of TM4 and TM5 in dopamine uptake *(52)*.

6.3. Site-Directed Mutagenesis

Identification of individual residues involved in substrate and inhibitor interactions has been accomplished through site-directed mutagenesis studies, either alone, or along with chimera studies that first identified a functionally significant region.[2] In TM1, a highly conserved aspartic acid in the biogenic amine transporters, $Asp_{1.45}$, was originally shown to be crucial for DAT function *(53)*. Mutation of this residue, $Asp79_{1.45}$ substantially decreased CFT binding and dramatically impaired uptake of dopamine and the neurotoxin, 1-methyl-4-phenyl pyridinium (MPP^+). In later work, mutation of $Asp_{1.45}$ to a variety of amino acids in SERT, severely reduced the uptake capacity of the transporter, and led to fivefold and 123-fold reductions in the potency of imipramine, and citalopram, respectively, without significantly impairing surface expression *(54)*.

Based on the model for catecholamine binding, in which a conserved aspartic acid in TM3 of the adrenergic receptor binds the amino group of catecholamines *(55)*, it was proposed that $Asp_{1.45}$ in the monoamine transporters is directly involved in binding the protonated amino group of monoamine substrates. Additional support for this hypothesis was provided in a structure-activity study in which $Asp98_{1.45}$ of SERT was mutated to glutamic acid, thereby extending the acidic side chain by one carbon *(54)*. Two serotonin analogs with shorter amine-containing side chains displayed increased affinity for $D98_{1.45}E$ over WT. Such evidence of complementarity between mutated sites in the transporter and functional substitutions within substrates bolsters the argument in favor of a direct interaction between the TM1 aspartic acid and the substrate amine moiety. Cysteines substituted for $Asp98_{1.45}$, $Gly100_{1.47}$, and $Asn101_{1.48}$ rendered hSERT sensitive to inhibition by MTSET with protection by 5-HT and/or cocaine *(55a)*. These findings suggest a possible contribution of these residues in TM1 to the substrate binding pocket.

The use of cross-species chimeras formed between human and *Drosophila melanogaster* SERT suggested that the primary site for species-selective recognition of mazindol and citalopram, two biogenic amine uptake inhibitors, lies within the TM1-2 region *(56)*. $Tyr95_{1.42}$ in TM1 of hSERT ($Phe90_{1.42}$ in dSERT), and the aligned $Phe72_{1.42}$ in hNET, were found to be responsible for this effect. The authors inferred that the interaction of the tyrosine hydroxyl with the hydroxyl in mazindol leads to a steric clash that results in the lower-affinity species-selective recognition of mazindol by hSERT. These studies also established a role for the hydroxyl of $Tyr95_{1.42}$ in determining the species-selectivity of indole-nitrogen substituted and 7-substituted tryptamines *(57)*. The proximity of $Tyr95_{1.42}$ and $Asp98_{1.45}$ further points to the possible participation of TM1 in the substrate permeation pathway, and moreover, the possibility that uptake blockers may work by directly obstructing the binding site for 5-HT.

Mutation to leucine in GAT of a highly conserved tryptophan, $W68_{1.50}$, disrupted the release of substrate to the intracellular medium, and the authors inferred from this that TM1 is involved with substrate interactions *(58)*. Mutation to alanine of $Pro87_{1.53}$ in TM1 or $Pro112_{2.50}$ in TM2 increased the K_m for [^{3}H]dopamine uptake 21-fold and 17-fold, respectively, consistent with either a direct or indirect effect on dopamine binding

[2] The criteria we used for inclusion of a mutation in this section are: (1) that binding or uptake was detected or the presence of transporter in the membrane was confirmed by immunoblotting or immunostaining, and (2) mutation must have produced a decrease in apparent affinity for substrates or affinity for antagonists of fivefold or greater.

and/or transport *(59)*. Mutation of Phe98$_{2.36}$ in TM2 resulted in decreased cocaine-analog affinity *(60)*.

SCAM studies in TM3 of SERT showed that replacement of the highly conserved Tyr176$_{3.50}$ with cysteine decreased the affinity for 5-HT (eightfold) and cocaine (five-fold), and that both 5-HT and cocaine protected I172$_{3.46}$C from inactivation by MTSET *(25)*. These data suggest that these two residues are in proximity to the 5-HT and cocaine binding site, and moreover, that the binding sites for substrate and blocker may overlap significantly. Protection by 5-HT from inactivation by MTSET at 3.46 was both temperature- and sodium-independent, and therefore inferred to result from direct steric block and not a conformational change *(61)*. In addition, cysteine substituted for Ile172$_{3.46}$ was accessible to both external and cytoplasmic reagents *(61)*, suggesting that this position may form a part of the substrate binding site that is alternately exposed to the extracellular and intracellular milieu, consistent with an alternate access model of transport *(62)*.

Substitution of TM3 from bovine DAT into human DAT abolished dopamine and MPP$^+$ transport and CFT binding *(63)*. Replacement into this chimera of the bovine Ile$_{3.46}$ by the hDAT residue Val152$_{3.46}$ restored transport and binding capacity to near wild-type hDAT values *(64)*. The corresponding Cys144$_{3.46}$ in TM3 of the creatine transporter is also thought to be close to the substrate binding site, as substrates inhibited MTSEA-induced inactivation *(65)*. Position 3.46, therefore, appears to play an important role in the interaction of ligand and transporter.

Tyrosine residues in TM3 of the GABA (Tyr140$_{3.50}$) and glycine (Tyr289$_{3.50}$) transporters, like SERT Tyr$_{3.50}$, were also shown to play a critical role in substrate recognition *(66,67)*. Species-scanning mutagenesis of bovine and human SERTs also suggested the involvement of Met180$_{3.54}$ in antidepressant recognition *(68)*. In addition, mutation of Phe155$_{3.49}$ in TM3 of rDAT resulted in a 30-fold decrease in the apparent affinity of dopamine uptake *(60)*. A phenylalanine-to-alanine mutation in TM3 of DAT, F154$_{3.48}$A, was shown to alter the stereospecificity of cocaine binding and lower its affinity 10-fold *(69)*. Mutation to alanine of Trp255$_{4.63}$ in TM4 of rDAT also resulted in a 16-fold decrease in the apparent affinity of dopamine uptake *(70)*.

In NET, TM6 and TM7 were inferred to be involved in the binding of the tricyclic antidepressants (TCA), desipramine and nortriptyline, based on chimeras constructed between NET and DAT *(71,72)*. Mutation of NET residues in this region to their DAT counterparts caused a significant loss of TCA binding affinity in mutants F316$_{6.52}$C (eightfold), and V356$_{7.44}$S (fivefold) *(73)*. Furthermore, mutation to alanine of Pro272$_{5.36}$, Trp310$_{6.44}$, and Phe364$_{7.50}$ of rDAT *(59,60,70)*, and to asparagine of Asp313$_{6.46}$ of hDAT *(74)* decreased the apparent affinity for dopamine uptake. In hDAT, D313$_{6.46}$N and W311$_{6.44}$L had five- and >100-fold, respectively, reduced affinity for dopamine *(74)*. Additionally, hDAT W311$_{6.44}$L, as well as rDAT Phe361$_{7.47}$A and Phe390$_{8.29}$A, had reduced affinity for cocaine *(60,74)*. In hNET, in S354$_{7.42}$A, the affinity of the inhibitor nisoxetine was decreased 69-fold, and the affinities of dopamine and *m*-tyramine for competition of nisoxetine binding in S354$_{7.42}$A were reduced as well *(75)*.

In NET, the naturally occurring A457$_{9.42}$P, associated with orthostatic intolerance, was shown to produce a 50-fold increase in K$_m$ for norepinephrine as compared with wild-type *(76)*. Species-scanning mutagenesis of bovine and human SERT revealed that Tyr495$_{10.28}$ and Ser513$_{10.46}$ (along with Met180$_{3.54}$) are essential for antidepressant recognition *(68)*. Positions 10.28 and 10.46 would be very far apart in an α-helix, and one (or both) of these mutations must exert its effect indirectly. Mutation to asparagine of hDAT Asp476$_{10.26}$ *(74)* and to alanine of rDAT Pro528$_{11.50}$ *(59)*, caused sevenfold and 44-fold reductions, respectively, in the apparent affinity for dopamine uptake *(70)*. In addition, mutation to alanine of rDAT Trp496$_{10.47}$ led to a sixfold reduction in cocaine-binding affinity *(70)*, and mutation to alanine of rSERT Ser545$_{11.49}$ reduced the affinities of imipramine and citalopram fivefold and sevenfold, respectively *(77)*.

To identify the residues responsible for the divergent TCA specificity in rat and human SERT, hSERT amino acids were substituted into each of four divergent aligned positions in rSERT, and Phe586$_{12.58}$ was shown to be responsible for the human selectivity of imipramine, desipramine, and nortriptyline, because valine is present at this position in rat SERT *(50)*. NET, however, binds desipramine and nortriptyline with very high affinity despite the fact that it contains Met586$_{12.58}$, arguing against the specific need for a phenylalanine at this position for TCA recognition.

Finally, loop regions have also been implicated in ligand interactions or their modulation. Chimeras generated between NET and DAT were assayed for ethanol sensitivity and revealed that Gly130$_{2.68}$ and Ile137$_{3.31}$ in IL1 are important for ethanol modulation of DAT activity *(78)*. In addition, mutation to alanine of Pro136$_{3.30}$ in IL1, as well as Pro553$_{12.43}$ in EL6, decreased the apparent affinity for dopamine uptake *(59)*.

7. CONFORMATIONAL CHANGES ASSOCIATED WITH TRANSPORT

Conformational changes associated with transport have been identified in both intracellular and extracellular loops as well as in transmembrane regions. In DAT, cocaine significantly retarded the reaction of MTS reagents with Cys135$_{3.29}$ in IL1, and Cys342$_{6.75}$ in IL3 *(13)*. Because mutation of these cysteines had no effect on cocaine's affinity for DAT, the protection most likely results from a cocaine-induced conformational change that reduces the accessibility of the cytoplasmic loop cysteines. Inward transport of the substrate, *m*-tyramine, enhanced the MTSEA-induced inactivation of dopamine uptake at Cys342$_{6.75}$ *(79)*. This effect of tyramine was temperature-sensitive and sodium-dependent, suggesting that tyramine exposes Cys342$_{6.75}$ during translocation, rather than simply by binding. Moreover, cocaine protected against tyramine's effect at Cys342$_{6.75}$, presumably by blocking transport and preventing exposure of this endogenous cysteine.

The corresponding position in SERT, Cys357$_{6.75}$, is also conformationally sensitive; the ability of 5-HT and cocaine to protect this site from MTSEA-induced inactivation presumably required a conformational change, as it was sodium- and temperature-dependent *(80)*. Moreover, reaction of Cys357$_{6.75}$ with MTSEA was increased in the presence of alkali cations, particularly K$^+$, suggesting that the accessibility of Cys357$_{6.75}$ varies with the conformation of the transporter. Consistent with a role of IL3 in conformational change, Tyr335$_{6.68}$ in DAT appears to play a role in regulating the distribution between different conformational states of the transport cycle *(81)*. Mutation of this residue to alanine caused conversion of the endogenous inhibitory Zn^{2+} switch to an activating one, in which binding of Zn^{2+} to inactive mutant transporter leads to uptake. The explanation offered for this observation was that mutation of Tyr335$_{6.68}$ alters the conformational equilibrium of the translocation cycle, with accumulation in conformational states following substrate binding, and that Zn^{2+} is able to reverse this change. The dramatic decrease (up to 150-fold) in affinity of $Y335_{6.68}A$ for cocaine-like inhibitors and the increased apparent affinity for substrates is consistent with an altered conformational equilibrium.

The accessibility to MTS reagents of $Cys399_{8.63}$ in IL4 of GAT-1 is dependent on transporter conformation *(17)*. In wild-type GAT-1, but not the $C399_{8.63}S$ mutant, binding of both sodium and chloride, as well as the nontransported GABA analog, SKF10030A, reduced the sensitivity to sulfhydryl modification; in contrast, GABA, in the presence of sodium chloride, increased the reactivity to MTS reagents. Finally, mutation of $Arg44_{1.26}$ in the cytoplasmic N-terminus of GAT was inferred to impair the reorientation of unloaded transporter after the release of substrate to the intracellular medium *(82)*.

Cocaine increased the rate of reaction of $Cys90_{1.56}$ in EL1 with MTS reagents, presumably by triggering a conformational change that increased the accessibility of this extracellular cysteine *(13)*. However, the DAT inhibitors benztropine and cocaine had different effects on the reaction of the extracellular $Cys90_{1.56}$ and the intracellular $Cys135_{3.29}$ *(83)*. Benztropine did not protect $Cys135_{3.29}$ in IL1 from MTSET inactivation, and also did not increase the reaction of $Cys90_{1.56}$ with MTSET. This finding suggests that inhibitors do not all stabilize the same conformational state. The same type of reactivity studies suggest that EL1 is also subject to ion-induced conformational changes: lithium caused a change in SERT $Cys109_{1.56}$ reactivity as well as cocaine binding and ion conductance *(84)*, whereas sodium and chloride protected against MTSET inactivation at a cysteine inserted at position 1.56 in GLYT2, in a temperature-dependent manner *(85)*.

A chimeric transporter, in which a short stretch of NET was substituted into the first half of EL2 in SERT, was substantially impaired for serotonin transport (< 10% wild-type SERT), whereas antagonist and substrate binding and surface expression were intact *(86)*. Chimeras in which the extracellular loops of SERT were replaced, one by one, with the corresponding sequence from NET, expressed and had normal SERT antagonist selectivity. In contrast, the EL4, EL5, and EL6 chimeras had dramatically impaired transport *(87)*. GAT EL4, EL5, and EL6 have been proposed to contribute to the substrate binding site, based on work that swapped external sequences between various GAT isoforms *(88)*. Substitution of three residues from EL5 of GAT-1 with the aligned residues in GAT-2, 3, and 4 imparted to GAT-1 sensitivity to β-alanine inhibition of GABA uptake, as seen with GAT-2, -3, and -4. Furthermore, Cao et al. used hSERT-rSERT chimeras and point mutations to identify position $490_{9.55}$ and neighboring residues in EL5 as determinants of the difference in pH-dependence of transport-associated currents between the hSERT and rSERT *(89)*. This region was inferred to participate in the external gating of SERT. The mutation $K448_{9.55}E$, in EL5 of GAT, was also shown to confer pH sensitivity and alter substrate interactions of the transporter *(90)*.

In the aforementioned SCAM study in TM3 of SERT, reaction of $I179_{3.53}C$ with MTSET inactivated transport but not binding *(25)*. Neither cocaine nor 5-HT protected against reaction with MTSET. These observations suggest that $Ile179_{3.53}$ is not associated with the binding site, but may be involved with some conformational event subsequent to substrate binding. Sodium enhanced the inactivation of $Ile179_{3.53}C$, as well as stimulated its reactivation by free cysteine, an effect that was enhanced by the presence of 5-HT *(61)*. These results suggest that because 5-HT can bind the inactivated

Ile179$_{3.53}$, the transporter is locked in an outward-facing form, and that Ile179$_{3.53}$ exists in a part of the transporter that changes conformation upon binding of sodium and 5-HT. In NET, MTSET inactivation of I155$_{3.53}$C was enhanced by cocaine but slowed by dopamine, presumably by transport-associated occlusion of 155$_{3.53}$. Thus, this region of the transporter has been proposed to form part of an extracellular gate, which is normally sequestered upon substrate translocation, and which is prevented from closing upon cysteine modification.

Glu101$_{2.55}$, at the intracellular end of TM2 in GAT, was inferred to be critical for the conformational changes GAT undergoes during its transport cycle *(91)*. Mutation of Glu101$_{2.55}$ to aspartate or other residues abolished transport despite unaltered surface expression. The transient sodium currents observed in wild-type, thought to represent sodium binding to the transporter and/or conformational changes in response to sodium binding, were not observed in these mutants. Thus, a defect in the binding or unbinding of sodium may result from mutation of Glu101$_{2.55}$.

8. CONCLUSIONS

The absence of a high-resolution structure of any transporter in the NT family greatly complicates the interpretation of structure–function studies. The recent identification of the bacterial and archaeal transporters and their potential suitability for direct structural studies raises the exciting prospect of being able to test specific structural hypotheses related to transport. In the meantime, the use of the numbering scheme introduced here in a context of predicted structural properties should facilitate communication among researchers on different members of the transporter family as the mechanistic details are being probed.

ACKNOWLEDGMENTS

This work was supported in part by National Institutes of Health Grants DA12408, DA11495, MH57324, DA00060, and the Lebovitz Fund.

REFERENCES

1. Norregaard, L. and Gether, U. (2001) The monoamine neurotransmitter transporters: structure, conformational changes and molecular gating. *Curr. Opin. Drug Discov. Dev.* **4,** 591–601.
2. Chen, N. and Reith, M. E. (2002) structure–function relationships for biogenic amine neurotransmitter transporters, in *Neurotransmitter Transporters* (Reith, M. E., ed.), Humana Press, Totowa. NJ, pp. 53–109.
3. Ballesteros, J. and Weinstein, H. (1995) Integrated methods for the construction of three-dimensional models of structure–function relations in G protein-coupled receptors. *Methods Neurosci.* **25,** 366–428.
4. Visiers, I., Ballesteros, J. A., and Weinstein, H. (2002) Computational methods for the construction and analysis of three dimensional representations of GPCR structures and mechanisms. *Methods Enzymol* **343,** 329–371.
5. Donnelly, D., Overington, J. P., Ruffle, S. V., Nugent, J. H., and Blundell, T. L. (1993) Modeling alpha-helical transmembrane domains: the calculation and use of substitution tables for lipid-facing residues. *Protein Sci.* **2,** 55–70.
6. Shi, L., Simpson, M. M., Ballesteros, J. A., and Javitch, J. A. (2001) The first transmembrane segment of the dopamine D2 receptor: accessibility in the binding-site crevice and position in the transmembrane bundle. *Biochemistry* **40,** 12339–12348.

7. Kyte, J. and Doolittle, R. F. (1982) A simple method for displaying the hydropathic character of a protein. *J. Mol. Biol.* **157,** 105–132.

8. Krogh, A., Larsson, B., von Heijne, G., and Sonnhammer, E. L. (2001) Predicting transmembrane protein topology with a hidden Markov model: application to complete genomes. *J. Mol. Biol.* **305,** 567–580.

9. Rost, B., Casadio, R., Fariselli, P., and Sander, C. (1995) Transmembrane helices predicted at 95% accuracy. *Protein Sci.* **4,** 521–533.

10. Roberts, D. D., Lewis, S. D., Ballou, D. P., Olson, S. T., and Shafer, J. A. (1986) Reactivity of small thiolate anions and cysteine-25 in papain toward methyl methanethiosulfonate. *Biochemistry* **25,** 5595–5601.

11. Karlin, A. and Akabas, M. H. (1998) Substituted-cysteine accessibility method. *Methods Enzymol.* **293,** 123–145.

12. Holmgren, M., Liu, Y., Xu, Y., and Yellen, G. (1996) On the use of thiol-modifying agents to determine channel topology. *Neuropharmacology* **35,** 797–804.

13. Ferrer, J. V. and Javitch, J. A. (1998) Cocaine alters the accessibility of endogenous cysteines in putative extracellular and intracellular loops of the human dopamine transporter. *Proc. Natl. Acad. Sci. USA* **95,** 9238–9243.

14. Chen, J. G., Liu-Chen, S., and Rudnick, G. (1997) External cysteine residues in the serotonin transporter. *Biochemistry* **36,** 1479–1486.

15. Yu, N., Cao, Y., Mager, S., and Lester, H. A. (1998) Topological localization of cysteine 74 in the GABA transporter, GAT1, and its importance in ion binding and permeation. *FEBS Lett.* **426,** 174–178.

16. Wang, J. B., Moriwaki, A., and Uhl, G. R. (1995) Dopamine transporter cysteine mutants: second extracellular loop cysteines are required for transporter expression. *J. Neurochem.* **64,** 1416–1419.

17. Golovanevsky, V. and Kanner, B. I. (1999) The reactivity of the gamma-aminobutyric acid transporter GAT-1 toward sulfhydryl reagents is conformationally sensitive. Identification of a major target residue. *J. Biol. Chem.* **274,** 23020–23026.

18. Kamdar, G., Penado, K. M., Rudnick, G., and Stephan, M. M. (2001) Functional role of critical stripe residues in transmembrane span 7 of the serotonin transporter. Effects of Na^+, Li^+, and methanethiosulfonate reagents. *J. Biol. Chem.* **276,** 4038–4045.

19. Chen, J. G., Liu-Chen, S., and Rudnick, G. (1998) Determination of external loop topology in the serotonin transporter by site-directed chemical labeling. *J. Biol. Chem.* **273,** 12675–12681.

20. Androutsellis-Theotokis, A. and Rudnick, G. (2002) Accessibility and conformational coupling in serotonin transporter predicted internal domains. *J. Neurosci.* **22,** 8370–8378.

21. Bruss, M., Hammermann, R., Brimijoin, S., and Bonisch, H. (1995) Antipeptide antibodies confirm the topology of the human norepinephrine transporter. *J. Biol. Chem.* **270,** 9197–9201.

22. Bennett, E. R. and Kanner, B. I. (1997) The membrane topology of GAT-1, a (Na^+ + Cl^-)-coupled gamma-aminobutyric acid transporter from rat brain. *J. Biol. Chem.* **272,** 1203–1210.

23. Clark, J. A. (1997) Analysis of the transmembrane topology and membrane assembly of the GAT-1 gamma-aminobutyric acid transporter. *J. Biol. Chem.* **272,** 14695–14704.

24. Olivares, L., Aragon, C., Gimenez, C., and Zafra, F. (1997) Analysis of the transmembrane topology of the glycine transporter GLYT1. *J. Biol. Chem.* **272,** 1211–1217.

25. Chen, J. G., Sachpatzidis, A., and Rudnick, G. (1997) The third transmembrane domain of the serotonin transporter contains residues associated with substrate and cocaine binding. *J. Biol. Chem.* **272,** 28321–28327.

26. Penado, K. M., Rudnick, G., and Stephan, M. M. (1998) Critical amino acid residues in transmembrane span 7 of the serotonin transporter identified by random mutagenesis. *J. Biol. Chem.* **273,** 28098–28106.

27. Norregaard, L., Frederiksen, D., Nielsen, E. O., and Gether, U. (1998) Delineation of an endogenous zinc-binding site in the human dopamine transporter. *Embo. J.* **17,** 4266–4273.
28. Loland, C. J., Norregaard, L., and Gether, U. (1999) Defining proximity relationships in the tertiary structure of the dopamine transporter. Identification of a conserved glutamic acid as a third coordinate in the endogenous Zn(2+)-binding site. *J. Biol. Chem.* **274,** 36928–36934.
29. Norregaard, L., Visiers, I., Loland, C. J., Ballesteros, J., Weinstein, H., and Gether, U. (2000) Structural probing of a microdomain in the dopamine transporter by engineering of artificial Zn^{2+} binding sites. *Biochemistry* **39,** 15836–15846.
30. Glusker, J. P. (1991) Structural aspects of metal liganding to functional groups in proteins. *Adv. Prot. Chem.* **42,** 1–76.
31. MacAulay, N., Bendahan, A., Loland, C. J., Zeuthen, T., Kanner, B. I., and Gether, U. (2001) Engineered Zn^{2+} switches in the gamma-aminobutyric acid (GABA) transporter-1. Differential effects on GABA uptake and currents. *J. Biol. Chem.* **276,** 40476–40485.
32. Kilic, F. and Rudnick, G. (2000) Oligomerization of serotonin transporter and its functional consequences. *Proc. Natl. Acad. Sci. USA* **97,** 3106–3111.
33. Schmid, J. A., Scholze, P., Kudlacek, O., Freissmuth, M., Singer, E. A., and Sitte, H. H. (2001) Oligomerization of the human serotonin transporter and of the rat GABA transporter 1 visualized by fluorescence resonance energy transfer microscopy in living cells. *J. Biol. Chem.* **276,** 3805–3810.
34. Berger, S. P., Farrell, K., Conant, D., Kempner, E. S., and Paul, S. M. (1994) Radiation inactivation studies of the dopamine reuptake transporter protein. *Mol. Pharmacol.* **46,** 726–731.
35. Milner, H. E., Beliveau, R., and Jarvis, S. M. (1994) The in situ size of the dopamine transporter is a tetramer as estimated by radiation inactivation. *Biochim. Biophys. Acta* **1190,** 185–187.
36. Hastrup, H., Karlin, A., and Javitch, J. A. (2001) Symmetrical dimer of the human dopamine transporter revealed by cross-linking Cys-306 at the extracellular end of the sixth transmembrane segment. *Proc. Natl. Acad. Sci. USA* **98,** 10055–10060.
37. Russ, W. P. and Engelman, D. M. (2000) The GxxxG motif: a framework for transmembrane helix-helix association. *J. Mol. Biol.* **296,** 911–919.
38. Senes, A., Gerstein, M., and Engelman, D. M. (2000) Statistical analysis of amino acid patterns in transmembrane helices: the GxxxG motif occurs frequently and in association with beta-branched residues at neighboring positions. *J. Mol. Biol.* **296,** 921–936.
39. Norgaard-Nielsen, K., Norregaard, L., Hastrup, H., Javitch, J. A., and Gether, U. (2002) Zn^{2+} site engineering at the oligomeric interface of the dopamine transporter. *FEBS Lett.* **524,** 87–91.
40. Torres, G. E., Carneiro, A., Seamans, K., Fiorentini, C., Sweeney, A., Yao, W. D., and Caron, M. G. (2002) Oligomerization and trafficking of the human dopamine transporter: mutational analysis identifies critical domains important for the functional expression of the transporter. *J. Biol. Chem.* **278,** 2731–2739.
41. Scholze, P., Freissmuth, M., and Sitte, H. H. (2002) Mutations within an intramembrane leucine heptad repeat disrupt oligomer formation of the rat gaba transporter 1. *J. Biol. Chem.* **277,** 43682–43690.
42. Horiuchi, M., Nicke, A., Gomeza, J., Aschrafi, A., Schmalzing, G. and Betz, H. (2001) Surface-localized glycine transporters 1 and 2 function as monomeric proteins in Xenopus oocytes. *Proc. Natl. Acad. Sci. USA* **98,** 1448–1453.
43. Vaughan, R. A. and Kuhar, M. J. (1996) Dopamine transporter ligand binding domains. Structural and functional properties revealed by limited proteolysis. *J. Biol. Chem.* **271,** 21672–21680.
44. Vaughan, R. A. (1995) Photoaffinity-labeled ligand binding domains on dopamine transporters identified by peptide mapping. *Mol. Pharmacol.* **47,** 956–964.
45. Vaughan, R. A., Agoston, G. E., Lever, J. R., and Newman, A. H. (1999) Differential binding of tropane-based photoaffinity ligands on the dopamine transporter. *J. Neurosci.* **19,** 630–636.

46. Vaughan, R. A., Gaffaney, J. D., Lever, J. R., Reith, M. E., and Dutta, A. K. (2001) Dual incorporation of photoaffinity ligands on dopamine transporters implicates proximity of labeled domains. *Mol. Pharmacol.* **59,** 1157–1164.

47. Newman, A. H. and Kulkarni, S. (2002) Probes for the dopamine transporter: new leads toward a cocaine-abuse therapeutic. *Med. Res. Rev.* **22,** 1–36.

48. Lee, S. H., Chang, M. Y., Jeon, D. J., Oh, D. Y., Son, H., Lee, C. H., and Lee, Y. S. (2002) The functional domains of dopamine transporter for cocaine analog, CFT binding. *Exp. Mol. Med.* **34,** 90–94.

49. Barker, E. L., Kimmel, H. L., and Blakely, R. D. (1994) Chimeric human and rat serotonin transporters reveal domains involved in recognition of transporter ligands. *Mol. Pharmacol.* **46,** 799–807.

50. Barker, E. L. and Blakely, R. D. (1996) Identification of a single amino acid, phenylalanine 586, that is responsible for high affinity interactions of tricyclic antidepressants with the human serotonin transporter. *Mol. Pharmacol.* **50,** 957–965.

51. Blakely, R. D., Moore, K. R., and Qian, Y. (1993) Tails of serotnin and norepinephrine transporters: deletions and chimeras retain function. *Soc. Gen. Phsiolog. Series* **48,** 283–300.

52. Itokawa, M., Lin, Z., Cai, N. S., Wu, C., Kitayama, S., Wang, J. B., and Uhl, G. R. (2000) Dopamine transporter transmembrane domain polar mutants: DeltaG and DeltaDeltaG values implicate regions important for transporter functions. *Mol. Pharmacol.* **57,** 1093–1103.

53. Kitayama, S., Shimada, S., Xu, H., Markham, L., Donovan, D. M., and Uhl, G. R. (1992) Dopamine transporter site-directed mutations differentially alter substrate transport and cocaine binding. *Proc. Natl. Acad. Sci. USA* **89,** 7782–7785.

54. Barker, E. L., Moore, K. R., Rakhshan, F., and Blakely, R. D. (1999) Transmembrane domain I contributes to the permeation pathway for serotonin and ions in the serotonin transporter. *J. Neurosci.* **19,** 4705–4717.

55. Strader, C. D., Gaffney, T., Sugg, E. E., Candelore, M. R., Keys, R., Patchett, A. A., and Dixon, R. A. (1991) Allele-specific activation of genetically engineered receptors. *J. Biol. Chem.* **266,** 5–8.

55a. Henry, L. K., Adkins, E. M., Han, Q., and Blakely, R. D. (2003). Serotonin and cocaine-sensitive inactivation of human serotonin transporters by methanethiosulfonates targeted to transmembrane domain 1. *J. Biol. Chem.* [Epub ahead of print].

56. Barker, E. L., Perlman, M. A., Adkins, E. M., Houlihan, W. J., Pristupa, Z. B., Niznik, H. B., and Blakely, R. D. (1998) High affinity recognition of serotonin transporter antagonists defined by species-scanning mutagenesis. An aromatic residue in transmembrane domain I dictates species-selective recognition of citalopram and mazindol. *J. Biol. Chem.* **273,** 19459–19468.

57. Adkins, E. M., Barker, E. L., and Blakely, R. D. (2001) Interactions of tryptamine derivatives with serotonin transporter species variants implicate transmembrane domain I in substrate recognition. *Mol. Pharmacol.* **59,** 514–523.

58. Mager, S., Kleinberger-Doron, N., Keshet, G. I., Davidson, N., Kanner, B. I., and Lester, H. A. (1996) Ion binding and permeation at the GABA transporter GAT1. *J. Neurosci.* **16,** 5405–5414.

59. Lin, Z., Itokawa, M., and Uhl, G. R. (2000) Dopamine transporter proline mutations influence dopamine uptake, cocaine analog recognition, and expression. *FASEB J.* **14,** 715–728.

60. Lin, Z., Wang, W., Kopajtic, T., Revay, R. S., and Uhl, G. R. (1999) Dopamine transporter: transmembrane phenylalanine mutations can selectively influence dopamine uptake and cocaine analog recognition. *Mol. Pharmacol.* **56,** 434–447.

61. Chen, J. G. and Rudnick, G. (2000) Permeation and gating residues in serotonin transporter. *Proc. Natl. Acad. Sci. USA* **97,** 1044–1049.

62. Hilgemann, D. W. and Lu, C. C. (1999) GAT1 (GABA:Na$^+$:Cl$^-$) cotransport function. Database reconstruction with an alternating access model. *J. Gen. Physiol.* **114,** 459–475.

63. Lee, S. H., Kang, S. S., Son, H., and Lee, Y. S. (1998) The region of dopamine transporter encompassing the 3rd transmembrane domain is crucial for function. *Biochem. Biophys. Res. Commun.* **246,** 347–352.

64. Lee, S. H., Chang, M. Y., Lee, K. H., Park, B. S., Lee, Y. S., and Chin, H. R. (2000) Importance of valine at position 152 for the substrate transport and 2beta-carbomethoxy-3beta-(4-flurophenyl)tropane binding of dopamine transporter. *Mol. Pharmacol.* **57,** 883–889.

65. Dodd, J. R. and Christie, D. L. (2001) Cysteine 144 in the third transmembrane domain of the creatine transporter is located close to a substrate-binding site. *J. Biol. Chem.* **276,** 46983–46988.

66. Bismuth, Y., Kavanaugh, M. P., and Kanner, B. I. (1997) Tyrosine 140 of the gamma-aminobutyric acid transporter GAT-1 plays a critical role in neurotransmitter recognition. *J. Biol. Chem.* **272,** 16096–16102.

67. Ponce, J., Biton, B., Benavides, J., Avenet, P., and Aragon, C. (2000) Transmembrane domain III plays an important role in ion binding and permeation in the glycine transporter GLYT2. *J. Biol. Chem.* **275,** 13856–13862.

68. Mortensen, O. V., Kristensen, A. S., and Wiborg, O. (2001) Species-scanning mutagenesis of the serotonin transporter reveals residues essential in selective, high-affinity recognition of antidepressants. *J. Neurochem.* **79,** 237–247.

69. Lin, Z. and Uhl, G. R. (2002) Dopamine transporter mutants with cocaine resistance and normal dopamine uptake provide targets for cocaine antagonism. *Mol. Pharmacol.* **61,** 885–891.

70. Lin, Z., Wang, W., and Uhl, G. R. (2000) Dopamine transporter tryptophan mutants highlight candidate dopamine- and cocaine-selective domains. *Mol. Pharmacol.* **58,** 1581–1592.

71. Buck, K. J. and Amara, S. G. (1995) Structural domains of catecholamine transporter chimeras involved in selective inhibition by antidepressants and psychomotor stimulants. *Mol. Pharmacol.* **48,** 1030–1037.

72. Giros, B., Wang, Y. M., Suter, S., McLeskey, S. B., Pifl, C., and Caron, M. G. (1994) Delineation of discrete domains for substrate, cocaine, and tricyclic antidepressant interactions using chimeric dopamine-norepinephrine transporters. *J. Biol. Chem.* **269,** 15985–15988.

73. Roubert, C., Cox, P. J., Bruss, M., Hamon, M., Bonisch, H., and Giros, B. (2001) Determination of residues in the nonrepineprhine transporter that are critical for tricyclic antidepressant affinity. *J. Biol. Chem.* **276,** 8254–8260.

74. Chen, N., Vaughan, R. A., and Reith, M. E. (2001) The role of conserved tryptophan and acidic residues in the human dopamine transporter as characterized by site-directed mutagenesis. *J. Neurochem.* **77,** 1116–1127.

75. Danek Burgess, K. S. and Justice, J. B., Jr. (1999) Effect of serine mutations in transmembrane domain 7 of the human norepinephrine transporter on substrate binding and transport. *J. Neurochem.* **73,** 656–664.

76. Paczkowski, F. A., Bonisch, H., and Bryan-Lluka, L. J. (2002) Pharmacological properties of the naturally occurring Ala(457)Pro variant of the human norepinephrine transporter. *Pharmacogenetics* **12,** 165–173.

77. Sur, C., Betz, H., and Schloss, P. (1997) A single serine residue controls the cation dependence of substrate transport by the rat serotonin transporter. *Proc. Natl. Acad. Sci. USA* **94,** 7639–7644.

78. Maiya, R., Buck, K. J., Harris, R. A., and Mayfield, R. D. (2002) Ethanol-sensitive sites on the human dopamine transporter. *J. Biol. Chem.* **277,** 30724–30729.

79. Chen, N., Ferrer, J. V., Javitch, J. A., and Justice, J. B., Jr. (2000) Transport-dependent accessibility of a cytoplasmic loop cysteine in the human dopamine transporter. *J. Biol. Chem.* **275,** 1608–1614.

80. Androutsellis-Theotokis, A., Ghassemi, F., and Rudnick, G. (2001) A conformationally sensitive residue on the cytoplasmic surface of serotonin transporter. *J. Biol. Chem.* **276,** 45933–45938.

81. Loland, C. J., Norregaard, L., Litman, T., and Gether, U. (2002) Generation of an activating Zn^{2+} switch in the dopamine transporter: mutation of an intracellular tyrosine constitutively alters the conformational equilibrium of the transport cycle. *Proc. Natl. Acad. Sci. USA* **99,** 1683–1688.

82. Bennett, E. R., Su, H., and Kanner, B. I. (2000) Mutation of arginine 44 of GAT-1, a (Na^+ + Cl^-)-coupled gamma-aminobutyric acid transporter from rat brain, impairs net flux but not exchange. *J. Biol. Chem.* **275,** 34106–34113.

83. Reith, M. E., Berfield, J. L., Wang, L. C., Ferrer, J. V., and Javitch, J. A. (2001) The uptake inhibitors cocaine and benztropine differentially alter the conformation of the human dopamine transporter. *J. Biol. Chem.* **276,** 29012–29018.

84. Ni, Y. G., Chen, J. G., Androutsellis-Theotokis, A., Huang, C. J., Moczydlowski, E., and Rudnick, G. (2001) A lithium-induced conformational change in serotonin transporter alters cocaine binding, ion conductance, and reactivity of Cys-109. *J. Biol. Chem.* **276,** 30942–30947.

85. Lopez-Corcuera, B., Nunez, E., Martinez-Maza, R., Geerlings, A., and Aragon, C. (2001) Substrate-induced conformational changes of extracellular loop 1 in the glycine transporter GLYT2. *J. Biol. Chem.* **276,** 43463–43470.

86. Stephan, M. M., Chen, M. A., Penado, K. M., and Rudnick, G. (1997) An extracellular loop region of the serotonin transporter may be involved in the translocation mechanism. *Biochemistry* **36,** 1322–1328.

87. Smicun, Y., Campbell, S. D., Chen, M. A., Gu, H., and Rudnick, G. (1999) The role of external loop regions in serotonin transport. Loop scanning mutagenesis of the serotonin transporter external domain. *J. Biol. Chem.* **274,** 36058–36064.

88. Tamura, S., Nelson, H., Tamura, A., and Nelson, N. (1995) Short external loops as potential substrate binding site of gamma-aminobutyric acid transporters. *J. Biol. Chem.* **270,** 28712–28715.

89. Cao, Y., Li, M., Mager, S., and Lester, H. A. (1998) Amino acid residues that control pH modulation of transport-associated current in mammalian serotonin transporters. *J. Neurosci.* **18,** 7739–7749.

90. Forlani, G., Bossi, E., Ghirardelli, R., et al. (2001) Mutation K448E in the external loop 5 of rat GABA transporter rGAT1 induces pH sensitivity and alters substrate interactions. *J. Physiol.* **536,** 479–494.

91. Keshet, G. I., Bendahan, A., Su, H., Mager, S., Lester, H. A., and Kanner, B. I. (1995) Glutamate-101 is critical for the function of the sodium and chloride-coupled GABA transporter GAT-1. *FEBS Lett.* **371,** 39–42.

13

Functional Mechanisms of G Protein-Coupled Receptors in a Structural Context

Marta Filizola, Irache Visiers, Lucy Skrabanek, Fabien Campagne, and Harel Weinstein

1. INTRODUCTION

The central role of G protein-coupled receptors (GPCRs) in most aspects of biological signal transduction has made them the object of extensive studies for a long period of time. These studies have revealed the key physiological roles of the many members of this family, and the manifold functions they have in the central nervous system (CNS) and the periphery (for recent reviews, *see* refs. *[1,2]*). Despite the abundance of information available in the literature, however, many of the fundamental questions regarding the molecular and structural requirements for GPCR function remain unanswered. A large number of reviews and compendia of results have been devoted to such fundamental elements in the biological mechanisms of GPCR *(3–9)*. For this reason, we review here only some of the key aspects of recent progress in the development and application of approaches aiming to elucidate functional mechanisms of GPCRs in a detailed structural context. A central aim of our own collaborative studies of GPCRs is to develop such a coherent structural context (e.g., *see [5,10–15]*) that can serve in the interpretation, as well as the integration into a mechanistic understanding, of the abundant data about these systems. For this reason, we focus here specifically on the following three aspects of recent developments in the field: (1) the management of the copious data accumulated from structure–function studies, including genomic information; (2) some novel insights about intramolecular mechanisms triggering the activation of GPCRs; and (3) the recently characterized oligomerization of the receptors. The concluding Perspective section points to the integration of the mechanistic insights at the level of GPCR function with the growing understanding of signal-transduction pathways in the cells.

2. MANAGING THE INFORMATION FROM STRUCTURE–FUNCTION STUDIES: MUTATION DATABASES

Directed mutagenesis experiments play a central role in the study of functional and structural features of GPCRs. Massive amounts of data from these extensive studies

From: *Molecular Neuropharmacology: Strategies and Methods*
Edited by: A. Schousboe and H. Bräuner-Osborne © Humana Press Inc., Totowa, NJ

describing mutations, phenotypes, and inferences generalizable to various GPCR families continue to accumulate at a very rapid pace *(16)*. These data are mostly recorded in articles published in a large number of journals, making it challenging for investigators to track mutations of a certain receptor family or of similar receptor families. To support research efforts aimed at studying structure and function of GPCRs, a few groups have developed electronic mutation databases. Early examples include MRS *(17)* and GRAP *(18)*. MRS was organized to facilitate structural studies, whereas GRAP also provided quantitative pharmacological data for mutants and enhanced searching capabilities. Neither MRS nor GRAP are maintained any longer. TinyGRAP *(19)* was introduced in 1995 as a replacement to GRAP. This "tiny" version of GRAP allowed the curators to cope with growing amounts of published mutant information. In contrast to GRAP, Tiny-GRAP does not store quantitative pharmacological data, but instead records a number of qualitative attributes that describe the experiments in which the mutant was used. Such attributes include the type of pharmacological studies that were performed with the mutant (e.g., agonist binding affinity measurements). With 10,500 mutants from 1,380 papers, TinyGRAP is the most prominent and useful source of GPCR mutant data. Mutant data from TinyGRAP can be visualized in snake-like diagrams produced by the Viseur program and the RbDe web service *(20–22)*. (RbDe allows the construction of custom diagrams, that show only specific mutations.) Such diagrams are conveniently offered by GPCRDB and referenced by SWISS-PROT entries *(23,24)*. Yet, at the time of writing, new data have not been entered in TinyGRAP since April 2001, so researchers should be cautious and check the time of the last update before relying on TinyGRAP to find more recent mutation information. An update of TinyGRAP with new mutants is planned (Edvardsen, Ø. and Beukers, M.W., personal communication).

The difficulty in maintaining up-to-date databases is one of the challenges facing researchers who are developing resources to manage GPCR mutation data. Most of the data available in TinyGRAP has been entered by its curators, who have manually extracted data from articles. Although TinyGRAP does not support direct submission of data, an experiment has been conducted at GPCRDB to offer submission forms that make possible the direct submission of mutation data to the database. Although the GPCRDB database is a high-traffic web site (with peaks at 100,000 page visits per month), no visitor submitted information about new mutants (Beukers, M.W. and Horn, F., personal communication). The apparent complexity of the submission page may have discouraged submitters, suggesting that future research in the development of Bioinformatics tools could focus on ways to improve the ergonomics of submission tools. Clearly, other factors were at play as well. For example, at the time that the experiment was conducted, journals and funding agencies did not request mutation data to be deposited electronically into a database.

Other challenges exist, in addition to the curation, in the construction of mutation databases for GPCRs (and other integral membrane proteins studied by a combination of directed mutagenesis and pharmacological approaches). For instance, information about putative genomic sequence variants in populations is becoming increasingly available (e.g., HUGO *[25]*, dbSNP *[26]*). It would therefore be very interesting to have a way to link data about sequence variants, in the coding region of a GPCR, with mutagenesis results.

In considering some of these challenges, we are developing the Arcadia database for GPCRs and neurotransmitter transporters. Some of the main features of this new resource are illustrated in Fig. 1, and described briefly below.

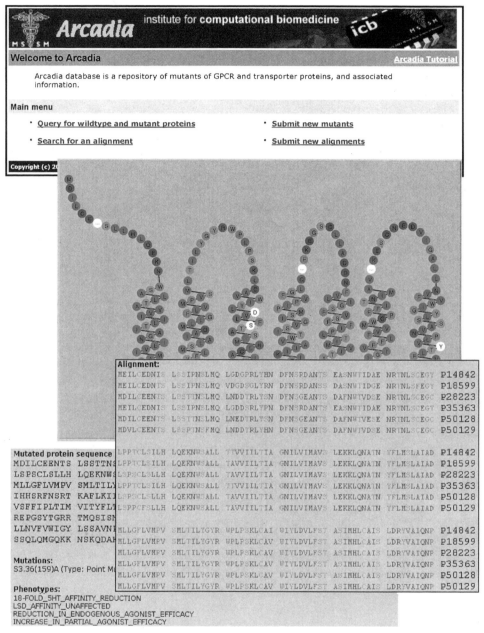

Fig. 1. Illustration of the major types of visualization available from the Arcadia database. The features (snake-like diagram, alignment and mutant details) are superimposed on the home page of the database.

2.1. Arcadia: Contents and Submissions

GPCR and transporter protein information from SWISS-PROT has been imported as background data into the Arcadia database. This background information is used in Arcadia to simplify mutant submissions (wild-type proteins are already in the database and therefore do not have to be added by the user; residue positions of the mutations can be checked against the sequence), and to link to other databases. To further facilitate data submission, we have implemented two possible methods. Mutant data can be entered into the database either via a wizard or by uploading an XML file. The wizard is an interactive method that takes the user step-by-step through the submission process, designed for investigators who will submit only a few mutants. The XML upload method is useful for submitting large numbers of mutants more efficiently.

In addition to mutations at the protein level, users can also submit genomic variants for specific entries. This is the first step to integrating sequence variations in the coding sequence of GPCRs to their pharmacological effects. Although it would be possible to import variation data from dbSNP directly, Small et al. recently reported 68% of false-positives when trying to confirm putative, nonsynonymous sequence variants in 25 GPCRs *(27)*. For this reason, only sequence variants that have been verified will be included in Arcadia.

2.2. The Use and Query of the Arcadia Database

A number of useful features from previous databases have been incorporated in this new tool. Thus, an interactive query retrieval system was implemented, to allow the user to search for wild-type and mutant proteins using keywords from many different data fields, such as keyword in protein name, protein family, or submitter. One of the more interesting query features is the ability to search by mutated position (such as specific positions in a transmembrane segment, e.g., TM3). The positions are identified both by the absolute sequence numbering and by the generic numbering system adopted for GPCRs. The "generic numbering system" for the positions in the GPCRs makes it possible to refer, comparatively, to structurally cognate receptors, including a structural template, e.g., rhodopsin *(3)*. The numbering system described originally in Ballesteros and Weinstein *(14)* has been used in many publications and in databases *(17)*. Briefly, in this generic numbering scheme, two numbers (N1.N2) are assigned to amino acid residues in TMs. N1 refers to the TM number. For N2, the numbering is relative to the most conserved residue in each TM, which is assigned 50; the other residues in the TM are numbered in relation to this conserved residue, with numbers decreasing towards the N-terminus and increasing toward the C-terminus. Thus, the proline in TM6 of rhodopsin would appear as P6.50, or P6.50(267) where the number in parentheses represents the sequence number in the individual receptor (rhodopsin, in this case). Similarly, the conserved arginine in the ERY motif in TM3 of rhodopsin would be R3.50(135), and the preceding glutamate would be E3.49(134).

Two-dimensional representations of the secondary structural elements of the proteins ("snake-like diagrams" *[20–22]*) have been integrated into the Arcadia database. Such diagrams provide a means of visualizing mutant data in the context of the sequence and topology of the protein. For wild-type proteins, the snake-like diagrams indicate all the positions that have been mutated in the protein; the mutant diagrams highlight only positions affected in the mutant. Notably, mutated residues are hyperlinked both in snake and in alignment views. Moreover, the views of alignments are configurable: Arcadia makes it possible for users to input custom alignments, for instance to view mutations in the context of different structural hypotheses. Arcadia also supports the export of alignments in FASTA format.

The data submitted to the Arcadia database will continue to be freely available, and can be exported for download as an XML file. In this way, the data in this database can be transferred to other systems or transformed and manipulated in a variety of ways by the user.

In summary, the new database, Arcadia, is an information management system for information from mutagenesis experiments. The features that set it apart from existing databases include: (1) the ability to view mutations in the context of different alignments; (2) information management features supporting user submission of both wild-type and mutant proteins, as well as user-defined alignments; (3) import and export of data using XML; and 4) free availability of both source code and data.

Arcadia is available at http://icb.mssm.edu/crt/Arcadia/

3. A STRUCTURAL CONTEXT FOR THE ACTIVATION MECHANISM OF GPCRS: INSIGHTS FROM MODELING AND COMPUTATIONAL SIMULATIONS

In the absence of detailed structural information about GPCRs, much of the efforts to interpret experimental results in a structural context has focused on creating molecular representations of these proteins that can incorporate directly and consistently the many types of function-related information (for a recent review, *see* ref. *[5]*). In turn, such molecular models serve as hypotheses-generators for experimental probing of functional inferences, and are continuously refined by the data obtained from such experiments. Listed below are some of the main advantages of such an iterative approach, as illustrated in this chapter:

1. It provides increasingly reliable and useful three-dimensional (3D) model constructs of specific members of the GPCR family (*see* below) for which complete structures are not available from experiments.
2. It yields generalizable structural templates for cognate families of these proteins, and identifies functionally important details about various subtypes *(3,5,7)*.
3. The 3D molecular constructs can be developed and modified to address characteristics of different states of the receptor molecules that relate to measurable functional characteristics such as activation, levels of constitutive activity, desensitization, and so forth (e.g., *see* refs. *[5,7,9–13,28–35]*).
4. The molecular models can be used in computational simulations of functional mechanisms to generate and/or probe specific mechanistic hypotheses for structural changes involved in the various states of the receptors, both wild-type and mutant constructs. The structural context makes these hypotheses testable in collaborative experiments designed to probe specific predictions and refine functional insights (for comprehensive review, *see* ref. *[5]*).

Together, all the inferences from both computational modeling and simulation (which can reveal novel aspects of the receptor mechanisms, based on the dynamic properties of the proteins) serve as mechanistic working hypotheses for new and more focused experiments. This mode of closely considered interactions and synergy between computational developments and experimental probing of the receptor systems has become a sustained characteristic of current studies of structure–function relations of GPCRs (for some recent examples and reviews, *see* refs. *[5,8,11,29,36–39]*).

3.1. Structural Motifs As Functional Microdomains

A key element in the development of a structure-based insight about the intramolecular mechanisms of GPCR activation has been the proposal to parse the structure into specific regions identified as Structural Motifs that act as (often conserved) Functional Microdomains (the "SM/FM"). The general concept of parsing the structure of GPCRs into identifiable SM/FM elements has been described in detail (e.g., *see* refs. *[10,13,40,41]*), and has been reviewed recently *(5,7,42)* in the context of the new crystal structure of rhodopsin *(3,43,44)*. Especially noteworthy in this context is the confirmation by the crystal structure of rhodopsin of structural predictions from the models, regarding some key SM/FM, and the corroboration of their functional properties inferred from the structural context provided by the molecular models of GPCRs in the rhodopsin-like family. This validation includes the following SM/FM:

- The predicted Arg-cage in TM3 *[13]* consisting of R3.50 interacting with D3.49 and reinforced by interactions with residues in TM6—especially E6.30—that constrains the receptor in the inactive form (*see [5,7,12,29,32,45–48]*);
- The specific interaction between TM2 and TM7 that determines some functional properties of the GPCRs (for recent reviews see *[5]*, and also *[49–52]*);
- The cluster of aromatic residues in TM6 composed of residues at positions 6.52/6.48/6.44 on a continuous face of the TM6 helix *(53)*, and their mutual orientation that appears to trigger activation *(5,8,38);* and
- The functional role of residue 7.53 in the conserved NPxxY motif in TM7 as a modulator of receptor activation, based on an interaction with Hx8, the segment C-terminal to TM7 that is known to fold into a helical structure *(10,54)*.

3.2. SM/FM in the Receptor Activation Trigger

Of the SM/FM identified in the structures of rhodopsin-like GPCRs, the cluster of aromatic residues in TM6 appears to be the most directly related to the triggering of GPCR activation by ligands. Thus, one of the components of this "aromatic cluster" in most neurotransmitter GPCRs, the conserved F6.52, has been shown to be accessible in the binding site *(55)* and has recently been suggested to serve as a sensor for the orientation of the ligand in the binding pocket *(11)*. Through the interactions of a rhodopsin-like GPCR ligand with this sensor (F6.52), the special structural properties of the aromatic cluster (*see* refs. *[5,53]*) allow it to respond to ligand binding through sterically determined conformational rearrangements of the three aromatic side chains, like a "toggle switch," to promote agonist-mediated receptor activation (for details of this hypothesis, *see [8,38,41]*). The apparent role of the orientation of the ligand in the binding pocket, in determining the measurable pharmacological response (e.g., full vs partial agonist) emphasizes the importance of this interaction with the sensor and provides a structural context for the definition of drug efficacy *(11)*. Notably, in some GPCRs (e.g., the muscarinic receptors), the role of the aromatic residue at position 6.52 seems to be subsumed by other residues that are more likely to engage in favorable interactions with the endogenous ligand, but preserve the steric properties that can trigger the interrelated rearrangement of the cluster owing to steric interactions. Thus, the recognition of the ligand by the 6.52 sensor has been suggested to give rise to a conformational change that is propagated into the aromatic cluster *(5,56)*. The functional implications of this conformational change are based on the involvement of the conserved W6.48 center of the aromatic cluster in the activation of a GPCR. This involvement was first proposed by Lin and Sakmar, based on the observation of a conformational change in the orientation of W6.48 in rhodopsin from an orientation "perpendicular" to the membrane in the inactive form, to a "parallel" orientation in activated rhodopsin *(57)*.

A mechanism by which binding of the ligand in the recognition pocket of a GPCR can trigger the receptor response through interaction with this cluster of aromatic residues was proposed earlier *(56)*, and illustrated for the serotonin 5-HT2A receptor (5-HT$_{2A}$R). The dynamics of the local rearrangements in the structure of the aromatic cluster that can be triggered by ligand binding in the pocket and interaction with the F6.52 sensor, were probed with computational simulations in a 3D model of the serotonin 5HT$_{2A}$R *(5,56)*. A molecular dynamics (MD) simulation of the effect of 5-HT binding to the receptor in the orientations described recently *(11)*, with an original distance of 4.2 Å from F6.52 as suggested in the literature *(55,58)*, was carried out utilizing the CHARMM program. Constraints were imposed to account for the effects of the surrounding environment, and to reflect the information obtained from experiment about the conformation of the activated form of a GPCR (e.g., *see* refs. *[5,33,38,39,41,52,59,60]*). Consequently, the structural context of the simulations is a model that accommodates the current information about the inactive form of the receptor based on the rhodopsin template *(3,43)*, as well as current data about the active state. The latter involves the increase in distance between the cytoplasmic ends of TM3 and TM6 upon activation, and the change in angle between W6.48 and the plane of the membrane (*see [5,52,57]*). The separation between TM3 and TM6 is achieved by imposing a NOE-type constraint (K = 1) on the distance between the Calpha of R3.50 and the Calpha of residues at positions 6.31 (min 15.8 max 19.8 Å), 6.32 (min 13.0, max 16.0 Å), and 6.34 (min 16.8, max 20.8 Å), in agreement with the changes reported by Hubbell et al. (*see* refs. *[52,59,60]*). The extracellular end of TM6 moves under initial constraints imposed on the position of the backbone of residues F6.53 to F6.59 with a K = 1, and on the dihedral angles of residues 6.30 to 6.44, so that the dynamic changes in the distance will be owing only to changes in the dihedral angles of the residues surrounding P6.50, and part of the aromatic cluster. Similarly, the position of the Calpha atoms of helices 1 to 5 and 7 are restrained with a force constant of 1. Note, however, that all the constraints imposed at the beginning of the simulation are released in small steps until a constraint-free production run is performed. The constraints are released in a total of seven steps, each followed by 50ps of equilibration. The total time spent in the heating (12ps) and subsequent equilibration steps is 350ps, followed by 1.5 nanoseconds of production run in the absence of constraints.

The simulations yielded the comparative behavior identified in Fig. 2 showing the angle between the indole plane of W6.48 and the plane of the membrane in the presence (Fig. 2A) and absence (Fig. 2B) of 5-HT interacting with F6.52 in the pocket. The energetics of the interaction affect the probability of the change in angle, thus pointing to the interaction between the aromatic moieties of F6.52 and 5HT as a trigger for the conformational change in the highly conserved W6.48 *(56)*. Other types of simulations,

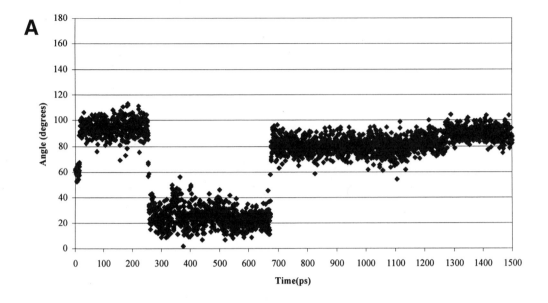

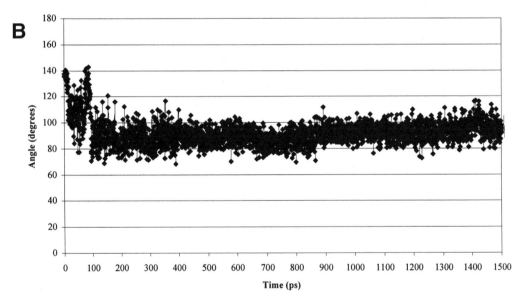

Fig. 2. Results from the MD simulations showing the values of the angle between the indole plane of W6.48 and the plane of the membrane in the presence (**A**) and absence (**B**) of the ligand, 5-HT, interacting with the F6.52 sensor in the binding pocket of the 5-HT2A receptor model.

based on Monte Carlo approaches, showed recently the same type of involvement of the aromatic cluster in the trigger of GPCR response to ligand binding *(38,39)*.

Notably, only part of the aromatic cluster is conserved in rhodopsin where the "ligand" (retinal) is covalently attached to the receptor. Thus, the 6.52 position is not needed to function as a ligand-sensor, as the conserved W6.48(265) is restrained in the "inactive" perpendicular position suggested from spectroscopy *(57)*, by interaction with the ring of retinal (see the crystal structure of the inactive form of rhodopsin *[3,43]*). It is likely, therefore, that the rearrangement in TM6 of the neurotransmitter receptor indicated from the simulations as being triggered by the interaction of the ligand with F6.52, is produced in rhodopsin by the conformational change of the retinal after isomerization around the double bond (see ref. *[38]* for a different trigger of the same motion, proposed for the cannabinoids receptor). This prediction for rhodopsin should be verified in a structure of an activated form of the receptor, where W6.48 should appear to have "escaped" the constraint of the retinal and have the ring positioned in parallel to the membrane plane. The concerted conformational changes propagated in this cluster of aromatic residues upon activation affects the kink produced in TM6 by the conserved P6.50(267). The change can transmit the binding stimulus registered by the aromatic cluster further down the TM6 helix, towards the cytoplasmic side of rhodopsin. This mechanism is likely to be conserved in the rhodopsin-like GPCRs.

The rearrangement of the aromatic cluster in response to ligand binding, and the subsequent conformational change induced at the P6.50 kink, are the likely triggers of the transition from the inactive form to the active state of the receptor *(5)*. The full dynamic rearrangement involves a series of TM helix movements described from experiments (e.g., *see* Farrens et al. *[59,61]*, Gether et al. *(15,33)*, and *(52)* for a review). These dynamic rearrangements take advantage of the special local flexibility conferred to a helix by the proline kink (e.g., see Sansom and Weinstein *[62]*). The dynamic mechanism involving the proline kink appears to be a generalized mechanism in membrane proteins (*see* refs. *[14,62-64]*), and it is not surprising to find that an absolutely conserved proline (P6.50) is likely to play a central role in the structural rearrangement of rhodopsin-like GPCRs upon transition from the inactive to the activated state. The nature of the rearrangement that identifies key differences between the structures of the active and inactive state models of these GPCRs (*see* above) are thus likely to relate to the geometry of the proline kink at P6.50, because this residue is in a special position relative to the aromatic cluster. The conserved proline is positioned between the W6.48 that has been implicated in activation *(55)*, and the "ligand sensor" F6.52 *[5,11]*. Like the MD simulations described earlier, computational simulations of using the Monte Carlo method have shown *(38,39)* that the side-chain conformation and backbone angles of the aromatic cluster region in TM6 are dynamically interrelated, and that the rearrangements also determine the preferred size and geometry of the kink produced by P6.50. Taken together, these findings point to a structure-based hypothesis involving a direct link between the recognition of the ligand by the sensor mechanism in the binding pocket, the reorientation of W6.48, and the change in the properties of the proline kink (*see* ref. *[65]*) that reposition the cytoplasmic end of TM6 and affect its neighbors. The changes in the relative position of the cytoplasmic ends of TM3 and TM6 observed experimentally *(52,60,62)* as well as computationally, require rearrangements of the hydrogen bonding network and ionic interactions that involve yet other SM/FM (most importantly, the arginine cage *[13,41]*). The role of this ionic SM/FM has been extensively probed and evaluated in the recent literature (e.g., *see* refs. *[12,15,29,32,33,35,42,47,48,66]*). The congruence of the findings from experiments and molecular modeling makes it all the more evident that molecular models must incorporate such structural rearrangements in representing the different states of

GPCRs in order to serve as templates and aid in the analysis of ligand-receptor complexes. Thus, the validated structural insights described earlier show that it is not appropriate to utilize the crystal structure of the inactive form of rhodopsin as a universal template if the modeled functional details pertain to an activated (e.g., agonist-bound) state of the GPCR.

4. THE OLIGOMERIZATION OF GPCRS

The possibility that functional mechanisms of GPCRs are constitutively regulated by oligomerization immediately brings to the fore the types of mechanistic questions that accompany most studies of protein interactions in biological systems. Among the principal considerations are the following: (1) are the GPCR oligomers preformed dimers (i.e., constitutive dimers) or dynamic structures promoted and/or modulated by regulators and/or ligands? (2) Is the interface of receptor oligomerization an element in the regulatory mechanism, or a fixed structural element? (3) Are there specific functional consequences to the 3D arrangement adopted by GPCR oligomers? The experimental exploration of these types of structure–function elements in the physiological mechanisms of GPCRs is only beginning. Especially at this early stage, both the analysis of results and the integration into mechanistic inferences can gain much from the application of biophysical principles and insights obtained from other studies of protein interaction in biological systems.

Therefore, in reviewing here the current insights concerning GPCR oligomerization, we emphasize the context provided by structural models of putative ligand–GPCR–GPCR complexes as a means to relate mechanistic hypotheses about GPCR function, to the wealth of data about structure–function relationships for individual GPCRs. In the absence of structural data at the atomic level of detail, special attention is given to: (1) inferences from computational studies regarding the likely oligomerization interfaces between TM helices of GPCRs; and (2) the possible 3D arrangements of the TM domains of GPCR dimers. Future computational simulations of these ligand–GPCR–GPCR complexes in their interaction with G proteins and in the presence of the correct environment are expected to provide a more complete characterization of the dynamic properties underlying the function of these complex biological systems. The types of GPCR interface models presented here, together with results from such simulations, will provide mechanistic working hypotheses for yet other experiments directed to the identification of specific requirements for protein interactions involved in GPCR function.

4.1. GPCR–Protein Interactions

Protein–protein interactions have often been recognized to play an essential role in protein function. Like many other signaling proteins, GPCRs have also been demonstrated to participate in such protein–protein interactions, the best known example being the interaction of these receptors with heterodimeric G proteins. Specifically, this interaction has been suggested to be the consequence of allosterism in monomeric GPCRs upon ligand activation (*see* above, and ref. *[67]* for a recent review). Ligand–GPCR complexes have also been shown to modulate various signaling pathways via direct, G protein-independent interactions with specific proteins (*see* refs. *[67–71]* for recent reviews). Among these proteins are: (1) ion channels; (2) β-arrestin and the SRC family

Table 1
Current Homo- and Hetero-Oligomers of GPCRs

Receptors	Reference	Receptors	Reference
Class A		MT$_1$-MT$_2$ Melatonin	*(115)*
A$_1$ Adenosine	*(202)*	NY1R neuropeptide Y	*(222)*
A$_1$ Adenosine-D$_1$ Dopamine	*(86)*	NY2R neuropeptide Y	*(222)*
A$_1$ Adenosine-Muscarinic mGluR$_1$	*(203)*	NY5R neuropeptide Y	*(222)*
A$_1$ Adenosine-P2Y$_1$ Purinergic	*(204)*	δ-Opioid	*(120,127)*
A$_{2A}$ Adenosine-D$_2$ Dopamine	*(205)*	μ-Opioid	*(97)*
AT$_1$ Angiotensin II	*(94)*	κ-Opioid	*(96)*
AT$_1$-AT$_2$ Angiotensin	*(206)*	δ-μ Opioid	*(97,123)*
AT$_1$ Angiotensin-B2 Bradykinin	*(94,207)*	δ-κ Opioid	*(96)*
β$_1$-Adrenoceptor	*(114)*	(δ,κ)-opioid-β$_2$-Adrenoceptor	*(125)*
β$_1$α$_{2a}$-Adrenoceptors	*(208)*	(δ,μ,κ)-CCR5	*(197)*
β$_2$-Adrenoceptor	*(117,126, 127)*	sst$_{2A}$ Somatostatin	*(124)*
		sst$_5$ Somatostatin	*(106)*
β$_1$-β$_2$-Adrenoceptor	*(114,209)*	sst$_3$ Somatostatin	*(124)*
B2 Bradykinin	*(134)*	sst$_1$ Somatostatin	*(106)*
CB1 cannabinoid	*(210)*	sst$_1$-sst$_5$ Somatostatin	*(106)*
CCR2 Chemokine	*(98,101)*	sst$_{2A}$-sst$_3$Somatostatin	*(124)*
CCR5 Chemokine	*(98,103)*	sst$_{2A}$ Somatostatin-μ-Opioid	*(223)*
CXCR4 Chemokine	*(104)*	sst$_5$ Somatostatin-D$_2$ Dopamine	*(95)*
CCR2-CCR5 Chemokine	*(98)*	Thyrotropin	*(87,116)*
D$_1$ Dopamine	*(138,211)*	TRHR1-TRHR2 thyrotropin	*(224)*
D$_2$ Dopamine	*(74,92, 118)*	V$_2$ Vasopressin	*(148)*
		Class B	
D$_3$ Dopamine	*(212)*	Gonadotropin-releasing hormone	*(129,225)*
D$_2$-D$_3$ Dopamine	*(213)*	Ig hepta	*(135)*
H$_2$ Histamine	*(214)*	**Class C**	
H$_4$ Histamine	*(215)*	Extracellular calcium-sensing	*(121)*
5-HT$_{1B}$ Serotonin	*(216)*	GABA$_B$(1)	*(226)*
5-HT$_{1D}$ Serotonin	*(144)*	GABA$_B$(2)	*(226)*
5-HT$_{1B}$-5-HT$_{1D}$ Serotonin	*(217)*	GABA$_B$(1)-GABA$_B$(2)	*(79–81)*
Luteinizing Hormone/hCGR	*(218,219)*	Metabotropic mGluR$_1$	*(133)*
M$_2$ Muscarinic Acetylcholine	*(85)*	Metabotropic mGluR$_5$	*(119,147)*
M$_3$ Muscarinic Acetylcholine	*(122)*	T1R1-T1R3 amino-acid taste	*(227)*
M$_2$-M$_3$ Muscarinic Acetylcholine	*(220,221)*	T1R2-T1R3 amino-acid taste	*(227,228)*
MT$_1$ Melatonin	*(115)*	***Class D***	
MT$_2$ Melatonin	*(115)*	Yeast α–factor receptor STE-2	*(128,229)*

of tyrosine kinases; (3) effector proteins containing PDZ domains; and (4) SH3-containing effector proteins (for recent reviews *see* refs. *[1,2]*). In light of these findings, the recently demonstrated ability of GPCRs to form homo- or hetero-oligomers (*see* refs. *[68,69,72]* for recent reviews) enriches the repertoire of GPCR-protein interactions, most likely leading to yet more signaling possibilities for these membrane proteins.

In the following paragraphs we summarize the current opinion on GPCR oligomerization and provide the details of the accumulated experimental and computational data on GPCR–GPCR interactions.

4.2. Understanding the Role of Oligomerization in GPCR Function

The increasingly recognized ability of GPCRs to aggregate into homomeric or complexes takes the generic name of oligomerization. The smallest oligomer is the "dimer." However, the interchangeable use of the terms "dimer" and "oligomer" in many reports on GPCR oligomerization has often confused the issue of the relevant oligomeric state, if any, for a GPCR to become functionally effective. It is still unclear whether all GPCRs can form high-order oligomers and how large they would be.

A variety of experimental approaches have produced evidence interpretable as homo- and/or hetero-oligomers of GPCRs (Table 1). Although the physiological relevance and/or functional significance of GPCR oligomerization remains to be elucidated, tantalizing hypotheses have been put forth for the potential functional roles of some of these homo- and hetero-oligomeric complexes of GPCRs. Specifically, oligomerization has been implicated in the identification of types of receptors that have been known only from their pharmacological properties, but not gene sequence (e.g., the δ1, δ2, μ1, μ2, κ1, κ2, and κ3 opioid receptor subtypes [73]), as well as in various processing and regulation procedures. Such procedures include: (1) chaperoning and intracellular transport of receptors [74–81]; (2) signal amplification and cooperativity [82–95]; and (3) the generation of novel pharmacological phenotypes that differ from the properties of the constituent monomers (e.g, the δ-μ opioid [96,97] and CCR2-CCR5[98] receptor heterodimers).

Given the lingering uncertainties about the details of cellular signal transduction involving GPCRs, the list of potential functional roles for GPCR homo- and hetero-oligomers is quite long [69,72,99]. This only reinforces the need for an understanding of the many central issues in GPCR homo- and hetero-oligomerization, preferably at a detailed structural level that would make possible the incorporation of these phenomena into physiologically relevant functional models for GPCRs. Clearly, a central issue in this respect is the role of the ligand in GPCR oligomerization.

Numerous examples of both ligand-induced [98,100–111] and ligand-independent (constitutive) oligomerization [79–81,86,94,96,97,106,112–130] have been reported in the literature over the last few years. However, inferences from these studies have not yet allowed a definitive conclusion about constitutive (i.e., GPCRs exist as stable preformed dimers) or ligand-induced oligomerization.

4.3. Structural Features of GPCR Oligomerization

Exploration of the structural features of GPCR oligomerization is at its very beginning. The experimental data collected thus far provide a general structural context for both homo- and hetero-dimerization of GPCRs. In reviewing the structural elements identified so far, we will pay special attention to: (1) GPCR regions implicated in oligomerization; (2) the nature of the interaction between GPCR monomers; and (3) suggested modes of association between GPCR monomers.

4.3.1. GPCR Regions Implicated in Oligomerization

Extracellular, transmembrane (TM), and/or C-terminal regions have been suggested to be involved in the oligomerization process of GPCRs.

The involvement of the extracellular domain in GPCR oligomerization has mostly been reported for members of Class C, such as the extracellular calcium-sensing receptor *(131)* and the metabotropic glutamate receptor *(132,133)*. However, evidence for the participation of the GPCR extracellular domain in oligomerization also exists for members of Class A rhodopsin-like, Class B secretin-like, and Class D fungal pheromone receptors, such as bradykinin B2 *(134)*, Ig hepta *(135)*, and the yeast alpha factor receptor STE-2 *(136)*, respectively. The high-resolution crystal structures of the extracellular ligand-binding region of the metabotropic glutamate receptor mGluR1 *(133)* suggested a mechanism by which the N-terminus could mediate receptor dimerization of this particular class of GPCRs. Specifically, three different crystal structures of the extracellular ligand-binding region of mGluR1 were determined, in a complex with glutamate and in two unliganded forms. All three different crystal structures appeared as disulphide-linked homodimers, in the presence or absence of the ligand. Examination of the conformations of the mGluR1 homodimer in the three crystal forms, combined with modeling studies, confirmed the involvement of several conformers of the mGluR1 extracellular ligand-binding region. "Active" and "resting" conformations were identified as the result of interdomain movement and relocation of the dimer interface. Binding of glutamate to the extracellular domain of mGluR1 was implicated in the selective stabilization of the activated domain conformations in equilibrium with all other states. The interdomain movements in the dimer related to glutamate binding were suggested to produce an allosteric effect on the TM or intracellular regions of the mGluR1 receptor, leading to its activation *(133)*.

The GPCR C-terminal region, with or without a coiled-coil motif, has been implicated in the heterodimerization process of $GABA_B(1)$-$GABA_B(2)$ *(78,80)*, as well as in the homodimerization of δ-opioid receptor *(120)*. However, the possibility cannot be ruled out that the indirect evidence for the involvement of the C-terminal domain results from changes produced by the deletion of the C-terminal tail, that alters the conformation of the receptors and causes decreased interactions between transmembrane helices and oligomer disruption.

Based on evidence from other membrane proteins for which more detailed structural information has become available *(137)*, it is likely that the GPCR oligomerization process involves not only the extracellular and/or C-terminal domains, but also their TM regions. Indeed, GPCR dimer interfaces involving TM regions have been proposed for several GPCR family members, such as β2-adrenergic *(117)*, D1 *(138)*, and D2 *(118)* dopamine, adrenergic-muscarinic *(139,140)*, CCK *(141)*, and yeast alpha factor receptors *(136)*. The presence of the glycophorin A dimerization motif LXXXGXXXGXXXL *(142)* in the TM6 of β2-adrenergic receptors suggested the involvement of this TM region in the receptor dimerization. Inhibition studies with a synthetic peptide based on the amino acid sequence of the sixth TM domain of β2-adrenergic receptor corroborated the hypothesis of dimeric interaction through TM6 for this GPCR family member. The use of synthetic peptides also allowed identification of the sixth and seventh TM domains as possible dimerization interfaces in D2 dopamine receptors *(118)*. On the other hand, a peptide based on the TM6 sequence of D1 dopamine receptor did not affect the level of receptor dimers *(138)*, suggesting that the dimerization interfaces of GPCRs may be subtype-specific. Very recently, fluorescence resonance energy transfer (FRET) and endocytosis assays of oligomerization in vivo have implicated TM1 and TM2 in the dimerization of the yeast alpha factor receptor *(136)*.

4.3.2. The Nature of the Interaction between GPCR Monomers

Both covalent (disulfide) and noncovalent (hydrophobic) interactions have been implicated in mediating the interaction between GPCR monomers.

The sensitivity of several GPCR homo-oligomers to reducing agents suggested the involvement of disulfide bonds in the mechanism of dimerization. Evidence of oligomeric dissociation by reducing agents has been reported for GPCR subtypes, including metabotropic glutamate 1 *(143)*, metabotropic glutamate 5 *(119)*, calcium-sensing *(121,132)*, muscarinic M3 *(122)*, δ-opioid *(120)*, κ-opioid *(96)*, dopamine D1 *(144)*, serotonin $5HT_{1B}$ *(144)*, serotonin $5HT_{1D}$ *(144)*, and vasopressin V2 *(77)* receptors. However, a resistance to dissociation has been observed even for these, as well as other GPCRs *(117,118,144)*, and recent evidence on calcium-sensing *(145)* and metabotropic glutamate receptors *(146,147)* suggests that disulfide bonds are not necessary for oligomerization of these proteins. Taken together, these experimental data favor the view that noncovalent rather than covalent interactions may play an essential role in the dimerization of GPCR monomers.

4.3.3. Suggested Modes of Association Between GPCR Monomers

An important question related to the functional role of the association of GPCRs in the cell membrane relates to the manner in which the oligomers are formed: either as interacting entities that retain their structural identity, or as interlaced units. Both modes of association of the TM helices of GPCR monomers into dimers have been proposed. One of them consists of simple 1:1 stoichiometric molecular complexes of the GPCR monomers (contact dimers) *(148)*. The other one involves the exchange between monomers of complementary TM domains, such as the N-terminal (TMs 1-5) and C-terminal (TMs 6-7) regions (domain swapping) *(149–152)*.

Although domain-swapped association has been demonstrated for truncated and mutant forms of adrenergic-muscarinic *(139)* and type 1 angiotensin II *(153)* receptors, more recent experimental evidence *(74,136,140,141,148)* suggests that this proposed mechanism of dimerization may represent only a mechanism of functional rescue in GPCRs. Accordingly, the fact that vasopressin V2 receptor mutants altered in the N-terminal folding domain (TMs 1-5) cannot be rescued by co-expression of N-terminal receptor fragments and mutated V2 receptors *(148)* suggests that oligomers of this receptor may form by contact rather than domain-swapping. Experiments with point mutants and truncation mutants of D2DR containing TMs 1-5 or TMs 6-7 also suggested that the mechanism of dimerization of GPCRs does not involve domain-swapped association *(74)*. Moreover, photoaffinity label experiments using a peptide agonist of the cholecystokinin receptor showed binding of this peptide to two regions (TM1 and TM7) of the same receptor *(141)*. The fact that this peptide did not link covalently two different cholecystokinin receptors, but binds to only one monomer also argues against the mechanism of domain swapping in the dimerization of GPCRs. The same conclusion emerges from mapping the proximities between the TM1 and TM7 domains of the M3 muscarinic acetylcholine receptor, using an *in situ* disulfide cross-linking strategy *[140]*. In fact, intramolecular cross-linking between the TM1 and TM7 domains of this GPCR did not change with ligand binding and, more importantly, it did not produce dimers. Finally, a very recent study on the yeast alpha factor receptor *(136)* also favors a dimer contact model over domain swapping because small receptor fragments, including one consisting only of the N-terminus plus TM1, can self-associate, unlike several receptor fragments that lack TM1.

The structural context of the interactions among GPCRs further supports the argument against domain swapping in the oligomerization of GPCRs. Based on the accepted topology of rhodopsin-like GPCRs, the two-dimensional diffraction maps of rhodopsin *(154–156)* offer information about the mode of association of these proteins. In these density maps, the TM domains are arranged in neighboring hydrophobic bundles such that lateral intermolecular interactions seem to occur between them, as in contact dimers rather than domain-swapped dimers. The recent atomic-force microscopy map of rhodopsin molecules *(157)*, which shows closely packed dimers in native membranes, also seems to support the hypothesis of contact dimers rather than domain-swapped dimers.

Even with the contact dimer geometry as the most likely form of GPCR association, the identity of the interface remains unknown. Yet, the nature and geometry of the interface(s) will be essential for understanding the role and implications of GPCR oligomerization in the functional mechanisms of the receptors. The use of computational methods combining modeling and Bioinformatics tools to identify the likely oligomerization interfaces is therefore described below as a basis for complementing and informing experimental explorations of GPCR dimerization.

4.4. Computational Methods for the Prediction of GPCR–GPCR Interactions

Although the character of interfaces in various known protein–protein complexes has been analyzed in some detail (*see* refs. *[158–160]* for reviews), the lingering difficulties in crystallizing functionally relevant complexes are holding back the identification of the underlying protein interaction networks. This problem is particularly acute for the GPCRs where structural information is scant even for the monomers. Consequently, analogy and homology serve in the analysis of putative interaction modes of GPCRs.

Complementing a variety of experimental techniques, such as analysis of isolated protein complexes by mass spectrometry *(161,162)*, protein microarrays *(163)*, and automated yeast two-hybrid-based methods *(164)*, several computational methods have been developed in order to help identify protein interaction networks (*see* refs. *[165]* and *[166]* for recent reviews). These computational methods can be divided into two categories, depending on the availability of the 3D structures of the interacting proteins in the complex. Specifically, if the structural information is available for the interacting proteins in the complex, the general approaches are in the "docking methods" category *(166)*, used in an attempt to identify the protein interfaces. Unfortunately, the case in which all interacting proteins in the complex have known 3D structures is unusual, and even in this case, the accuracy of the predictions derived from computational techniques for physical docking is limited *(167)*. On the other hand, several new computational methods based on sequence and genomic information have emerged to identify protein-protein interactions in the absence of complete structural information about the interacting proteins in a complex. The main limitations and predictive power of these methods that are based on sequence and genomic information, have been discussed in a recent review *(165)*.

Among the computational techniques for predicting protein-protein interactions, only a few methods have been applied to GPCRs. For example, a data-mining approach based on pattern discovery and membrane topology prediction was applied to GPCRs in an attempt to identify patterns of amino acid residues in the cytoplasmic domains of these receptors that are specific for coupling to a particular class of GPCRs *(168)*. Co-evolutionary analysis has recently been used to predict interactions between the GPCR/G-α receptor families *(169)*. Predictions of residues of GPCRs that are responsible for G protein coupling were also proposed using correlated mutation analysis (CMA)-based techniques *(170–175)*. Based on the principle that residues involved in a common function tend to mutate together, the CMA has often been used to detect intramolecular *(176–178)* and intermolecular *(179,180)* contacts between protein residues.

In the particular case of GPCRs, the application of CMA techniques, including the evolutionary trace method *(181–183)*, has been aimed at the prediction of putative functionally important residues involved in GPCR–G protein interactions *(170–175)*, GPCR–GPCR interactions *(151,152,175,184–189)*, ligand–protein interactions *(23,190,191)*, as well as signal transduction *(170,192,193)*.

4.4.1. The Prediction of Likely Interfaces of Oligomerization Between TM Regions of GPCRs

To obtain a 3D model of a GPCR oligomerizing through the TM domains of the monomers, the key predictive steps involve the identification of the TM interfaces. The number of possibilities is extremely large, as even the model of a GPCR dimer would present at least 49 (= 7 × 7) different alternative configurations for a heterodimer, and at least 28 (= 7(7 + 1)/2) for a homodimer. These alternatives are independent, in principle, because there is no evidence regarding the possibility that the homo- and hetero-oligomerization interfaces between the TM regions of GPCRs coincide.

A combination of the structural information of GPCR monomers derived from homology modeling, using the rhodopsin crystal structure *(43)* as a template, with either correlated mutation or evolutionary trace analyses, has been used to obtain insights that could reduce the number of different configurations in which TM bundles can be packed next to each other to form homo- and/or hetero-oligomers of GPCRs *(151,152,175,184–189)*.

The earlier computational studies *(151,152,175,184,185)* considered both domain-swapped and contact dimers as equally possible mechanisms of GPCR oligomerization. In contrast, the later computational studies on GPCR oligomerization *(186–189)* take into account only the hypothesis of contact dimers, supported by the more recent experimental evidence. For the prediction of heterodimer interfaces, the recent studies use a modified CMA methodology, termed subtractive correlated mutation (SCM) analysis *(187,188)*. A similar method for the identification of physically interacting protein pairs has recently been reported in the literature *(180)*.

Briefly, the SCM method first provides the complete list of intra- and intermolecular pairs of interacting GPCR monomers by predicting correlated mutations from the multiple sequence alignment of concatenated monomeric sequences from equal species. Subsequently, the intramolecular pairs of correlated residues within each GPCR monomer, which can be calculated using correlated mutation analysis of separated multiple sequence alignments, are filtered out from this complete list of intra- and intermolecular pairs. Finally, the remaining pairs of residues are evaluated based on solvent accessibility values calculated from the atomic coordinates of the 3D structures of each GPCR monomer. According to this criterion, the intermolecular pairs where either one or both residues are considered completely or partially inaccessible to the solvent (based on the structural template of the monomer) are eliminated from the list. The remaining residues for each GPCR monomer are candidates for the heterodimerization interface. Additional criteria for the strengthening of the predictive value of the approach include considerations of "interaction neighborhoods" on the specific helix face proposed as an interface.

Application of the SCM method to the opioid receptor subtypes *(187,188)* supported the experimental evidence of subtype specificity in the process of heterodimerization of these receptors *(96)*. Although likely heterodimerization interfaces of the δ-μ and δ-κ opioid receptor complexes were predicted using the SCM method, the application of this procedure to explore the putative interface between μ and κ opioid receptors resulted in a null set for residues that indicates the absence of a predicted dimerization interface. This result is in full agreement with previous experimental observations *(96)*.

Although the evolutionary trace method applied to GPCRs *(175,185)* fails to detect any residues responsible for the subtype-specific heterodimerization that has recently been demonstrated for opioid *(96)*, somatostatin *(106)*, and chemokine *(98)* receptors, correlated mutation analysis had already been demonstrated to be able to identify useful details of molecular specificity *(184)*. Thus, the molecular basis of specificity was hypothesized to reside in outward (i.e., lipid) facing residues of TM5 and TM6 that exhibited evolutionarily correlated mutations and differed between receptor subtypes *(184)* in the case of dimerization. In the case of oligomers, the key interface between different subtypes was suggested to be the 2,3-interface *(152)* rather than the 5,6-interface.

The hypothesis that oligomerization may have a specific functional role in GPCR activity raises the possibility that various receptor subtypes may use different oligomerization interfaces to exhibit functional selectivity. Therefore, using a combination of structural information and correlated mutation analysis, we have explored the hypothesis that the oligomerization interfaces might vary among GPCRs. The first set of GPCRs for which this analysis was carried out in detail included the rhodopsin-like GPCR subtypes for which homodimerization has been demonstrated experimentally (manuscript in preparation; *see* ref. *[186]*). The approach was validated using the structural information available from the electron-density maps of opsins in different species. The results suggested that at least for this group of GPCRs, different dimerization interfaces are involved in homodimerization of different GPCR subtypes.

4.4.2. Putative Configurations of GPCR Dimers

Correlated mutation analysis does not yield a precise and exhaustive enumeration of all the interacting residues in a complex. Rather, tests have shown *(177)* that the result is a definition of the neighborhood of interacting regions. This makes the careful construction of explicit 3D models of GPCR homodimers, using the information derived from correlated mutation analysis, all the more important in the effort to reveal additional loci of interaction between monomeric interfaces, and to eliminate false-positives among the predicted correlated mutations.

Models of the geometrically feasible configurations of GPCR dimers that incorporate the information derived from correlated mutation analysis satisfy a number of criteria. These criteria represent working hypotheses about the nature of the dimers, and are used in the construction of the models. The first criterion is that the orientation of the 7TM interacting bundles of the GPCR dimer in the membrane is maintained as in the original monomer template. Second, all combinations of the likely interfaces of dimerization predicted by correlated mutation analysis are considered as equally possible. Finally, the 3D models of GPCR dimers maximize the number of interactions between the predicted correlated residues on the appropriate lipid-facing surface of the TMs in each monomer.

These criteria were applied to the construction of 3D molecular models of δ, μ, and κ opioid receptor homodimers *(189)*. The most likely dimerization interfaces obtained from the correlated mutation analysis yielded three equally possible 3D models (TM4-TM4, TM4-TM5, and TM5-TM5) for the δ-opioid homodimer, but only one configuration was possible for μ-opioid (TM1-TM1) and κ-opioid (TM5-TM5) homodimers. By listing the interface residues together with those at distances within a 5 Å range between the closest heavy atoms of the side chains of the energy minimized 3D models of the δ, μ, and κ opioid receptor homodimers, explicit predictions are obtained for testing with specific experiments designed to not only validate the structural predictions but also to probe the functional implications of the dimerization mechanism of the opioid receptor subtypes. The same approach is being used to build 3D molecular models of other GPCR dimers that have been shown to exist by experiments, and to guide pointed experiments designed to identify specific structural elements responsible for dimerization and specificity. Mutant constructs based on inferences from such studies should be valuable in the elucidation of functional implications of the formation of GPCR homodimers and heterodimers, and its physiological consequences.

5. PERSPECTIVE

As indicated in this chapter, there is a rich and constantly growing literature describing detailed studies of structure–function relations for GPCRs in the various classes. This is a most powerful testimonial to the importance of these proteins for understanding fundamental aspects of physiological processes based on signal transduction. It also demonstrates the continued high interest in GPCRs as valuable targets for drug design (for a recent review, *see* ref. *[37]*). The main characteristics that make GPCRs so important in this respect are that: (1) they constitute mostly validated targets for drug design, both pharmacologically and physiologically; (2) they are pharmacologically "complete," in that ligands can be designed to elicit responses in the full spectrum, from antagonism to partial and full agonism; and (3) they offer a very high "genomic leverage" for the designed ligands, given the large number of GPCRs in the genome, the functional importance they have, and the structural similarity within the classes of GPCRs. With the growing ability to relate the functional properties of the GPCRs to specific structural elements, such as the SM/FM illustrated here, for their ability to provide a defined structural context to the intramolecular mechanisms (e.g.,

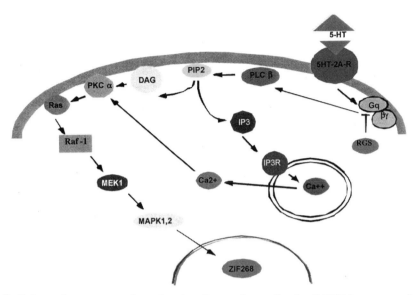

Fig. 3. Schematic representation of a signaling pathway for the 5-HT2A receptor, in which activation by 5-HT leads to the regulation of expression of an immediate early gene, Zif268.

specific ligand recognition, response triggering, signal propagation to interaction with G proteins), there is increasing confidence that new types of ligands can be discovered in a structure-guided approach. Importantly, such approaches are expected to yield ligands that can selectively address the different forms of the receptors, e.g., constitutively active (for a review, *see* ref. *[36]*), or in the oligomeric forms discussed earlier (for a review, *see* ref. *[69]*).

It is important to note, however, that the emerging view of GPCRs as interconnected elements in signaling ensembles regulated by protein oligomerization (*see* refs. *[73,99,194,195]* for recent reviews) raises the question of the mode of integration of this mechanism in the pharmacological classification of drug action. If the relevant unit defining the pharmacological characteristics is not the specific GPCR, then the classical elements of drug classification, definitions of selectivity, and measurements of efficacy must be reconsidered (*see* ref. *[69]* for a recent review, and refs. *[125,128,144,148,175,189,196,197]*). The conceptual difficulties that might be created by this new perspective are perhaps mitigated by the promise for discovering new types of functional modulation of receptor activity, and novel approaches to drug development. In particular, such novel approaches are likely to emerge from the recognition that the physiological consequences of the signal-transduction processes, in which the GPCRs play a key role, depend on a series of intracellular interactions in the signal-transduction pathways. Consequently, we and others have been developing new types of Bioinformatics approaches to study the structure–function relations of GPCRs in a context of their signaling pathways (an example for the 5-HT2A receptor is schematically illustrated in Fig. 3). Understanding signal flow from the cell surface to the nucleus requires detailed knowledge of signaling events at the plasma membrane, coupled with movement of the signal into the nucleus to regulate gene expression. Initial models of such complex signaling pathways, in which the interactions shown schematically in Fig. 3 are accounted for quantitatively, have been demonstrated *(198–201).* Application to GPCR signaling will allow us to determine refined approaches for the realistic representation of signaling in dynamic compartments within the cell, and to discover how the movement of signaling components between compartments regulates the flow of the signal triggered by GPCR activation. Such models are likely to be increasingly necessary for development of an understanding of how the multiple functions of a cell are coordinated and regulated, and to realize the specific role of the GPCRs in these functions.

Implementation of quantitative methods of computational modeling and simulation, made possible by the combination of discrete molecular models and systems level simulations, should eventually enable an understanding of the regulation of cellular functions by the ligands of GPCRs with various pharmacological profiles. These approaches will utilize the new type of data collected from quantitative biochemical approaches, as well as high-throughput genomic and proteomic methods. Based on such data, the methods of Bioinformatics, molecular biophysics and modeling at mesoscale and macroscopic levels will serve to attain a quantitative, heretofore unachievable understanding of the mechanisms underlying systems behavior at the cellular level, and its modulation by the actions of various agents acting on GPCRs.

ACKNOWLEDGMENTS

The work was supported by NIH grants P01 DA-12923, P20 HL69733, and K05 DA-00060 (to HW). MF is supported by NRSA T32 DA07135. Computational support was provided by the National Science Foundation Terascale Computing System at the Pittsburgh Supercomputing Center and by the computer facilities of the Institute for Computational Biomedicine of the Mount Sinai School of Medicine.

REFERENCES

1. Pierce, K. L., Premont, R. T., and Lefkowitz, R. J. (2002) Seven-transmembrane receptors. *Nature Rev. Mol. Cell Biol.* **3,** 639–650.
2. Rockman, H. A., Koch, W. J., and Lefkowitz, R. J. (2002) Seven-transmembrane-spanning receptors and heart function. *Nature* **415,** 206–212.
3. Stenkamp, R. E., Teller, D. C., and Palczewski, K. (2002) Crystal structure of rhodopsin: a G protein-coupled receptor. *Chem. BioChem.* **3,** 963–967.
4. Oussama, E. F. and Heinrich, B. (2002) G protein-coupled receptors for neurotransmitter amino acids: C-terminal tails, crowded signalosomes. *Biochem. J.* **365,** 329–336.
5. Visiers, I., Ballesteros, J. A., and Weinstein, H. (2002) Three-dimensional representations of G protein-coupled receptor structures and mechanisms. *Methods Enzymol* **343,** 329–371.
6. Kroeze, W. K., Kristiansen, K., and Roth, B. L. (2002) Molecular biology of serotonin receptors structure and function at the molecular level. *Curr. Top. Med. Chem.* **2,** 507–528.
7. Ballesteros, J. A., Shi, L., and Javitch, J. A. (2001) Structural mimicry in G protein-coupled receptors: implications of the high-resolution structure of rhodopsin for structure–function analysis of rhodopsin-like receptors. *Mol. Pharmacol.* **60,** 1–19.
8. Shi, L. and Javitch, J. A. (2002) The binding site of aminergic G protein-coupled receptors. *Annu. Rev. Pharmacol. Toxicol.* **42,** 437–467.
9. Javitch, J. A., Ballesteros, J. A., Chen, J., Chiappa, V., and Simpson, M. M. (1999) Electrostatic and aromatic microdomains within the binding-site crevice of the D2 receptor: contributions of the second membrane-spanning segment. *Biochemistry* **38,** 7961–7968.

10. Prioleau, C., Visiers, I., Ebersole, B. J., Weinstein, H., and Sealfon, S. (2002) Conserved helix7 tyrosine acts as a multistate conformational switch in the 5-HT2C receptor:Identification of a novel "locked-on" phenotype and double revertant mutations. *J. Biol. Chem.* **277,** 36577–36584.

11. Ebersole, B. J., Visiers, I., Weinstein, H., and Sealfon, S. (2003) Molecular basis of partial agonism: orientation of indolamine ligands in the binding pocket of the human serotonin 5-HT2A receptor determines relative efficacy. *Mol. Pharmacol.* **63,** 36–43.

12. Huang, P., Li, J., Visiers, I., Weinstein, H., and Liu-Chen, L.-Y. (2001) Functional role of a conserved motif in TM6 of the rat m opioid receptor: constitutively active and inactive receptors result from substitutions of Thr6.34(279) with Lys and Asp. *Biochemistry* **40,** 13501–13509.

13. Ballesteros, J. A., Kitanovic, S., Guarnieri, F., et al. (1998) Functional microdomains in G protein-coupled receptors: The conserved arginine cage motif in the gonadotropin-releasing hormone receptor. *J. Biol. Chem.* **273,** 10445–10453.

14. Ballesteros, J. A. and Weinstein, H. (1995) Integrated methods for the construction of three-dimensional models and computational probing of structure–function relations in G protein-coupled receptors. *Methods Neurosci.* **25,** 366–428.

15. Gether, U., Lin, S., Ghanouni, P., Ballesteros, J. A., Weinstein, H., and Kobilka, B. K. (1997) Agonists induce conformational changes in transmembrane domains III and VI of the b2 adrenergic receptor. *EMBO J.* **16,** 6737–6747.

16. Anastasiadis-Pool, A. and Schwalbe, H. (2002) Web sites about G protein-coupled receptors. *ChemBioChem.* **3,** 1029–1030.

17. van Rhee, A. M. and Jacobson, K. A. (1996) Molecular architecture of G protein coupled receptors. *Drug Dev. Res.* **37,** 1–38.

18. Kristiansen, K., Dahl, S. G., and Edvardsen, O. (1996) A database of mutants and effects of site-directed mutagenesis experiments on GPCRs. *Prot. Struct. Func. Genet.* **26,** 81–94.

19. Beukers, M. W., Kristiansen, I., AP, I. J., and Edvardsen, I. (1999) TinyGRAP database: a bioinformatics tool to mine G protein-coupled receptor mutant data. *Trends Pharmacol. Sci.* **20,** 475–477.

20. Campagne, F., Jestin, R., Reversat, J. L., Bernassau, J. M., and Maigret, B. (1999) Visualisation and integration of G protein-coupled receptor related information help the modelling: description and applications of the Viseur program. *J. Comput, Aided Mol. Des.* **13,** 625–643.

21. Campagne, F. and Weinstein, H. (1999) Schematic representation of residue-based protein context-dependent data: an application to transmembrane proteins. *J. Mol. Graph. Model* **17,** 207–213.

22. Konvicka, K., Campagne, F., and Weinstein, H. (2000) Interactive construction of residue-based diagrams of proteins: the RbDe web service. *Prot. Eng.* **13,** 395–396.

23. Horn, F., Bywater, R., Krause, G., et al. (1998) The interaction of class B G protein-coupled receptors with their hormones. *Recep. Channels* **5,** 305–314.

24. Bairoch, A. and Apweiler, R. (2000) The SWISS-PROT protein sequence database and its supplement TrEMBL in 2000. *Nucleic Acids Res.* **28,** 45–48.

25. Cotton, R. G. and Horaitis, O. (2002) The HUGO Mutation Database Initiative. Human Genome Organization. *Pharmacogen. J.* **2,** 16–19.

26. Sherry, S. T., Ward, M. H., Kholodov, M., Baker, J., Phan, L., Smigielski, E. M., and Sirotkin, K. (2001) dbSNP: the NCBI database of genetic variation. *Nucleic Acids Res.* **29,** 308–311.

27. Small, K. M., Seman, C. A., Castator, A., Brown, K. M., and Liggett, S. B. (2002) False positive non-synonymous polymorphisms of G protein coupled receptor genes. *FEBS Lett.* **516,** 253–256.

28. Visiers, I., Hassan, S. A., and Weinstein, H. (2001) Differences in conformational properties of the second intracellular loop (IL2) in 5HT(2C) receptors modified by RNA editing can account for G protein coupling efficiency. *Prot. Eng.* **14,** 409–414.

29. Shapiro, D. A., Kristiansen, K., Weiner, D. M., Kroeze, W. K., and Roth, B. L. (2002) Evidence for a model of agonist-induced activation of 5-HT2A serotonin receptors that involves the disruption of a strong ionic interaction between helices 3 and 6. *J. Biol. Chem.* **277,** 11441–11449.

30. Fanelli, F., Menziani, C., Scheer, A., Cotecchia, S., and DeBenedetti, P. G. (1999) Theoretical study on receptor-G protein recognition: new insights into the mechanism of the α_{1b}-adrenergic receptor activation. *Intl. J. Quant. Chem.* **73,** 71–83.

31. Fanelli, F. (2000) Theoretical study on mutation-induced activation of the luteinizing hormone receptor. *J. Mol. Biol.* **296,** 1333–1351.

32. Huang, P., Visiers, I., Weinstein, H., and Liu-Chen, L.-Y. (2002) The local environment at the cytoplasmic end of TM6 of the m opioid receptor differs from those of rhodopsin and monoamine receptors: introduction of an ionic lock between the cytoplasmic ends of helices 3 and 6 by a L6.30(275)E mutation inactivates the m opioid receptor and reduces the constitutive activity of its Thr6.34(279)K mutant. *Biochemistry* **41,** 11972–11980.

33. Jensen, A. D., Guarnieri, F., Rasmussen, S. G., Asmar, F., Ballesteros, J. A., and Gether, U. (2001) Agonist-induced conformational changes at the cytoplasmic side of transmembrane segment 6 in the beta 2 adrenergic receptor mapped by site-selective fluorescent labeling. *J. Biol. Chem.* **276,** 9279–9290.

34. Ghanouni, P., Gryszynski, Z., Steeinhuis, J. J., Lee, T. W., Farrens, D. L., Lakowitz, J. R., and Kobilka, B. K. (2001) Functionally different agonists induce distinct conformations in the G protein coupling domain of the β2 adrenergic receptor. *J. Biol. Chem.* **276,** 24433–24436.

35. Ghanouni, P., Steenhuis, J. J., Farrens, D. L., and Kobilka, B. K. (2001) Agonist-induced conformational changes in the G protein-coupling domain of the beta 2 adrenergic receptor. *Proc. Natl. Acad. Sci. USA* **98,** 5997–6002.

36. Chalmers, D. T. and Behan, D. P. (2002) The use of constitutively active GPCRs in drug discovery and functional genomics. *Nat. Rev. Drug Dis.* **1,** 599–608.

37. Klabunde, T. and Hessler, G. (2002) Drug design strategies for targeting G protein-coupled receptors. *ChemBioChem* **3,** 928–944.

38. Singh, R., Hurst, D. P., Barnett-Norris, J., Lynch, D. L., Reggio, P. H., and Guarnieri, F. (2002) Activation of the cannabinoid CB1 receptor may involve a W6.48/F3.36 rotamer toggle switch. *J. Peptide Res.* **60,** 357–370.

39. Shi, L., Liapakis, G., Xu, R., Guarnieri, F., Ballesteros, J. A., and Javitch, J. A. (2002) beta2 Adrenergic receptor activation. *J. Biol. Chem.* **277,** 40989–40996.

40. Sealfon, S. C., Chi, L., Ebersole, B. J., Rodic, V., Zhang, D., Ballesteros, J. A., and Weinstein, H. (1995) Related contribution of specific helix 2 and 7 residues to conformational activation of the serotonin 5-HT$_{2A}$ receptor. *J. Biol. Chem.* **270,** 16683–16688.

41. Visiers, I., Ebersole, B. J., Dracheva, S., Ballesteros, J., Sealfon, S. C., and Weinstein, H. (2002) Structural motifs as functional microdomains in GPCRs: energetic considerations in the mechanism of activation of the serotonin 5-HT2A receptor by disruption of the ionic lock of the arginine cage. *Intl. J. Quant. Chem.* **88,** 65–75.

42. Ballesteros, J. A., Jensen, A. D., Liapakis, G., Rasmussen, S. G., Shi, L., Gether, U., and Javitch, J. A. (2001) Activation of the {beta}2 adrenergic receptor involves disruption of an ionic lock between the cytoplasmic ends of transmembrane segments 3 and 6. *J. Biol. Chem.* **276,** 29171–29177.

43. Palczewski, K., Kumasaka, T., Hori, T., et al. (2000) Crystal structure of rhodopsin: a G protein-coupled receptor. *Science* **289,** 739–745.

44. Teller, D. C., Okada, T., Behnke, C. A., Palczewski, K., and Stenkamp, R. E. (2001) Advances in determination of a high-resolution three-dimensional structure of rhodopsin, a model of G protein-coupled receptors (GPCRs). *Biochemistry* **40,** 7761–7772.

45. Angelova, K., Fanelli, F., and Puett, D. (2002) A model for constitutive lutropin receptor activation based on molecular simulation and engineered mutations in transmembrane helices 6 and 7. *J. Biol. Chem.* **277,** 32202–32213.

46. Greasley, P. J., Fanelli, F., Scheer, A., Abuin, L., Nenniger-Tosato, M., DeBenedetti, P. G., and Cotecchia, S. (2001) Mutational and computational analysis of the alpha(1b)-adrenergic receptor. Involvement of basic and hydrophobic residues in receptor activation and G protein coupling. *J. Biol. Chem.* **276,** 46485–46495.
47. Greasley, P. J., Fanelli, F., Rossier, O., Abuin, L., and Cotecchia, S. (2002) Mutagenesis and modelling of the alpha(1b)-adrenergic receptor highlight the role of the helix 3/helix 6 interface in receptor activation. *Mol. Pharmacol.* **61,** 1025–1032.
48. Cotecchia, S., Rossier, O., Fanelli, F., Leonardi, A., and De Benedetti, P. G. (2000) The alpha 1a and alpha 1b-adrenergic receptor subtypes: molecular mechanisms of receptor activation and of drug action. *Pharm. Acta Helv.* **74,** 173–179.
49. Zhou, W., Flanagan, C., Ballesteros, J. A., et al. (1994) A reciprocal mutation supports helix 2 and helix 7 proximity in the gonadotropin-releasing hormone receptor. *Mol. Pharmacol.* **45,** 165–170.
50. Xu, W., Ozdener, F., Li, J. G., Chen, C., de Riel, J. K., Weinstein, H., and Liu Chen, L.-Y. (1999) Functional role of the spatial proximity of Asp2.50(114) in TMH2 and Asn7.49(332) in TMH7 of the μ opioid receptor. *FEBS Lett.* **447,** 318–324.
51. Sealfon, S. C., Chi, L., Ebersole, B. J., Rodic, V., Zhang, D., Ballesteros, J. A., and Weinstein, H. (1995) Related contribution of specific helix 2 and 7 residues to conformational activation of the serotonin 5-HT2A receptor. *J. Biol. Chem.* **270,** 16683–16688.
52. Meng, E. C. and Bourne, H. R. (2001) Receptor activation: what does the rhodopsin structure tell us? *Trends Pharmacol. Sci.* **22,** 587–593.
53. Javitch, J. A., Ballesteros, J. A., Weinstein, H., and Chen, J. (1998) A cluster of aromatic residues in the sixth membrane-spanning segment of the dopamine D2 receptor is accessible in the binding site crevice. *Biochemistry* **37,** 998–1006.
54. Krishna, A. G., Menon, S. T., Terry, T. J., and Sakmar, T. P. (2002) Evidence that helix 8 of rhodopsin acts as a membrane-dependent conformational switch. *Biochemistry* **41,** 8298–8309.
55. Choudhary, M. S., Craigo, S., and Roth, B. L. (1993) A single point mutation (Phe[340] to Leu[340]) of a conserved phenylalanin abolishes 4-[^{125}I]Iodo-(2,5-dimethoxy) phenylisopropylamine and [^{3}H]Mesulergine but not [^{3}H]Ketanserin binding to 5-hydroxytryptamine[2] receptors. *Mol. Pharmacol.* **43,** 755–761.
56. Visiers, I., Ballesteros, J. A., and Weinstein, H. (2000) A spatially ordered sequence of intramolecular rearrangements observed from simulations of agonist-related activation of 5HT(2C) receptors. *Biophys. J.* **78,** 394.
57. Lin, S. W. and Sakmar, T. P. (1996) Specific tryptophan UV-absorbance changes are probes of the transition of rhodopsin to its active state. *Biochemistry* **35,** 11149–11159.
58. Choudhary, M. S., Craigo, S., and Roth, B. L. (1992) Identification of receptor domains that modify ligand binding to 5-hydroxytryptamine2 and 5-hydroxytryptamine1C serotonin receptors. *Mol. Pharmacol.* **42,** 627–633.
59. Farrens, D. L., Altenbach, C., Yang, K., Hubbell, W. L., and Khorana, H. G. (1996) Requirement of rigid-body motion of transmembrane helices for light activation of rhodopsin. *Science* **274,** 768–770.
60. Klein-Seetharaman, J. (2002) Dynamics in rhodopsin. *ChemBioChem* **3,** 981–986.
61. Sheikh, S. P., Zvyaga, T. A., Lichtarge, O., Sakmar, T. P., and Bourne, H. R. (1996) Rhodopsin activation blocked by metal-ion-binding sites linking transmembrane helices C and F. *Nature* **383,** 347–350.
62. Sansom, M. S. P. and Weinstein, H. (2000) Hinges, swivels and switches: the role of prolines in signalling via transmembrane alpha-helices. *Trends Pharmacol. Sci.* **21,** 445–451.
63. Tieleman, D. P., Shrivastava, I. H., Ulmschneider, M. R., and Sansom, M. S. (2001) Proline-induced hinges in transmembrane helices: possible roles in ion channel gating. *Proteins* **44,** 63–72.

64. Ballesteros, J. A. and Weinstein, H. (1992) The role of Pro/Hyp-kinks in determining the transmembrane helix length and gating mechanism of a Leu-zervamicin channel. *Biophys. J.* **62,** 110–111.

65. Visiers, I., Braunheim, B. B., and Weinstein, H. (2000) Prokink: a protocol for numerical evaluation of helix distortions by proline. *Prot. Eng.* **13,** 603–606.

66. Scheer, A., Costa, T., Fanelli, F., et al. (2000) Mutational analysis of the highly conserved arginine within the Glu/Asp-Arg-Tyr motif of the alpha(1b)-adrenergic receptor: effects on receptor isomerization and activation. *Mol. Pharmacol.* **57,** 219–231.

67. Christopoulos, A. and Kenakin, T. (2002) G protein-coupled receptor allosterism and complexing. *Pharmacol. Rev.* **54,** 323–374.

68. Brady, A. E. and Limbird, L. E. (2002) G protein coupled receptor interacting proteins: emerging roles in localization and signal transduction. *Cell. Signal.* **14,** 297–309.

69. George, S. R., O'Dowd, B. F., and Lee, S. P. (2002) Oligomerization and its potential for drug discovery. *Nat. Rev. Drug Disc.* **1,** 808–820.

70. Kenakin, T. (2002) Efficacy at G protein-coupled receptors. *Nat. Rev. Drug Discov.* **1,** 103–110.

71. Hur, E. M. and Kim, K. T. (2002) G protein-coupled receptor signalling and cross-talk: achieving rapidity and specificity. *Cell. Signal.* **14,** 397–405.

72. Angers, S., Salahpour, A., and Bouvier, M. (2002) Dimerization: an emerging concept for G protein coupled receptor ontogeny and function. *Annu. Rev. Pharmacol. Toxicol.* **42,** 409–435.

73. Levac, B. A., O'Dowd, B. F., and George, S. R. (2002) Oligomerization of opioid receptors: generation of novel signaling units. *Curr. Opin. Pharmacol.* **2,** 76–81.

74. Lee, S. P., O'Dowd, B. F., Ng, G. Y., et al. (2000) Inhibition of cell surface expression by mutant receptors demonstrates that D2 dopamine receptors exist as oligomers in the cell. *Mol. Pharmacol.* **58,** 120–128.

75. Karpa, K. D., Lin, R., Kabbani, N., and Levenson, R. (2000) The dopamine D3 receptor interacts with itself and the truncated D3 splice variant d3nf: D3-D3nf interaction causes mislocalization of D3 receptors. *Mol. Pharmacol.* **58,** 677–683.

76. Benkirane, M., Jin, D. Y., Chun, R. F., Koup, R. A., and Jeang, K. T. (1997) Mechanism of transdominant inhibition of CCR5-mediated HIV-1 infection by ccr5delta32. *J. Biol. Chem.* **272,** 30603–30606.

77. Zhu, X. and Wess, J. (1998) Truncated V2 vasopressin receptors as negative regulators of wild-type V2 receptor function. *Biochemistry* **37,** 15773–15784.

78. Kuner, R., Kohr, G., Grunewald, S., Eisenhardt, G., Bach, A., and Kornau, H. C. (1999) Role of heteromer formation in GABAB receptor function. *Science* **283,** 74–77.

79. Jones, K. A., Borowsky, B., Tamm, J. A., et al. (1998) GABA(B) receptors function as a heteromeric assembly of the subunits GABA(B)R1 and GABA(B)R2. *Nature* **396,** 674–679.

80. White, J. H., Wise, A., Main, M. J., et al. (1998) Heterodimerization is required for the formation of a functional GABA(B) receptor. *Nature* **396,** 679–682.

81. Kaupmann, K., Malitschek, B., Schuler, V., et al. (1998) GABA(B)-receptor subtypes assemble into functional heteromeric complexes. *Nature* **396,** 683–687.

82. Limbird, L. E., Meyts, P. D., and Lefkowitz, R. J. (1975) Beta-adrenergic receptors: evidence for negative cooperativity. *Biochem. Biophys. Res. Commun.* **64,** 1160–1168.

83. Henis, Y. I. and Sokolovsky, M. (1983) Muscarinic antagonists induce different receptor conformations in rat adenohypophysis. *Mol. Pharmacol.* **24,** 357–365.

84. Mattera, R., Pitts, B. J., Entman, M. L., and Birnbaumer, L. (1985) Guanine nucleotide regulation of a mammalian myocardial muscarinic receptor system. Evidence for homo- and heterotropic cooperativity in ligand binding analyzed by computer-assisted curve fitting. *J. Biol. Chem.* **260,** 7410–7421.

85. Wreggett, K. A. and Wells, J. W. (1995) Cooperativity manifest in the binding properties of purified cardiac muscarinic receptors. *J. Biol. Chem.* **270,** 22488–22499.

86. Gines, S., Hillion, J., Torvinen, M., et al. (2000) Dopamine D1 and adenosine A1 receptors form functionally interacting heteromeric complexes. *Proc. Natl. Acad. Sci. USA* **97,** 8606–8611.

87. Chazenbalk, G. D., Kakinuma, A., Jaume, J. C., McLachlan, S. M., and Rapoport, B. (1996) Evidence for negative cooperativity among human thyrotropin receptors overexpressed in mammalian cells. *Endocrinology* **137,** 4586–4591.

88. Boyer, J. L., Martinez-Carcamo, M., Monroy-Sanchez, J. A., Posadas, C., and Garcia-Sainz, J. A. (1986) Guanine nucleotide-induced positive cooperativity in muscarinic-cholinergic antagonist binding. *Biochem. Biophys. Res. Commun.* **134,** 172–177.

89. Potter, L. T., Ballesteros, L. A., Bichajian, L. H., Ferrendelli, C. A., Fisher, A., Hanchett, H. E., and Zhang, R. (1991) Evidence of paired M2 muscarinic receptors. *Mol. Pharmacol.* **39,** 211–221.

90. Sinkins, W. G., Kandel, M., Kandel, S. I., Schunack, W., and Wells, J. W. (1993) G protein-linked receptors labeled by [³H]histamine in guinea pig cerebral cortex. I. Pharmacological characterization [corrected]. *Mol. Pharmacol.* **43,** 583–594.

91. Hirschberg, B. T. and Schimerlik, M. I. (1994) A kinetic model for oxotremorine M binding to recombinant porcine m2 muscarinic receptors expressed in Chinese hamster ovary cells. *J. Biol. Chem.* **269,** 26127–26135.

92. Armstrong, D. and Strange, P. G. (2001) Dopamine d2 receptor dimer formation. evidence from ligand binding. *J. Biol. Chem.* **276,** 22621–22629.

93. Pizard, A., Marchetti, J., Allegrini, J., Alhenc-Gelas, F., and Rajerison, R. M. (1998) Negative cooperativity in the human bradykinin B2 receptor. *J. Biol. Chem.* **273,** 1309–1315.

94. AbdAlla, S., Lother, H., and Quitterer, U. (2000) AT1-receptor heterodimers show enhanced G protein activation and altered receptor sequestration. *Nature* **407,** 94–98.

95. Rocheville, M., Lange, D. C., Kumar, U., Patel, S. C., Patel, R. C., and Patel, Y. C. (2000) Receptors for dopamine and somatostatin: formation of hetero-oligomers with enhanced functional activity. *Science* **288,** 154–157.

96. Jordan, B. A. and Devi, L. A. (1999) G protein-coupled receptor heterodimerization modulates receptor function. *Nature* **399,** 697–700.

97. George, S. R., Fan, T., Xie, Z., Tse, R., Tam, V., Varghese, G., and O'Dowd, B. F. (2000) Oligomerization of mu- and delta-opioid receptors. Generation of novel functional properties. *J. Biol. Chem.* **275,** 26128–26135.

98. Mellado, M., Rodriguez-Frade, J. M., Vila-Coro, A. J., Fernandez, S., Martin De Ana, A., Jones, D. R., et al. (2001) Chemokine receptor homo- or heterodimerization activates distinct signaling pathways. *Embo J,* **20,** 2497–2507.

99. Rios, C. D., Jordan, B. A., Gomes, I., and Devi, L. A. (2001) G protein coupled receptor dimerization: modulation of receptor function. *Pharmacol. Ther.* **92,** 71–87.

100. Rodriguez-Frade, J. M., Mellado, M., and Martinez-A, C. (2001) Chemokine receptor dimerization: two are better than one. *Trends Immunol.* **22,** 612–617.

101. Rodriguez-Frade, J. M., Vila-Coro, A. J., de Ana, A. M., Albar, J. P., Martinez-A, C., and Mellado, M. (1999) The chemokine monocyte chemoattractant protein-1 induces functional responses through dimerization of its receptor CCR2. *Proc. Natl. Acad. Sci. USA* **96,** 3628–3633.

102. Mellado, M., Rodriguez-Frade, J. M., Manes, S., and Martinez, A. C. (2001) Chemokine signaling and functional responses: the role of receptor dimerization and tk pathway activation. *Annu. Rev. Immunol.* **19,** 397–421.

103. Vila-Coro, A. J., Mellado, M., Martin de Ana, A., Lucas, P., del Real, G., Martinez-A, C., and Rodriguez-Frade, J. M. (2000) HIV-1 infection through the CCR5 receptor is blocked by receptor dimerization. *Proc. Natl. Acad. Sci. USA* **97,** 3388–3393.

104. Vila-Coro, A. J., Rodriguez-Frade, J. M., Martin De Ana, A., Moreno-Ortiz, M. C., Martinez-A, C., and Mellado, M. (1999) The chemokine SDF-1 alpha triggers CXCR4 receptor dimerization and activates the JAK/STAT pathway. *FASEB J.* **13,** 1699–1710.

105. Cheng, Z. J. and Miller, L. J. (2001) Agonist-dependent dissociation of oligomeric complexes of G protein-coupled cholecystokinin receptors demonstrated in living cells using bioluminescence resonance energy transfer. *J. Biol. Chem.* **276,** 48040–48047.

106. Rocheville, M., Lange, D. C., Kumar, U., Sasi, R., Patel, R. C., and Patel, Y. C. (2000) Subtypes of the somatostatin receptor assemble as functional homo-and heterodimers. *J. Biol. Chem.* **275,** 7862–7869.

107. Patel, R. C., Kumar, U., Lamb, D. C., et al. (2002) Ligand binding to somatostatin receptors induces receptors-specific oligomer formation in live cells. *Proc. Natl. Acad. Sci. USA* **99,** 3294–3299.

108. Horvat, R. D., Roess, D. A., Nelson, S. E., Barisas, B. G., and Clay, C. M. (2001) Binding of agonist but not antagonist leads to fluorescence resonance energy transfer between intrinsically fluorescent gonadotropin-releasing hormone receptors. *Mol. Endocrinol* **15,** 695–703.

109. Margeta-Mitrovic, M., Jan, Y. N., and Jan, L. Y. (2001) Ligand-induced signal transduction within heterodimeric GABA(B) receptor. *Proc. Natl. Acad. Sci. USA* **98,** 14643–14648.

110. Zhu, C. C., Cook, L. B., and Hinkle, P. M. (2002) Dimerization and phosphorylation of thyrotropin-releasing hormone receptors are modulated by agonist stimulation. *J. Biol. Chem.* **277,** 28228–28237.

111. Torvinen, M., Gines, S., Hillion, J., Latini, S., Canals, M., Ciruela, F., et al. (2002) Interactions among adenosine deaminase, adenosine A(1) receptors and dopamine D(1) receptors in stably cotransfected fibroblast cells and neurons. *Neuroscience* **113,** 709–719.

112. Kniazeff, J., Galvez, T., Labesse, G., and Pin, J. P. (2002) No ligand binding in the GB2 subunit of the GABA(B) receptor is required for activation and allosteric interaction between the subunits. *J. Neurosci.* **22,** 7352–7361.

113. Issafras, H., Angers, S., Bulenger, S., et al. (2002) Constitutive agonist-independent CCR5 oligomerization and antibody-mediated clustering occurring at physiological levels of receptors. *J. Biol. Chem.* **277,** 34666–34673.

114. Mercier, J. F., Salahpour, A., Angers, S., Breit, A., and Bouvier, M. (2002) Quantitative assessment of beta 1 and beta 2-adrenergic receptor homo and hetero-dimerization by bioluminescence resonance energy transfer. *J. Biol. Chem.* **277,** 44925–44931.

115. Ayoub, M. A., Couturier, C., Lucas-Meunier, E., Angers, S., Fossier, P., Bouvier, M., and Jockers, R. (2002) Monitoring of ligand-independent dimerization and ligand-induced conformational changes of melatonin receptors in living cells by bioluminescence resonance energy transfer. *J. Biol. Chem* **277,** 21522–21528.

116. Kroeger, K. M., Hanyaloglu, A. C., Seeber, R. M., Miles, L. E., and Eidne, K. A. (2001) Constitutive and agonist-dependent homo-oligomerization of the thyrotropin-releasing hormone receptor. Detection in living cells using bioluminescence resonance energy transfer. *J. Biol. Chem.* **276,** 12736–12743.

117. Hebert, T. E., Moffett, S., Morello, J. P., Loisel, T. P., Bichet, D. G., Barret, C., and Bouvier, M. (1996) A peptide derived from a beta2-adrenergic receptor transmembrane domain inhibits both receptor dimerization and activation. *J. Biol. Chem.* **271,** 16384–16392.

118. Ng, G. Y., O'Dowd, B. F., Lee, S. P., Chung, H. T., Brann, M. R., Seeman, P., and George, S. R. (1996) Dopamine D2 receptor dimers and receptor-blocking peptides. *Biochem. Biophys. Res. Commun.* **227,** 200–204.

119. Romano, C., Yang, W. L., and O'Malley, K. L. (1996) Metabotropic glutamate receptor 5 is a disulfide-linked dimer. *J. Biol. Chem.* **271,** 28612–28616.

120. Cvejic, S. and Devi, L. A. (1997) Dimerization of the delta opioid receptor: implication for a role in receptor internalization. *J. Biol. Chem.* **272,** 26959–26964.

121. Bai, M., Trivedi, S., and Brown, E. M. (1998) Dimerization of the extracellular calcium-sensing receptor (CaR) on the cell surface of CaR-transfected HEK293 cells. *J. Biol. Chem.* **273,** 23605–23610.

122. Zeng, F. Y. and Wess, J. (1999) Identification and molecular characterization of m3 muscarinic receptor dimers. *J. Biol. Chem.* **274,** 19487–19497.

123. Gomes, I., Jordan, B. A., Gupta, A., Trapaidze, N., Nagy, V., and Devi, L. A. (2000) Heterodimerization of mu and delta opioid receptors: a role in opiate synergy. *J. Neurosci.* **20,** RC110.

124. Pfeiffer, M., Koch, T., Schroder, H., et al. (2001) Homo- and heterodimerization of somatostatin receptor subtypes. Inactivation of sst(3) receptor function by heterodimerization with sst(2A). *J. Biol. Chem.* **276,** 14027–14036.

125. Jordan, B. A., Trapaidze, N., Gomes, I., Nivarthi, R., and Devi, L. A. (2001) Oligomerization of opioid receptors with beta 2-adrenergic receptors: a role in trafficking and mitogen-activated protein kinase activation. *Proc. Natl. Acad. Sci. USA* **98,** 343–348.

126. Angers, S., Salahpour, A., Joly, E., Hilairet, S., Chelsky, D., Dennis, M., and Bouvier, M. (2000) Detection of beta 2-adrenergic receptor dimerization in living cells using bioluminescence resonance energy transfer (BRET). *Proc. Natl. Acad. Sci. USA* **97,** 3684–3689.

127. McVey, M., Ramsay, D., Kellett, E., Rees, S., Wilson, S., Pope, A. J., and Milligan, G. (2001) Monitoring receptor oligomerization using time-resolved fluorescence resonance energy transfer and bioluminescence resonance energy transfer. The human delta -opioid receptor displays constitutive oligomerization at the cell surface, which is not regulated by receptor occupancy. *J. Biol. Chem.* **276,** 14092–14099.

128. Overton, M. C. and Blumer, K. J. (2000) G protein-coupled receptors function as oligomers in vivo. *Curr. Biol.* **10,** 341–344.

129. Cornea, A., Janovick, J. A., Maya-Nunez, G., and Conn, P. M. (2001) Gonadotropinreleasing hormone receptor microaggregation. Rate monitored by fluorescence resonance energy transfer. *J. Biol. Chem.* **276,** 2153–2158.

130. Babcock, G. J., Farzan, M., and Sodroski, J. (2002) Ligand-independent dimerization of CXCR4, a principal HIV-1 coreceptor. *J. Biol. Chem.* Nov 13; [epub ahead of print].

131. Goldsmith, P. K., Fan, G. F., Ray, K., Shiloach, J., McPhie, P., Rogers, K. V., and Spiegel, A. M. (1999) Expression, purification, and biochemical characterization of the aminoterminal extracellular domain of the human calcium receptor. *J. Biol. Chem.* **274,** 11303–11309.

132. Pace, A. J., Gama, L., and Breitwieser, G. E. (1999) Dimerization of the calcium-sensing receptor occurs within the extracellular domain and is eliminated by Cys → Ser mutations at Cys101 and Cys236. *J. Biol. Chem.* **274,** 11629–11634.

133. Kunishima, N., Shimada, Y., Tsuji, Y., et al. (2000) Structural basis of glutamate recognition by a dimeric metabotropic glutamate receptor. *Nature* **407,** 971–977.

134. AbdAlla, S., Zaki, E., Lother, H., and Quitterer, U. (1999) Involvement of the amino terminus of the B(2) receptor in agonist-induced receptor dimerization. *J. Biol. Chem.* **274,** 26079–26084.

135. Abe, J., Suzuki, H., Notoya, M., Yamamoto, T., and Hirose, S. (1999) Ig-hepta, a novel member of the G protein-coupled hepta-helical receptor (GPCR) family that has immunoglobulin-like repeats in a long N-terminal extracellular domain and defines a new subfamily of GPCRs. *J. Biol. Chem.* **274,** 19957–19964.

136. Overton, M. C. and Blumer, K. J. (2002) The extracellular N-terminal domain and transmembrane domains 1 and 2 mediate oligomerization of a yeast G protein-coupled receptor. *J. Biol. Chem.* **277,** 41463–41472.

137. Fleming, K. G. and Engelman, D. M. (2001) Specificity in transmembrane helix-helix interactions can define a hierarchy of stability for sequence variants. *Proc. Natl. Acad. Sci. USA* **98,** 14340–14344.

138. George, S. R., Lee, S. P., Varghese, G., Zeman, P. R., Seeman, P., Ng, G. Y., and O'Dowd, B. F. (1998) A transmembrane domain-derived peptide inhibits D1 dopamine receptor function without affecting receptor oligomerization. *J. Biol. Chem.* **273,** 30244–30248.

139. Maggio, R., Vogel, Z., and Wess, J. (1993) Coexpression studies with mutant muscarinic/adrenergic receptors provide evidence for intermolecular "cross-talk" between G protein-linked receptors. *Proc. Natl. Acad. Sci. USA* **90,** 3103–3107.

140. Hamdan, F. F., Ward, S. D., Siddiqui, N. A., Bloodworth, L. M., and Wess, J. (2002) Use of an in situ disulfide cross-linking strategy to map proximities between amino acid residues

in transmembrane domains I and VII of the M3 muscarinic acetylcholine receptor. *Biochemistry* **41,** 7647–7658.

141. Hadac, E. M., Ji, Z., Pinon, D. I., Henne, R. M., Lybrand, T. P., and Miller, L. J. (1999) A peptide agonist acts by occupation of a monomeric G protein-coupled receptor: dual sites of covalent attachment to domains near TM1 and TM7 of the same molecule make biologically significant domain-swapped dimerization unlikely. *J. Med. Chem.* **42,** 2105–2111.

142. Lemmon, M. A., Flanagan, J. M., Treutlein, H. R., Zhang, J., and Engelman, D. M. (1992) Sequence specificity in the dimerization of transmembrane α-helices. *Biochemistry* **31,** 12719–12725.

143. Ray, K. and Hauschild, B. C. (2000) Cys-140 is critical for metabotropic glutamate receptor-1 dimerization. *J. Biol. Chem.* **275,** 34245–34251.

144. Lee, S. P., Xie, Z., Varghese, G., Nguyen, T., O'Dowd, B. F., and George, S. R. (2000) Oligomerization of dopamine and serotonin receptors. *Neuropsychopharmacology* **23,** S32–40.

145. Zhang, Z., Sun, S., Quinn, S. J., Brown, E. M., and Bai, M. (2001) The extracellular calcium-sensing receptor dimerizes through multiple types of intermolecular interactions. *J. Biol. Chem.* **276,** 5316–5322.

146. Tsuji, Y., Shimada, Y., Takeshita, T., et al. (2000) Cryptic dimer interface and domain organization of the extracellular region of metabotropic glutamate receptor subtype 1. *J. Biol. Chem.* **275,** 28144–28151.

147. Romano, C., Miller, J. K., Hyrc, K., et al. (2001) Covalent and noncovalent interactions mediate metabotropic glutamate receptor mGlu5 dimerization. *Mol. Pharmacol.* **59,** 46–53.

148. Schulz, A., Grosse, R., Schultz, G., Gudermann, T., and Schoneberg, T. (2000) structural implication for receptor oligomerization from functional reconstitution studies of mutant V2 vasopressin receptors. *J. Biol. Chem.* **275,** 2381–2389.

149. Gouldson, P. R. and Reynolds, C. A. (1997) Simulations on dimeric peptides: evidence for domain swapping in G protein-coupled receptors? *Biochem. Soc. Trans.* **25,** 1066–1071.

150. Gouldson, P. R., Snell, C. R., and Reynolds, C. A. (1997) A new approach to docking in the beta 2-adrenergic receptor that exploits the domain structure of G protein-coupled receptors. *J. Med. Chem.* **40,** 3871–3886.

151. Gouldson, P. R., Snell, C. R., Bywater, R. P., Higgs, C., and Reynolds, C. A. (1998) Domain swapping in G protein coupled receptor dimers. *Prot. Eng.* **11,** 1181–1193.

152. Gouldson, P. R., Higgs, C., Smith, R. E., Dean, M. K., Gkoutos, G. V., and Reynolds, C. A. (2000) Dimerization and domain swapping in G protein-coupled receptors: a computational study. *Neuropsychopharmacology* **23,** S60–S77.

153. Monnot, C., Bihoreau, C., Conchon, S., Curnow, K. M., Corvol, P., and Clauser, E. (1996) Polar residues in the transmembrane domains of the type 1 angiotensin II receptor are required for binding and coupling. Reconstitution of the binding site by co-expression of two deficient mutants. *J. Biol. Chem.* **271,** 1507–1513.

154. Davies, A., Gowen, B. E., Krebs, A. M., Schertler, G. F., and Saibil, H. R. (2001) Three-dimensional structure of an invertebrate rhodopsin and basis for ordered alignment in the photoreceptor membrane. *J. Mol. Biol.* **314,** 455–463.

155. Krebs, A., Villa, C., Edwards, P. C., and Schertler, G. F. (1998) Characterisation of an improved two-dimensional p22121 crystal from bovine rhodopsin. *J. Mol. Biol.* **282,** 991–1003.

156. Schertler, G. F. and Hargrave, P. A. (1995) Projection structure of frog rhodopsin in two crystal forms. *Proc. Natl. Acad. Sci. USA* **92,** 11578–11582.

157. Fotiadis, D., Liang, Y., Filipek, S., Saperstein, D. A., Engel, A., and Palczewski, K. (2003) Atomic-force microscopy: rhodopsin dimers in native disc membranes. *Nature* **421,** 127–128.

158. Jones, S. and Thornton, J. M. (1996) Principles of protein-protein interactions. *Proc. Natl. Acad. Sci. USA* **93,** 13–20.

159. Norel, R., Petrey, D., Wolfson, H. J., and Nussinov, R. (1999) Examination of shape complementarity in docking of unbound proteins. *Proteins* **36,** 307–317.

160. Lo Conte, L., Chothia, C., and Janin, J. (1999) The atomic structure of protein-protein recognition sites. *J. Mol. Biol.* **285,** 2177–2198.

161. Gavin, A. C., Bosche, M., Krause, R., et al. (2002) Functional organization of the yeast proteome by systematic analysis of protein complexes. *Nature* **415,** 141–147.

162. Ho, Y., Gruhler, A., Heilbut, A., et al. (2002) Systematic identification of protein complexes in Saccharomyces cerevisiae by mass spectrometry. *Nature* **415,** 180–183.

163. Zhu, H., Bilgin, M., Bangham, R., et al. (2001) Global analysis of protein activities using proteome chips. *Science* **293,** 2101–2105.

164. Fields, S. and Song, O. (1989) A novel genetic system to detect protein-protein interactions. *Nature* **340,** 245–246.

165. Valencia, A. and Pazos, F. (2002) Computational methods for the prediction of protein interactions. *Curr. Opin. Struct. Biol.* **12,** 368–373.

166. Smith, G. R. and Sternberg, M. J. (2002) Prediction of protein-protein interactions by docking methods. *Curr. Opin. Struct. Biol.* **12,** 28–35.

167. Dixon, J. S. (1997) Evaluation of the CASP2 docking section. *Proteins* **Suppl. 1,** 198–204.

168. Moller, S., Vilo, J., and Croning, M. D. (2001) Prediction of the coupling specificity of G protein coupled receptors to their G proteins. *Bioinformatics* **17,** S174–181.

169. Goh, C. S. and Cohen, F. E. (2002) Co-evolutionary analysis reveals insights into protein-protein interactions. *J. Mol. Biol.* **324,** 177–192.

170. Oliveira, L., Paiva, A. C., and Vriend, G. (2002) Correlated mutation analysis on very large sequence families. *Chembiochem* **3,** 1010–1017.

171. Horn, F., van der Wenden, E. M., Oliveira, L., AP, I. J., and Vriend, G. (2000) Receptors coupling to G proteins: is there a signal behind the sequence? *Proteins* **41,** 448–459.

172. Oliveira, L., Paiva, A. C., and Vriend, G. (1999) A low resolution model for the interaction of G proteins with G protein-coupled receptors. *Prot. Eng.* **12,** 1087–1095.

173. Lichtarge, O., Sowa, M. E., and Philippi, A. (2002) Evolutionary traces of functional surfaces along G protein signaling pathway. *Methods Enzymol.* **344,** 536–556.

174. Lichtarge, O., Bourne, H. R., and Cohen, F. E. (1996) Evolutionarily conserved Galphabetagamma binding surfaces support a model of the G protein-receptor complex. *Proc. Natl. Acad. Sci. USA* **93,** 7507–7511.

175. Dean, M. K., Higgs, C., Smith, R. E., et al. (2001) Dimerization of G protein-coupled receptors. *J. Med. Chem.* **44,** 4595–4614.

176. Gobel, U., Sander, C., Schneider, R., and Valencia, A. (1994) Correlated mutations and residue contacts in proteins. *Proteins* **18,** 309–317.

177. Olmea, O. and Valencia, A. (1997) Improving contact predictions by the combination of correlated mutations and other sources of sequence information. *Fold Des.* **2,** S25–32.

178. Pazos, F., Olmea, O., and Valencia, A. (1997) A graphical interface for correlated mutations and other protein structure prediction methods. *Comput. Appl. Biosci.* **13,** 319–321.

179. Pazos, F., Helmer-Citterich, M., Ausiello, G., and Valencia, A. (1997) Correlated mutations contain information about protein-protein interaction. *J. Mol. Biol.* **271,** 511–523.

180. Pazos, F. and Valencia, A. (2002) In silico two-hybrid system for the selection of physically interacting protein pairs. *Proteins* **47,** 219–227.

181. Lichtarge, O., Bourne, H. R., and Cohen, F. E. (1996) An evolutionary trace method defines binding surfaces common to protein families. *J. Mol. Biol.* **257,** 342–358.

182. Madabushi, S., Yao, H., Marsh, M., Kristensen, D. M., Philippi, A., Sowa, M. E., and Lichtarge, O. (2002) Structural clusters of evolutionary trace residues are statistically significant and common in proteins. *J. Mol. Biol.* **316,** 139–154.

183. Lichtarge, O. and Sowa, M. E. (2002) Evolutionary predictions of binding surfaces and interactions. *Curr. Opin. Struct. Biol.* **12,** 21–27.

184. Gouldson, P. R., Dean, M. K., Snell, C. R., Bywater, R. P., Gkoutos, G., and Reynolds, C. A. (2001) Lipid-facing correlated mutations and dimerization in G protein coupled receptors. *Prot. Eng.* **14,** 759–767.

185. Gkoutos, G. V., Higgs, C., Bywater, R. P., Gouldson, P. R., and Reynolds, C. A. (1999) Evidence for dimerization in the β2-adrenergic receptor from the evolutionary trace method. *Intl. J. Quantum. Chem. Biophys. Q.* **74,** 371–379.

186. Filizola, M., Guo, W., Javitch, J. A., and Weinstein, H. (2003) Dimerization in G protein coupled receptors: Correlation analysis and electron density maps of rhodopsin from different species suggest subtype-specific interfaces. *Biophys. J.* **84(2),** 1309 Pos Part 2.

187. Filizola, M., Olmea, O., and Weinstein, H. (2002) Prediction of heterodimerization interfaces of G protein coupled receptors with a new subtractive correlated mutation method. *Prot. Eng.* **15,** 881–885.

188. Filizola, M., Olmea, O., and Weinstein, H. (2002) Using correlated mutation analysis to predict the heterodimerization interface of GPCRs. *Biophys. J.* **82,** 2307. (Part 2302 Jan 2002.)

189. Filizola, M. and Weinstein, H. (2002) Structural models for dimerization of G protein coupled receptors: the opioid receptor homodimers. *Biopolymers (Peptide Sci.)* **66,** 317–325.

190. Kuipers, W., Oliveira, L., Vriend, G., and Ijzerman, A. P. (1997) Identification of class-determining residues in G protein-coupled receptors by sequence analysis. *Recept. Channels* **5,**159–174.

191. Singer, M. S., Oliveira, L., Vriend, G., and Shepherd, G. M. (1995) Potential ligand-binding residues in rat olfactory receptors identified by correlated mutation analysis. *Recept. Channels* **3,** 89–95.

192. Geva, A., Lassere, T. B., Lichtarge, O., Pollitt, S. K., and Baranski, T. J. (2000) Genetic mapping of the human C5a receptor. Identification of transmembrane amino acids critical for receptor function. *J. Biol. Chem.* **275,** 35393–35401.

193. Baranski, T. J., Herzmark, P., Lichtarge, O., et al. (1999) C5a receptor activation. Genetic identification of critical residues in four transmembrane helices. *J. Biol. Chem.* **274,** 15757–15765.

194. Jordan, B. A., Cvejic, S., and Devi, L. A. (2000) Opioids and their complicated receptor complexes. *Neuropsychopharmacology* **23,** S5–S18.

195. Devi, L. A. (2000) G protein-coupled receptor dimers in the lime light. *Trends Pharmacol. Sci.* **21,** 324–326.

196. Ramsay, D., Kellett, E., McVey, M., Rees, S., and Milligan, G. (2002) Homo- and heterooligomeric interactions between G protein-coupled receptors in living cells monitored by two variants of bioluminescene resonance energy transfer. *Biochem. J.* **365,** 429–440.

197. Suzuki, S., Chuang, L. F., Yau, P., Doi, R. H., and Chuang, R. Y. (2002) Interactions of opioid and chemokine receptors: oligomerization of mu, kappa, and delta with CCR5 on immune cells. *Exp. Cell. Res.* **280,** 192–200.

198. Hoffmann, A., Levchenko, A., Scott, M. L., and Baltimore, D. (2002) The IkappaB-NF-KappaB signaling module: temporal control and selective gene activation. *Science* **298,** 1241–1245.

199. Bhalla, U. S., Ram, P. T., and Iyengar, R. (2002) MAP kinase phosphatase as a locus of flexibility in a mitogen-activated protein kinase signaling network. *Science* **297,** 1018–1023.

200. Neves, S. R. and Iyengar, R. (2002) Modeling of signaling networks. *Bioessays* **24,** 1110–1117.

201. Neves, S. R., Ram, P. T., and Iyengar, R. (2002) G protein pathways. *Science* **296,** 1636–1639.

202. Ciruela, F., Casado, V., Mallol, J., Canela, E. I., Lluis, C., and Franco, R. (1995) Immunological identification of A1 adenosine receptors in brain cortex. *J. Neurosci. Res.* **42,** 818–828.

203. Cirula, F., Escriche, M., Burgueno, J., et al. (2001) Metabotropic glutamate 1alpha and adenosine A1 receptors assemble into functionally interacting complexes. *J. Biol. Chem.* **276,** 18345–18351.

204. Yoshioka, K., Saitoh, O., and Nakata, H. (2001) Heteromeric association creates a P2Y-like adenosine receptor. *Proc. Natl. Acad. Sci. USA* **98,** 7617–7622.

205. Hillion, J., Canals, M., Torvinen, M., et al. (2002) Coaggregation, cointernalization, and codesensitization of adenosine A2A receptors and dopamine D2 receptors. *J. Biol. Chem.* **277,** 18091–18097.

206. AbdAlla, S., Lother, H., Abdel-tawab, A. M., and Quitterer, U. (2001) The angiotensin II AT2 receptor is an AT1 receptor antagonist. *J. Biol. Chem.* **276,** 39721–39726.

207. AbdAlla, S., Lother, H., el Massiery, A., and Quitterer, U. (2001) Increased AT(1) receptor heterodimers in preeclampsia mediate enhanced angiotensin II responsiveness. *Nat. Med.* **7,** 1003–1009.

208. Xu, J., He, J., Castleberry, A., Balasubramanian, S., Lau, A. G., and Hall, R. A. (2003) Heterodimerization of alpha-2A- and beta-1-adrenergic receptors. *J. Biol. Chem.* **278,** 10770–10777.

209. Lavoie, C., Mercier, J. F., Salahpour, A., Umapathy, D., Breit, A., Villeneuve, L. R., et al. (2002) Beta 1/beta 2-adrenergic receptor heterodimerization regulates beta 2-adrenergic receptor internalization and ERK signaling efficacy. *J. Biol. Chem.* **277,** 35402–35410.

210. Mukhopadhyay, S., McIntosh, H. H., Houston, D. B., and Howlett, A. C. (2000) The CB(1) cannabinoid receptor juxtamembrane C-terminal peptide confers activation to specific G proteins in brain. *Mol. Pharmacol.* **57,** 162–170.

211. Ng, G. Y., Mouillac, B., George, S. R., Caron, M., Dennis, M., Bouvier, M., and O'Dowd, B. F. (1994) Desensitization, phosphorylation and palmitoylation of the human dopamine D1 receptor. *Eur. J. Pharmacol.* **267,** 7–19.

212. Nimchinsky, E. A., Hof, P. R., Janssen, W. G., Morrison, J. H., and Schmauss, C. (1997) Expression of dopamine D3 receptor dimers and tetramers in brain and in transfected cells. *J. Biol. Chem.* **272,** 29229–29237.

213. Scarselli, M., Novi, F., Schallmach, E., et al. (2001) D2/D3 dopamine receptor heterodimers exhibit unique functional properties. *J. Biol. Chem.* **276,** 30308–30314.

214. Fukushima, Y., Asano, T., Satioh, T., et al. (1997) Oligomer formation of histamine H2 receptors expressed in Sf9 and COS7 cells. *FEBS Lett.* **409,** 283–286.

215. Nguyen, T., Shapiro, D. A., George, S. R., et al. (2001) Discovery of a novel member of the histamine receptor family. *Mol. Pharmacol.* **59,** 427–433.

216. Ng, G. Y., George, S. R., Zastawny, R. L., Caron, M., Bouvier, M., Dennis, M., and O'Dowd, B. F. (1993) Human serotonin1B receptor expression in Sf9 cells: phosphorylation, palmitoylation, and adenylyl cyclase inhibition. *Biochemistry* **32,** 11727–11733.

217. Xie, Z., Lee, S. P., O'Dowd, B. F., and George, S. R. (1999) Serotonin 5-HT1B and 5-HT1D receptors form homodimers when expressed alone and heterodimers when coexpressed. *FEBS Lett.* **456,** 63–67.

218. Indrapichate, K., Meehan, D., Lane, T. A., et al. (1992) Biological actions of monoclonal luteinizing hormone/human chorionic gonadotropin receptor antibodies. *Biol. Reprod.* **46,** 265–278.

219. Roess, D. A., Horvat, R. D., Munnelly, H., and Barisas, B. G. (2000) Luteinizing hormone receptors are self-associated in the plasma membrane. *Endocrinology* **141,** 4518–4523.

220. Maggio, R., Barbier, P., Colelli, A., Salvadori, F., Demontis, G., and Corsini, G. U. (1999) G protein-linked receptors: pharmacological evidence for the formation of heterodimers. *J. Pharmacol. Exp. Ther.* **291,** 251–257.

221. Sawyer, G. W. and Ehlert, F. J. (1999) Muscarinic M3 receptor inactivation reveals a pertussis toxin-sensitive contractile response in the guinea pig colon: evidence for M2/M3 receptor interactions. *J. Pharmacol. Exp. Ther.* **289,** 464–476.

222. Dinger, M. C., Bader, J. E., Kobor, A. D., Kretzschmar, A. K., and Beck-Sickinger, A. G. (2003) Homodimerization of neuropeptide Y receptors investigated by fluorescence resonance energy transfer in living cells. *J. Biol. Chem.* **10,** 10.

223. Pfeiffer, M., Koch, T., Schroder, H., Laugsch, M., Hollt, V., and Schulz, S. (2002) Heterodimerization of somatostatin and opioid receptors cross-modulates phosphorylation, internalization, and desensitization. *J. Biol. Chem.* **277,** 19762–19772.

224. Hanyaloglu, A. C., Seeber, R. M., Kohout, T. A., Lefkowitz, R. J., and Eidne, K. A. (2002) Homo and hetero-oligomerization of thyrotropin releasing hormone (TRH) receptor subtypes: differential regulation of -arrestins 1 and 2. *J. Biol. Chem.* **277,** 50422–50430.

225. Cornea, A. and Michael Conn, P. (2002) Measurement of changes in fluorescence resonance energy transfer between gonadotropin-releasing hormone receptors in response to agonists. *Methods* **27,** 333.

226. Ng, G. Y., Clark, J., Coulombe, N., et al. (1999) Identification of a GABAB receptor subunit, gb2, required for functional GABAB receptor activity. *J. Biol. Chem.* **274,** 7607–7610.

227. Nelson, G., Chandrashekar, J., Hoon, M. A., Feng, L., Zhao, G., Ryba, N. J., and Zuker, C. S. (2002) An amino-acid taste receptor. *Nature* **416,** 199–202.

228. Nelson, G., Hoon, M. A., Chandrashekar, J., Zhang, Y., Ryba, N. J., and Zuker, C. S. (2001) Mammalian sweet taste receptors. *Cell* **106,** 381–390.

229. Yesilaltay, A. and Jenness, D. D. (2000) Homo-oligomeric complexes of the yeast alpha-factor pheromone receptor are functional units of endocytosis. *Mol. Biol. Cell.* **11,** 2873–2884.